Lecture Notes in Computer S

Edited by G. Goos and J. Hartmanis

Advisory Board: W. Brauer   D. Gries   J. Stoer

K. U. Schulz (Ed.)

Word Equations and Related Topics

1st International Workshop, IWWERT '90
Tübingen, Germany, October 1-3, 1990
Proceedings

Springer-Verlag

Berlin Heidelberg New York
London Paris Tokyo
Hong Kong Barcelona
Budapest

Series Editors

Gerhard Goos
Universität Karlsruhe
Postfach 69 80
Vincenz-Priessnitz-Straße 1
W-7500 Karlsruhe, FRG

Juris Hartmanis
Department of Computer Science
Cornell University
5148 Upson Hall
Ithaca, NY 14853, USA

Volume Editor

Klaus Ulrich Schulz
Center for Information and Language Processing (CIS)
Ludwig-Maximilian University Munich
Leopoldstraße 139, W-8000 München 40, FRG

CR Subject Classification (1991): F.4.1-3, E.1, I.2.3

ISBN 3-540-55124-7 Springer-Verlag Berlin Heidelberg New York
ISBN 0-387-55124-7 Springer-Verlag New York Berlin Heidelberg

© Springer-Verlag Berlin Heidelberg 1992
Printed in Germany

Typesetting: Camera ready by author
Printing and binding: Druckhaus Beltz, Hemsbach/Bergstr.
45/3140-543210 - Printed on acid-free paper

Preface

This volume contains papers presented at the first international workshop on word equations and related topics which was held at the SNS Institute of the University of Tübingen, Germany, in October 1990.

One year earlier, in 1989, several people were working in the field of associative unification without much contact among each other. At that time the idea arose of organizing a workshop where researchers with special interest in this topic — or, almost equivalently, in word equations — could present and discuss actual results. It soon became clear that the central topic, word equations, lay at the intersection of several broader areas of computer science which are not similarly related elsewhere, such as unification theory, combinatorics on words, list processing and constraint logic programming.

Thus the second, more general idea arose of using the workshop as a forum where researchers from these different domains could exchange ideas, concentrating on a common aspect, thereby supporting interaction and cross-fertilization between those fields with a practical orientation and those with a more theoretical one. The first workshop showed that this concept was rather attractive, in the meantime a second workshop was held in Rouen and a third is planned.

The papers in this volume represent a preliminary result of these intentions. The topics cover investigations on free groups (Makanin, Koscielski), associative unification and Makanin's algorithm to decide the solvability of equations in free semi-groups (Abdulrab, Schulz), general unification theory (Baader, Ohsuga and Sakai) and its relationship to algebra and model theory (Bockmayr), Thue systems (Book, Wrathall), and finitely presented groups (Makanina).

We would like to express our gratitude to the "Seminar für natürlich-sprachliche Systeme" (SNS), the "Wilhelm-Schickard-Institut für Informatik" (both University of Tübingen), the "Centrum für Informations- und Sprachverarbeitung" (University of Munich), the "Münchener Universitätsgesellschaft e.V." and, more personally, Prof. Guenthner and Prof. Loos without whose support the workshop would not have been possible. Thanks are also due to the speakers, authors and referees and to Springer-Verlag for the good collaboration.

Munich, October 1991   Klaus U. Schulz

Contents

Contents

Investigations on Equations
in a Free Group

G.S. Makanin
Steklov Institute of Mathematics
Academy of Science of the USSR
Vavilov Str. 42, Moscow 117966, GSP-1, USSR

Suppose $\mathcal{G}$ is a free group with a finite alphabet of generators $a_1, \ldots, a_\omega$. By an *equation* in $\mathcal{G}$ with unknowns $x_1, \ldots, x_\lambda$, we mean an equality of the form

$$W(x_1, \ldots, x_\lambda, a_1, \ldots, a_\omega) = 1, \tag{1.1}$$

where W is a word formed from the letters $x_1, \ldots, x_\lambda, a_1, \ldots, a_\omega$ and their inverse. A list of words

$$X_1, \ldots, X_\lambda \tag{1.2}$$

in the alphabet $a_1, \ldots, a_w, a_1^{-1}, \ldots, a_\omega^{-1}$ is called a *solution* of equation (1.1) if the word $W(X_1, \ldots, X_\lambda, a_1, \ldots, a_\omega)$ is equal to 1 in G.

In 1918 Nielsen [1], in studying the automorphisms of the free group on two generators a, b, arrived at the equation $xyx^{-1}y^{-1} = aba^{-1}b^{-1}$ in the free group and described the set of solutions of this equation. In 1959 Lyndon [2] proved the following: If elements A, B, C, of the free group $\mathcal{G}$ satisfy the equality $AABB = CC$, then they belong to a cyclic subgroup of $\mathcal{G}$. In 1960 Lyndon [3], [4] considered equations with one unknown in a free group and proved that all solutions of such equations can be represented in terms of a finite number of parametric words. In 1968 Lorents [5] simplified the structure of Lyndon's parametric words and obtained the definitive result: the set of solutions of any equation with one unknown in a free group consists of a finite set of words of the form $AB^\gamma C$, where A, B, C are elements of the free group and γ is a natural parameter. From the results of Nielsen [1] and Whitehead [6] one can obtain an algorithm for recognizing the solvability of equations with two unknowns of the form $f(x_1, x_2) = A$, where A is an element of the free group and the left-hand side is a word formed from the unknowns $x_1, x_2, x_1^{-1}, x_2^{-1}$. In 1971 Hmelevskii [7],[8] constructed an algorithm for recognizing the solvability of equations with two separated unknowns, $\varphi(x_1, a_1, \ldots, a_m) = \psi(x_2, a_1, \ldots, a_m)$, and systems of equations involving equations of the form $f(x_1, x_2) = A$ and equations with two separated unknowns.

In 1982 Makanin [9], [10] obtained a complete solution of the problem of solvability of equations in a free group. Makanin proved that if a given equation of the notational length δ has a solution in a free group, then the length of every component of a minimal (counting the length of the maximal component) solution does not exceed a number $\Upsilon(\delta)$, where $\Upsilon(\delta)$ is a recursive function. This statement gives a simple brute-force search algorithm recognizing the solvability of an arbitrary equation in a free group. The main notations and results of the papers [9] and [10] will be presented below.

Suppose $\mathcal{D}$ is the free semigroup on the paired alphabet of generators

$$a_1, \ldots, a_\omega, a_1^{-1}, \ldots, a_\omega^{-1}. \tag{1.3.}$$

By a generalized equation Ω in the paired alphabet (1.3) we mean any system consisting of the following seven parts.

1. **Coefficients.** A generalized equation Ω contains a list of nonempty noncontractible words in the alphabet (1.3):

$$A_1, A_2, \ldots, A_m \qquad (m \geq 0). \tag{2.1}$$

These words are called the *coefficients* of the generalized equation Ω.

2. **Word variables.** A generalized equation Ω contains a table of variables

$$\begin{aligned} &t, x_1, \ldots, x_n, x_{n+1}, \ldots, x_{2n}, l_1, \ldots, l_\rho, r_1, \ldots, r_\rho \\ &h_1, \ldots, h_{\rho-1}, b_1, \ldots, b_k, c_1, \ldots, c_k \qquad (n \geq 0, \rho \geq 1, k \geq 0) \end{aligned} \tag{2.2}$$

the values of which are words in the alphabet (1.3). To each variable x_i, b_i, c_i of table (2.2) there corresponds a nonemptiness condition

$$x_i > 0 \quad (i = 1, \ldots, 2n), \quad b_j > 0, c_j > 0 \quad (j = 1, \ldots, k) \tag{2.3}$$

To each pair of variables x_i, x_{i+n} there corresponds an equality of variables

$$x_i = x_{i+n} \quad (i = 1, \ldots, n) \tag{2.4}$$

3. **Bases.** A generalized equation Ω contains a function ε with the following domain and range:

$$\varepsilon : \qquad \{1, \ldots, 2n\} \rightarrow \{+1, -1\}.$$

The variables

$$x_1^{\varepsilon(1)}, \ldots, x_n^{\varepsilon(n)}, x_{n+1}^{\varepsilon(n+1)}, \ldots, x_{2n}^{\varepsilon(2n)} \tag{2.5}$$

are called the *bases* of the generalized equation. We define

$$\Delta(i) = \begin{cases} i + n & \text{if } 1 \leq i \leq n, \\ i - n & \text{if } 1 + n \leq i \leq 2n. \end{cases}$$

For each $i = 1, \ldots, 2n$ we call the base $x_i^{\varepsilon(i)}$ the *dual* of the base $x_{\Delta(i)}^{\varepsilon(\Delta(i))}$.

4. **Boundaries.** The variables

$$l_1, \ldots, l_\rho \tag{2.6}$$

in (2.2) are called the *boundaries* of the generalized equation Ω . To each boundary l_i there corresponds a boundary equality

$$l_i r_i = t \qquad (i = 1, \ldots, \rho). \tag{2.7}$$

A system of conditions

$$1 = l_1 < l_2 < \ldots < l_{\rho-1} < l_\rho = t \tag{2.8}$$

is called a *boundary order table*.

5. Objects. A generalized equation contains functions α and β with $\alpha(i) < \beta(i)$ for all i and the following domain and range:

$$\{1, \ldots, 2_n, 2_{n+1}, \ldots, 2_{n+m}\} \to \{1, \ldots, \rho\}. \tag{2.9}$$

The bases $x_1^{\epsilon(1)}, \ldots, x_n^{\epsilon(n)}, x_n^{\epsilon(n+1)}, \ldots, x_{2n}^{\epsilon(2n)}$ and the coefficients $A_1, \ldots, A_m$ are called the *objects* of the generalized equation Ω and are denoted respectively by

$$w_1, \ldots, w_n, w_{n+1}, \ldots, w_{2n}, w_{2n+1}, \ldots, w_{2n+m}. \tag{2.10}$$

To each object w_i in the list (2.10) there corresponds a situation equality

$$l_{\alpha(i)} w_i r_{\beta(i)} = t, \qquad \alpha(i) < \beta(i) \qquad (i = 1, \ldots, 2n + m). \tag{2.11}$$

We call $l_{\alpha(i)}$ the *left boundary* of the object w_i and $l_{\beta(i)}$ the *right boundary*. We say that the boundary l_j touches the left end of w_i if $j = \alpha(i)$ and touches the right end of w_i if $j = \beta(i)$. We say l_j cuts w_i if $\alpha(i) < j < \beta(i)$.

A boundary that cuts at least one object is called an *open* boundary. A boundary that does not cut any object is called a *closed* boundary. A generalized equation Ω contains ν ($\nu \geq 0$) open boundaries and σ ($\sigma \geq 1$) closed boundaries, so that

$$\rho = \nu + \sigma. \tag{2.12.}$$

6. Intervals. The variables

$$h_1, h_2, \ldots, h_{\rho-1} \tag{2.13}$$

in (2.2) are called the intervals of the generalized equation Ω. To each interval h_λ there corresponds an *interval situation equality*

$$l_\lambda h_\lambda r_{\lambda+1} = t \qquad (\lambda = 1, \ldots, \rho - 1). \tag{2.14}$$

A product of consecutive intervals

$$h_i h_{i+1}, \ldots, h_{j-2} h_{j-1} \qquad (i < j)$$

is called the $[i, j]$ section of the generalized equation. We say that the $[i, j]$ section belongs to the object w_p (or that w_p contains the $[i, j]$ section) if

$$\alpha(p) \leq i < j \leq \beta(p). \tag{2.15}$$

We say that the object w_p belongs to the $[i,j]$ section if

$$i \leq \alpha(p) < \beta(p) \leq j. \tag{2.16}$$

The $[i,j]$ section is called a closed section if the boundaries l_i and l_j are closed and all of the boundaries $l_{i+1}, \ldots, l_{j-1}$ are open. We say that the object w_i and w_j intersect if there is some interval h_u that belongs both w_i and w_j and to w_j.

7. **Boundary connections.** A generalized equation Ω contains k *boundary connections* $\Sigma_1, \ldots, \Sigma_k$. The number k here is equal to the number of variables b_i in (2.2). Each boundary connection Σ_i $(i = 1, \ldots, k)$ is a system of equalities

$$x_{\lambda_i}^{\epsilon(\lambda_i)} = b_i c_i, \tag{2.17.1}$$

$$l_{p_i} = l_{\alpha(\lambda_i)} b_i, \tag{2.17.2}$$

$$l_{q_i} = \begin{cases} l_{\alpha(\Delta(\lambda_i))} b_i & \text{if } \epsilon(\lambda_i) = \epsilon(\Delta(\lambda_i)), \\ l_{\alpha(\Delta(\lambda_i))} c_i^{-1} & \text{if } \epsilon(\lambda_i) = -\epsilon(\Delta(\lambda_i)), \end{cases} \tag{2.17.3}$$

which connects the boundaries l_{p_i} and l_{q_i} $(p_i \neq q_i)$ by means of a pair of dual bases $x_{\lambda_i}^{\epsilon(\lambda_i)}$ and $x_{\Delta(\lambda_i)}^{\epsilon(\Delta(\lambda_i))}$ such that

$$\alpha(\lambda_i) < p_i < \beta(\lambda_i), \quad \alpha(\Delta(\lambda_i)) < q_i < \beta(\Delta(\lambda_i)). \tag{2.17.4}$$

We will write the boundary connection Σ_i as a sequence

$$l_{p_i}, x_{\lambda_i}^{\epsilon(\lambda_i)}, x_{\Delta(\lambda_i)}^{\epsilon(\Delta(\lambda_i))}, l_{q_i} \tag{2.18.1}$$

or as a defining integral vector

$$(p_i, \lambda_i, q_i). \tag{2.18.2}$$

Boundary connections are distinct if their defining integral vectors are distinct.

In a generalized equation Ω all boundary connections $\Sigma_1, \ldots, \Sigma_\kappa$ are distinct.

A generalized equation containing only one boundary is called *trivial*. A trivial equation contains no coefficients, bases, intervals or boundary connections. It contains variables t, l_1, r_1, a closed boundary l_1, a boundary equality $l_1 r_1 = t$, and a boundary order table $1 = l_1 = t$.

Let us turn to the definition of a *solution* of a generalized equation.

A table of words in the alphabet (1.3)

$$T, X_1, \ldots, X_n, X_{n+1}, \ldots, X_{2n}, L_1, \ldots, L_\rho, R_1, \ldots, R_\rho,$$
$$H_1, \ldots, H_{\rho-1}, B_1, \ldots, B_k, C_1, \ldots, C_k \tag{2.19}$$

where n, ρ and k are equal to the corresponding numbers in the table of word variables (2.2) and the words

$$X_1, \ldots, X_n, X_{n+1}, \ldots, X_{2n} \tag{2.20}$$

are noncontractible, will be called a *form* of the generalized equation Ω. We will say that the form (2.19) satisfies a given word equality in the alphabet of variables (2.2) if upon substituting for each small Roman letter the word denoted in (2.19) by the corresponding capital letter we obtain a graphical equality. We will say that the form (2.19) satisfies a given word inequality in the alphabet of variables (2.2) if upon substituting for each small Roman letter the length of the word denoted in (2.19) by the corresponding capital letter we obtain a true numerical inequality.

A form (2.19) will be called a solution of the generalized equation Ω if it satisfies:

the nonemptiness conditions (2.3),

the equalities of variables (2.4),

the boundary equalities (2.7),

the boundary order table (2.8),

the object situation equalities (2.11),

the interval situation equalities (2.14), and

all boundary connections of Ω.

By the *length* of solution (2.19) of the equation Ω we mean the number $\delta(T)$. A solution $\mathcal{R}$ of Ω will be called minimal if no solution of Ω has length less than that of $\mathcal{R}$. By the *exponent of periodicity* of solution (2.19) we mean the exponent of periodicity of the list of words (2.20). A trivial equation has a unique solution of length 0.

It is clear from the definition in the preceding section that any generalized equation in a paired alphabet consists of an unchangeable structural part and a changeable parametric part. Any list of nonempty noncontractible words $A_1, \ldots, A_m$ in the alphabet (1.3.) uniquely determines the coefficients of a generalized equation, the numbers n, ρ and k uniquely determine the table of word variables, the function ε uniquely determines the list of bases, and the functions α and β uniquely determine the situation of objects. Each boundary connection Σ_i of the form (2.18.1) is uniquely determined by a vector of positive integers

$$(p_i, \lambda_i, q_i). \tag{3.1}$$

Thus, a set containing a list of words, $A_1, \ldots, A_m$, numbers n, ρ, k, functions ε, α, β, and k numerical vectors of type (3.1) uniquely determines a generalized equation. A generalized equation can be regarded as a function of this set.

The object situation equalities (2.11) uniquely determine the number ν of open boundaries and the number σ of closed boundaries. A generalized equation containing σ closed boundaries contains, by definition, $\sigma - 1$ closed sections. The number

$$\xi = n - (\sigma - 1), \tag{3.2}$$

where n is the number of pairs of bases of the bases of the equation and $\sigma - 1$ is the number of closed sections, will be called the *complexity* of the generalized equation. The number

$$d = \delta(A_1 \ldots A_m) + 1, \tag{3.3}$$

where $A_1, \ldots, A_m$ is the list of coefficients of the equation, will be called the *coefficient parameter* of the generalized equation. Let m be the number of coefficients and ρ the number of boundaries of a generalized equation.

By the *principal parameters* of a generalized equation we mean the parameters

$$A_1, \ldots, A_m \ ; \ m \ ; \ d \ ; \ n \ ; \ \rho \ ; \ \nu \ ; \ \sigma \ ; \ \xi.$$

We construct the function $\Upsilon(\delta)$. We remark $\nu \leq \rho$;

$$\sigma \leq \rho \leq \delta \quad ; \quad m \leq \delta \quad ; \quad \xi \leq n+1 \leq \delta+1 \quad ; \quad \leq \delta+1.$$

$$f_1(n, m, \rho) = (n+1)(2n\rho^2 + 1)\rho^{4n+2m+1}2^{2n(\rho^2+1)}.$$
$$f_2(\xi, m, \nu) = f_1(m + 3\xi, m, 2\xi + m + \nu + 1).$$
$$f_3(\xi, m, \nu) = 3^{(2\xi+m+2\nu+2)f_2(\xi, m+\nu, \nu)[(m+\nu)!](m+2\nu)}$$
$$f_2(\xi, m+\nu, \nu)[(m+\gamma)!](m+2\nu)(2\xi + m + 2\nu + 1)^2.$$
$$f_4(n, m, \rho) = \rho 2^{f_1(n,m,\rho)[m!](\rho n+m)}.$$
$$f_5(\xi, m, \nu) = f_4(3\xi + m, m, 2\xi + m + \nu + 1).$$
$$f_6(n, m, \rho) = f_4(3(\rho + 4n + 2m)^2, 5(\rho + 4n + 2m)^4, 2(\rho + 4n + 2m)^2).$$
$$f_7(m, \xi, n, \nu, s) = 8n^2(s+2)(n+m)(2n+m-2)f_6(n, m, 2\xi + m + \nu + 1).$$
$$f_8(\xi, m, \nu) = 3^{2\xi+m+\nu+1}(2\xi + m + \nu + 1)^2 f_3(\xi, m + 2\nu + 2, \nu).$$

The function $F_1(n, \xi, m, \nu, s, \pi)$ is given recursively by $F_1(0, \xi, m, \nu, s, \pi) = 1$ and for $n \geq 1$ the function $F_1(n, \xi, m, \nu, s, \pi)$ is expressible in terms of the function $F_1(n-1, \xi, m, \nu, s, \pi)$ with the help of the functions $F_2, F_3, F_4, F_5, F_6, F_7, F_8, F_9$.

$$F_6(n-1, \xi, m, \nu, s, \pi) = \pi(n-1)^2 3^{(2\xi+m+\nu+3)}(2\xi + m + \nu + 1)^2$$
$$F_1(n-1, \xi, m + 2(\nu + 1)(n-1), \nu, s, \pi 3^{(2\xi+m+\nu+3)}(2\xi + m + \nu + 1)^2).$$
$$F_7(n-1, \xi, m, \nu, s, \pi) = \pi 3^{12(\xi+m+\nu+n+1)^2}(\xi + m + \nu + n)^4$$
$$(F_6(n-1, \xi, m, \nu, s, \pi) + 1).$$
$$F_2(n-1, \xi, m, \nu, s, \pi) = 2n^2(s+2)(n-1)^2\pi 3^{(2\xi+m+\nu+4)}(2\xi + m + \nu + 1)^2$$
$$F_7(n-1, \xi, m, \nu, s, \pi) \, F_1^2(n-1, \xi,$$
$$m + 24(n-1)(\xi + m + \nu + n)^2, 12(\xi + m + \nu + n)^2, s, F_7(n-1, \xi, m, \nu, s, \pi)).$$

Let us set

$$F_9(n-1, \xi, m, \nu, s, \pi) = n + \sum_{i=1}^{\pi f_7(m, \xi, n, \nu, s)+1} (q(i) + 1),$$

where the function $q(\alpha_1, \ldots \alpha_k)$ $(\alpha_i \geq 1)$, along with $P(\alpha_1, \ldots \alpha_k)$ and $\Psi(\alpha_1, \ldots \alpha_k)$ $(\alpha_i \geq 1)$, is defined by a simultaneous recursion on the basis of the following equations:

$$q(\alpha_1, \ldots \alpha_{n-1}) = s + 1,$$

$$q(\alpha_1, \ldots, \alpha_k) = \sum_{i=1}^{\psi(\alpha_1, \ldots, \alpha_k)+1} (q(\alpha_1, \ldots, \alpha_k, i) + 1) \quad (1 \leq k \leq n-2),$$

$$P(\alpha_1) = \sum_{i=1}^{\alpha_1-1}(q(i)+1),$$

$$P(\alpha_1,\ldots,\alpha_k) = P(\alpha_1,\ldots,\alpha_{k-1}) + \sum_{i=1}^{\alpha_k-1}(q(\alpha_1,\ldots,\alpha_{k-1},i)+1) \quad (2 \le k \le n-2),$$

$$\Psi(\alpha_1,\ldots,\alpha_k) = F_2(n-1,\xi,m,\nu 2^{P(\alpha_1,\ldots,\alpha_k)},s,\pi 2^{P(\alpha_1,\ldots,\alpha_k)}) \quad (1 \le k \le n-2)$$

Here the function $q(\alpha_1,\ldots\alpha_k)$ is defined by a recursion with respect to a well-ordering $(\alpha_1,\ldots,\alpha_s) < (\beta_1,\ldots,\beta_r)$ among all vectors with $1 \le s,r \le n-1$, which is defined as a lexicographical ordering if we represent these vectors symbolically in the form of $(n-1)$-dimensional vectors $(\alpha_1,\ldots,\alpha_s,\infty,\ldots,\infty)$ and $(\beta_1,\ldots,\beta_r,\infty,\ldots,\infty)$. The functions $F_3(n-1,\xi,m,\nu,s,\pi)$, $F_8(n-1,\xi,m,\nu,s,\pi)$ are defined by the following recursions:

$$A(1) = F_9(n-1,\xi,m,\nu,s,\pi),$$

$$A(i+1) = F_9(n-1,\xi,m,\nu 2^{A(i)+1},s,\pi 2^{A(i)+1}) + A(i)+1 \quad (i=1,\ldots,n-1),$$

$$F_3(n-1,\xi,m,\nu,s,\pi) = A(n)+n,$$

$$F_8(n-1,\xi,m,\nu,s,\pi) = 2^{F_3(n-1,\xi,m,\nu,s,\pi)}$$

$$F_2(n-1,\xi,m,\nu 2^{F_3(n-1,\xi,m,\nu,s,\pi)},s,\pi 2^{F_3(n-1,\xi,m,\nu,s,\pi)}).$$

$$F_4(n-1,\xi,m,\nu,s,\pi) = 3^{(2\xi+m+\nu+2)}(2\xi+m+\nu+1)^2$$

$$F_1(n-1,\xi,m+2\nu+2,s,\pi 3^{(2\xi+m+\nu+3)}(2\xi+m+\nu+1)^2).$$

$$F_5(n-1,\xi,m,\nu,s,\pi) = 2^{m\,F_3(n-1,\xi,m,\nu,s,\pi)}$$

$$f_8(\xi,m+3\nu 2^{m\,F_3(n-1,\xi,m,\nu,s,\pi)}+2,$$

$$\nu 2^{m\,F_3(n-1,\xi,m,\nu,s,\pi)}+2).$$

$$F_1(n,\xi,m,\nu,s,\pi) = F_5(n-1,\xi,m,\nu,s,\pi)$$

$$F_4(n-1,\xi,3\nu 2^{m\,F_3(n-1,\xi,m,\nu,s,\pi)}+2,$$

$$\nu 2^{m\,F_2(n-1,\xi,m,\nu,s,\pi)},s,8F_5(n-1,\xi,m,\nu,s,\pi)) + F_8(n-1,\xi,m,\nu,s,\pi).$$

Moreover $F_1(n,\xi,m,\nu,s) = F_1(n,\xi,m,\nu,s,1)$.

Theorem. *If a generalized equation $\Omega(A_1,\ldots,A_m,n,\rho,d)$ has a minimal solution of length J and exponent of periodicity s, then*

$$J \le 3^p(\Psi_1(n,n,m,\rho-1,s)+\rho^2)d,$$

where

$$\Psi_1(n,\xi,m,\nu,s) = f_3(\xi,m,\nu)[F_1(n,\xi,m,\nu,s)+f_5(\xi,m,\nu)+1].$$

A proof of this theorem and of the following may be found in [9] and [10]. The latter follows from the former. The following result is a consequence of the fact that the solvability problem for equations in a free group can be reduced to the solvability problem for generalized equations.

Theorem. *If an equation with notational length δ in a free group has a minimal solution $X_1, \ldots, X_\lambda$, then for each $i = 1, \ldots, \lambda$ we have $\delta(X_1) \leq \Upsilon(\delta)$, where*

$$\Upsilon(\delta) = \Psi_2(\delta^3, \delta^3, \delta^3, (6\delta^3)^{2^{(2\delta^{12})}} + 2)(\delta + 1),$$

where

$$\Psi_2(m, n, \rho, s) = 3^\rho(\Psi_1(n, n, m, \rho - 1, s) + \rho^2).$$

The elementary theory of the free group $\mathcal{G}$ is the collection of true closed formulas of the restricted predicate calculus with equality in $\mathcal{G}$. It is known that every closed formula of the restricted predicate calculus with equality in $\mathcal{G}$ can be represented in prenex disjunctive normal form, that is, in the form

$$Q_1 x_1 \ldots Q_\lambda x_\lambda \left(\bigvee_{i=1}^q \left(\bigwedge_{j=1}^{r_j} W_{i,j}(x_1, \ldots, x_\lambda, a_1, \ldots, a_w) =\neq 1 \right) \right).$$

Here $Q_i x_i$ is a universal or existential quantifier, $x_1, \ldots, x_\lambda$ are variables, $a_1, \ldots, a_w$ are constants, and each atomic formula has the form $W_{i,j}(x_1, \ldots, x_\lambda, a_1, \ldots, a_w) = 1$ or $W_{i,j}(x1, \ldots, x_\lambda, a_1, \ldots, a_w) \neq 1$, where $W_{i,j}$ is a group-theoretic word in the indicated alphabet.

The problem of decidability of the elementary theory of free groups, raised by Tarski [12], is still an open question. The two theorems presented above are the basis for partial solutions which will be presented below.

The universal theory of a free group G is the collection of true formulas in the class of formulas of the form

$$\exists x_1 \ldots \exists x_\lambda \left(\bigvee_{i=1}^q \left(\bigwedge_{j=1}^{r_i} W_{i,j}(x_1, \ldots, x_\lambda, a_1, \ldots, a_w) =\neq 1 \right) \right).$$

The positive theory of a free group is the collection of true formulas in the class of formulas of the form

$$Q_1 x_1 \ldots Q_\lambda x_\lambda \left(\bigvee_{i=1}^q \left(\bigwedge_{j=1}^{r_i} W_{i,j}(x_1, \ldots, x_\lambda, a_1, \ldots, a_w) = 1 \right) \right).$$

In 1984 Makanin [10] proved

Theorem. *The universal theory of the free group $\mathcal{G}$ is decidable.*

It follows from the theorem given below that each positive formula is equivalent in free groups to a formula of the simpler form (4.1).

Theorem (A.I. Mal'tsev). *The equation*

$$x_1^2 a_1 x_1^2 a_1^{-1} = (x_2 a_2 x_2 a_2^{-1})^2$$

in the free group $\mathcal{G}$ is equivalent to the conjunction of two equations:

$$x_1 = 1 \ \& \ x_2 = 1.$$

Proof. Let X_1, X_2 be a solution. Any three elements of the free group $\mathcal{G}$ satisfying the equation $A^2 B^2 = C^2$ belong to a cyclic subgroup of $\mathcal{G}$. Hence, $X_1^2 a_1$ and a_1 belong to a cyclic subgroup, and therefore $X_1 = a_1^p$ for some p. If $p \neq 0$, then $a_1^{4p} = (X_2 a_2 X_2 a_2^{-1})^2$ and $a_1^{2p} = (X_2 a_2)^2 a_2^{-2}$, whence it would follow that the generators a_1 and a_2 belong to a cyclic subgroup. Hence $p = 0, X_1 = 1, X_2 a_2 X_2 a_2^{-1} = 1$, and $X_2 = 1$.

Theorem (G.A. Gurevich). *The conjunction of the four equations*

$$(x_1 a_1^\varepsilon x_1 a_1^{-\varepsilon})(x_1 a_1^\delta x_1 a_1^{-\delta}) = (x_2 a_2^\delta x_2 a_2^{-\delta})(x_2 a_2^\varepsilon x_2 a_2^{-\varepsilon}) \quad (\varepsilon, \delta = \pm 1)$$

in the free group $\mathcal{G}$ is equivalent to the disjunction of two equations: $x_1 = 1 \vee x_2 = 1$.

Proof. Assume that X_1, X_2, is a solution of the system and assume that $X_1 \neq 1, X_2 \neq 1$. Then $X_1 a_1^\varepsilon X_1 a_1^{-\varepsilon} \neq 1$ and $X_2 a_2^\delta X_2 a_2^{-\delta} \neq 1$ for all $\varepsilon, \delta = \pm 1$. Hence, there exists a nonidentity element S in G such that $X_1 a_1^\varepsilon X_1 a_1^{-\varepsilon}$ and $a_2^\delta X_2 a_2^{-\delta}$ are nonzero powers of S for all $\varepsilon, \delta = \pm 1$. Therefore,

$$(X_1 a_1 X_1 a_1^{-1})^p = (X_1 a_1^{-1} X_1 a_1)^q,$$
$$(X_2 a_2 X_2 a_2^{-1})^r = (X_2 a_2^{-1} X_2 a_2)^t$$

for certain nonzero p, q, r and t. It follows that $S = a_1^g$ for some $g \neq 0$, and it follows that $S = a_2^h$ for some $h \neq 0$. This is a contradiction.

Assume that a formula with arbitrary quantifier prefix

$$\exists x_1 \ldots \exists x_{p_1} \forall y_1 \exists x_{p_1+1} \ldots \exists x_{p_2} \forall y_2 \ldots \exists x_{p_{s-1}+1} \ldots \exists x_{p_s} \forall y_s$$
$$\exists x_{p_s+1} \ldots \exists x_n (W(x_1, \ldots, x_n, y_1, \ldots, y_s, a_1, \ldots, a_w) = 1) \tag{4.1}$$

contains in its quantifier-free part one equation in the free group, and assume that $x_1, \ldots, x_n$ and $y_1, \ldots, y_s$ are unknowns and $a_1, \ldots, a_w$ are coefficients in this equation. We construct the formula

$$\exists x_1 \ldots \exists x_{p_1} \exists x_{p_1+1} \ldots \exists x_{p_2} \ldots \exists x_{p_{s-1}+1} \ldots \exists x_{p_s}$$
$$\exists x_{p_s+1} \ldots \exists x_n (W(x_1, \ldots, x_n, b_1, \ldots, b_s, a_1, \ldots, a_w) = 1) \tag{4.2}$$

which contains one equation in the free group with unknowns $x_1, \ldots, x_n$ and coefficients $a_1, \ldots, a_w$ and $b_1, \ldots, b_s$, where the values of the unknowns can only be words in the alphabets indicated in the following conditions:

$$
\begin{aligned}
&x_1, \ldots, x_{p1} \in \{a_1, \ldots, a_w\}, \\
&x_{p_1+1}, \ldots, x_{p_2} \in \{a_1, \ldots, a_w, b_1\}, \\
&\ldots \\
&x_{p_{s-1}+1}, \ldots, x_{p_s} \in \{a_1, \ldots, a_w, b_1, \ldots, b_{s-1}\}, \\
&x_{p_{s+1}+1}, \ldots, x_n \in \{a_1, \ldots, a_w, b_1, \ldots, b_{s-1}, b_s\}.
\end{aligned}
\tag{4.3}
$$

Theorem (Yu.I.Merzlyakov). *Formula (4.1) is true if and only if formula (4.2) with conditions (4.3) on the unknowns, is true.*

The proof is contained in the proof of Theorem 1 in [11].

In 1984 Makanin [10] proved

Theorem. *The positive theory of the free group $\mathcal{G}$ is decidable.*

While proving the decidability of the positive theory, Makanin used a result of Merzlyakov [11] on the elimination of universal quantifiers in positive formulas.

In 1984 Razborov [13] constructed an algorithm which, given a coefficient-free equation in a free group, found the rank of the equation. (The rank of coefficient-free equation in a free group is the largest r such equation has a solution which generates a subgroup of rank r.) In constructing the algorithm he used Makanin's method of generalized equations for double-alphabet semigroups. Razborov also obtained a description of the set of solutions of bounded periodicity index for any equation in a free group by means of arbitrary paths in a certain finite graph constructed from the given equation.

BIBLIOGRAPHY

1. J. Nielsen, *Die Isomorphismen der allgemeinen, unendlichen Gruppe mit zwei Erzeugenden*, Math. Ann. 78 (1918), 385-397.

2. R. C. Lyndon, *The equation $a^2b^2 = c^2$ in free groups*, Michigan Math. J. 6 (1959), 89-95.

3. R. C. Lyndon, *Equations in free groups*, Trans. Amer. Math. soc. (1960), 445-457.

4. R. C. Lyndon, *Groups with parametrical exponents*, Trans. Amer. Math. Soc. 96 (1960), 518-533.

5. A. A. Lorents, *Representation of solution sets of systems of equations with one unknown in free groups*, Dokl. Akad. Nauk SSSR 178 (1968), 290-292: English transl. in Soviet Math. Dokl. 9 (1968).

6. J. H. C. Whitehead, *On equivalent sets of elements in a free group*, Proc. London Math. Soc. (2) 37 (1936), 782-800.

7. Yu. I. Hhmelevskii, *Systems of equations in a free group I*, Izv. Akad. Nauk SSR Ser. Mat. 35 (1971), 1237-1268; English transl. in Math. USSR Izv. 5 (1971).

8. Yu. I. Hhmelevskii, *Systems of equations in a free group II*, Izv. Akad. Nauk SSR Ser. Mat. 36 (1972), pp. 110-179; English transl. in Math. USSR Izv. 6 (1972).

9. G. S. Makanin, *Equations in a free group*, Izv. Akad. Nauk SSR Ser. Math. 46 (1982), 1199-1273; English transl. in Math. USSR Izv. 21 (1983).

10. G. S. Makanin, *Decidability of the universal and positive theories of a free group*, Izv. Akad. Nauk SSSR Ser. Math. 48 (1984), 735-749; English transl. in Math. USSR Izv. 25 (1985).

11. Yu. i. Merzlyakov, *Positive formulas on free groups*, Algebra i Logika 5 (1966), no.4, 25-42.

12. A. Tarskii, *Basic Research in the foundations of mathematics*, Report for the period January 15, 1957 - June 30, 1959.

13. A. A. Razborov, *On systems of equations in a free group*, Izv. Akad. Nauk SSSR Ser. Mat. 48 (1984), 779-832; English transl. in Math. USSR Izv. 25 (1985).

An analysis of Makanin's algorithm deciding solvability
of equations in free groups

Antoni Kościelski

University of Wrocław, Institute of Computer Science,

ul. Przesmyckiego 20, 51-151 Wrocław

Abstract

We give slightly simplified version of the proof of Makanin's theorem on decidability of solvability problem for equations in free groups. We also provide an analysis of the Makanin's proof and point out why the estimates on the complexity of Makanin's algorithm and on the length of a minimal solution of an equation in a free group are not primitive recursive.

I. Generalized equations

Let A be a fixed non-empty finite set. We call A an alphabet. Assume that for each $a \in A$, a unique a^{-1} is given such that a^{-1} does not belong to A and for $a, b \in A$, if $a \neq b$, then $a^{-1} \neq b^{-1}$. Let $\mathcal{A}$ be the union of A and $\{a^{-1} : a \in A\}$. The set $\mathcal{A}$ is called a double alphabet. By $\mathcal{A}^*$ we denote the set of all words in $\mathcal{A}$. If $w_1, w_2 \in \mathcal{A}^*$, then $w_1 w_2$ is the concatenation of w_1 and w_2. If $w \in \mathcal{A}^*$ and p is a non negative integer, then w^p is a concatenation of p copies of w. We denote the length of a word w by $lh(w)$. The symbol ε denotes the empty word. A word w is irreducible if it does not contain any subword of the form aa^{-1} and $a^{-1}a$ for any $a \in A$. We define the operation $^{-1}$ on $\mathcal{A}^*$ putting

1) $\varepsilon^{-1} = \varepsilon$,
2) $(a)^{-1} = a^{-1}$ and $(a^{-1})^{-1} = a$ for all $a \in A$,
3) $(w_1 w_2)^{-1} = w_2^{-1} w_1^{-1}$ for all $w_1, w_2 \in \mathcal{A}^*$.

The set $\mathcal{A}^*$ with the operation of concatenation and the operation $^{-1}$ is called a semigroup over the double alphabet $\mathcal{A}$.

Let $\leq$ be a linear order on a non-empty finite set X. The smallest and the greatest element of X will be denoted respectively by $\varnothing$ and ∞. pred(x) and succ(x) denote respectively the predecessor of x and the successor of x, if they exist. We shall use the standard notations to denote segments of X:

$[x, y] = \{z \in X: x \leq z \leq y\}$,

$[x, y) = \{z \in X: x \leq z < y\}$,

$(x, y) = \{z \in X: x < z < y\}$

for x, y $\in$ X. For a given linearly ordered set X we define the set IS(X) of all increasing sequences $\langle x_1, x_2, \ldots, x_t \rangle$ of elements of X with t > 1. A sequence $\langle x_1, x_2, \ldots, x_t \rangle \in$ IS(X) codes the partition of $[x_1, x_t)$ into sets $[x_1, x_2), [x_2, x_3), \ldots, [x_{t-1}, x_t)$. Also it codes the set of segments of the form $[x_1, x_i)$, where $1 < i \leq t$.

Now, we define a generalized equation in a semigroup over a double alphabet $\mathcal{A}$. A generalized equation is a quintuple $E = \langle L, D^+, D^-, F, W \rangle$ such that

1) L is a finite, non-empty, linearly ordered set; elements of L are called boundaries,

2) D^+ and D^- are two weakly antisymmetric disjoint relations in IS(L) such that the lengths of x and y are equal whenever $\langle x, y \rangle \in D^+ \cup D^-$; elements of $D^+ \cup D^-$ are called dependencies,

3) F is a function which maps $D^+ \cup D^-$ into the cartesian product $\mathcal{A} \times \mathcal{A}$,

4) W is a partial function, which maps the set $\{\langle x, y \rangle \in L \times L: x < y\}$ into the set of non-empty words in $\mathcal{A}$.

Given a function $R: L \setminus \{\infty\} \longrightarrow \mathcal{A}^*$ and x, y $\in$ L such that $x \leq y$. We put $R[x, y) = \varepsilon$ if $x = y$ and $R[x, y) = R(x)R[succ(x), y)$ if $x < y$. A function $R: L \setminus \{\infty\} \longrightarrow \mathcal{A}^*$ is a solution of a generalized equation $E = \langle L, D^+, D^-, F, W \rangle$ if

1) the values of R are non-empty words in $\mathcal{A}$,

2) if $\langle\langle x_1, x_2, \ldots, x_t \rangle, \langle y_1, y_2, \ldots, y_t \rangle\rangle \in D^+$, then

    a) $R[x_1, x_t)$ is irreducible,

    b) $R[x_1, x_i) = R[y_1, y_i)$ for i = 2,...,t,

3) if $\langle\langle x_1, x_2, \ldots, x_t \rangle, \langle y_1, y_2, \ldots, y_t \rangle\rangle \in D^-$, then

    a) $R[x_1, x_t)$ is irreducible,

    b) $(R[x_1, x_i))^{-1} = R[y_{t-i+1}, y_t)$ for i = 2,...,t,

4) if $d = \langle\langle x_1, x_2, \ldots, x_t \rangle, \langle y_1, y_2, \ldots, y_t \rangle\rangle \in D^+ \cup D^-$ and F(d) = $\langle a, b \rangle$, then a is the first and b is the last letter of $R[x_1, x_t)$,

5) if $\langle x, y \rangle$ belongs to the domain of W, then $W(\langle x,y \rangle) = R[x, y)$.

Assume that $E = \langle L, D^+, D^-, F, W \rangle$ is a generalized equation, $x \in L$ and $y \notin L$. Let $L_1 = (L \setminus \{x\}) \cup \{y\}$. We can define an ordering of L_1 and D_1^+, D_1^-, F_1, W_1 replacing x by y in an ordering of L and in D^+, D^-, F, W. It is obvious, that $E_1 = \langle L_1, D_1^+, D_1^-, F_1, W_1 \rangle$ so obtained is also a generalized equation and E has a solution if and only if E_1 has a solution. Similarly, if we replace a dependence $\langle i, j \rangle$ by $\langle j, i \rangle$ and replace F by F_1 such that F and F_1 have the same values for arguments different from $\langle i, j \rangle$, $\langle j, i \rangle$ and

$$F_1(\langle j, i \rangle) = \langle a, b \rangle \text{ if } \langle i, j \rangle \in D^+,$$
$$F_1(\langle j, i \rangle) = \langle b^{-1}, a^{-1} \rangle \text{ if } \langle i, j \rangle \in D^-,$$

where $\langle a, b \rangle = F(\langle i, j \rangle)$, then we get a generalized equation, which has a solution if and only if the original one had. So, if we study the problem of the existence of a solution of a generalized equation, then we identify two generalized equations, which have different sets of boundaries of the same cardinality and have corresponding sets of dependencies and the functions. Also we identify generalized equations, which differ only on dependencies $\langle i, j \rangle$ and $\langle j, i \rangle$ as described above.

Lemma 1.1. Given a set X of words in $\mathcal{A}$ there exist at most $S_1(1)$ generalized equations $E = \langle L, D^+, D^-, F, W \rangle$ such that X is the set of values of W and L has l boundaries, where

$$S_1(l) = 2^{2^{2^{2l + 1}}} \cdot l^{1^2} \cdot a^{2^{l + 1}} \quad \text{and a is the cardinality of } \mathcal{A}.$$

Proof. There are at most

- 2^l increasing sequences of elements of L (elements of $IS(L)$),
- 2^{2l} possible dependencies,
- $m_1 = 2^{2^{2l}}$ weakly antisymmetric relations in $IS(L)$,
- $1/2 \cdot (1^2 - 1)$ elements, which belong to dom W or belong to X,
- $[1/2 \cdot (1^2 - 1) + 1]^{1/2 \cdot (1^2 - 1)} \leq m_2 = 1^{1^2}$ partial functions which map $\{\langle x, y \rangle \in L^2 : x < y\}$ into X,
- $m_3 = a^{2^{l + 1}}$ functions which map $IS(L)$ into $\mathcal{A} \times \mathcal{A}$.

The number of generalized equations with a given range of W and with l boundaries is not greater than $m_1^2 \cdot m_2 \cdot m_3$. $\blacksquare$

Lemma 1.2. Given a set of words X, which has x elements there exist at most $S_2(1, x)$ generalized equations $E = \langle L, D^+, D^-, F, W \rangle$ such that values of W are elements of X and L has at most l boundaries, where

$$S_2(1, x) = 1 \cdot 2^x \cdot S_1(1).$$

Proof. There exists no more than 2^X possible sets of values of W. Now, the lemma follows from Lemma 1.1. ∎

II. INTRODUCTION

This paper concerns the solvability of equations in finitely generated free groups. These groups can be constructed by an identification of all subwords of the form aa^{-1} and $a^{-1}a$ with the empty word in semigroups over a double alfabet. In [MA2] and [MA3] Makanin proved the decidability of the solvability problem for equations in such groups. His proof consists of three parts. In the first part he proves that the solvability problem for equations in free groups can be reduced to the problem of existence a non empty irreducible solution of systems of equations in semigroups over a double alfabet. Next Makanin introduces the generalized equations and proves that the problem of existence of a solution for such systems can be reduced to the problem of solvability of generalized equations. A very large part of [MA2] and [MA3] contains a description and a proof of correctness of an algorithm deciding the solvability of a generalized equation. It should be noticed that the notion of a generalized equation in [MA1] is different from the notion which is used in [MA2]. The problem of generating solutions of equations in free group have been studied in [RAZ].

We give a description and a proof of correctness of a slightly simplified version of the Makanin's algorithm deciding solvability of generalized equations which are defined in Chapter I. We hope that the paper is sufficiently completed. The main part of the proof is presented in Chapter X.

Chapter III contains a large part of all definitions and notations which are used in the paper. The next chapter contains descriptions of simple methods of a transforming generalized equations. These methods are called elementary. The Makanin's algorithm is based on repeated execution of the reducing transformation RTR or the main transformation MTR in order to decrease the complexity (cpl, see Chapter III) of a generalized equation. The definitions of these transformations are presented in Chapter V and Chapter VI. There are also given proofs that these transformations do not increase cpl. In Chapter V also we give bounds on the number of consecutive executions of RTR successively. Next we study properties of generalized equations which we get as results of RTR. We prove that MTR preserves these properties. In Chapter VIII we consider linear generalized equations. We prove there that the solvability problem for such equations is decidable and we estimate the length of a minimal solution R of a generalized equation E by the excess exc(E, R) (see Chapter III). This result is used in Chapter IX. Chapter IX contains important combinatorial results, which are used in the proof of the Makanin's theorem.

The final chapter contains a proof that the function which estimates the number of steps of Makanin's algorithm is not primitive recursive. In [KP2] was proved that this result is not a consequence of insufficiently precise estimations. Rather, it is a consequence of a small knowledge on the Makanin's algorithm and on equations in free groups. It is also possible that the solvability problem for free groups is one of the most complicated natural problems. The results of [KP1] and [KP3] gave better estimates on the number of steps in the Makanin's algorithm deciding solvability of equations in free semigroup. Combined with an observation make in [MA2] these results could also be applied to improve bound in the case of groups. However, in this case the improvement is not a substantial one.

I would like to express my great gratitude to Leszek Pacholski for encouraging me to write this paper and especially for helping me with the redaction.

The notion of a generalized equation which is defined in Chapter I is similar but different to one which is defined in [MA2]. So, it is necessary to study a connection between systems of equations in semigroups over a double alfabet and generalized equations. Similarly as in [MA2] we can prove the following theorem.

Theorem 2.1. For every system of equations $\mathcal{E}$ in a semigroup over a double alfabet $\mathcal{A}$ there exists a sequence of generalized equations $E_1, \ldots, E_n$ such that $\mathcal{E}$ has a solution composed of non-empty irreducible words if and only if some equation among $E_1, \ldots, E_n$ has a solution.
Proof. We give only a sketch of a proof without given a precise definition of some notions, which apply to semigroups over a double alfabet. Let $\mathcal{E}$ be a system which contains equations

$$l_1 = r_1, \ldots, l_m = r_m.$$

Every equation in a semigroup over a double alfabet is equivalent to one in which the operation $^{-1}$ is applied only to variables. So, we can assume that $l_1, \ldots, l_m, r_1, \ldots, r_m$ are words on a double alfabet, which includes $\mathcal{A}$ and all variables. Let X be the set of words of the form

$l_1 \ldots l_{i-1} w$, where w is a prefix of l_i or

$r_1 \ldots r_{i-1} w$, where w is a prefix of r_i.

Let x be the cardinality of X. Let f be a function maps X into non negative integers less than x such that

$f(l_1 \ldots l_i) = f(r_1 \ldots r_i)$ for $i \leq m$ and

w_1, $w_2 \in X$, $w_1 \neq w_2$ and w_1 is a prefix of w_2 implies $f(w_1) < f(w_2)$.

For such f and for a given function F we define $E_{f,F} = \langle L_f, D_f^+, D_f^-, F, W_f \rangle$ as follows:

L_f is the set of values of f with the natural ordering,

D_f^+ is the set of pairs of the form $\langle\langle f(w_1), f(w_1 z)\rangle, \langle f(w_2), f(w_2 z)\rangle\rangle$ or $\langle\langle f(w_1), f(w_1 z^{-1})\rangle, \langle f(w_2), f(w_2 z^{-1})\rangle\rangle$, where z is a variable, given arguments are in X and $f(w_1) \leq f(w_2)$,

D_f^- is the set of pairs of the form $\langle\langle f(w_1), f(w_1 z)\rangle, \langle f(w_2), f(w_2 z^{-1})\rangle\rangle$, where z is a variable and $w_1 z$, $w_2 z^{-1} \in X$,

dom W_f is the set of pairs $\langle f(w), f(wa)\rangle$, where wa $\in$ X and a $\in$ $\mathcal{A}$,

$W_f(\langle f(w), f(wa)\rangle) = a$.

The sequence E_1, ..., E_n consists of all $E_{f,F}$, which are generalized equations.

Assume that R is a solution of $\mathcal{E}$. Let f be a function on X such that f(w) is the cardinality of $\{lh(R(w_1)) : w_1 \in X$ and $lh(R(w_1)) < lh(R(w))\}$. Let $R_f(f(w))$ be the word such that $R(w)R_f(f(w)) = R(w_1)$, where $w_1 \in X$ satisfies $f(w_1) = f(w) + 1$. It is not difficult to see that we can define F such that $E_{f,F}$ is a generalized equation and R_f is a solution of $E_{f,F}$. On the other hand if R_f is a solution of $E_{f,F}$ and

$R(z) = R_f[f(w), f(wz))$

or $R(z) = (R_f[f(w), f(wz^{-1})))^{-1}$,

where z is a variable and respectively wz $\in$ X or $wz^{-1} \in$ X, then the function R is a solution of $\mathcal{E}$. ∎

III. Basic notions and notations

Chapter III contains most definitions of main notions or notations which are used in this paper. We hope this solution will make reading the paper easier. To find the definition of a notion or a notation one should first read this chapter and Chapter I. If a definitions is not here it should be found near the place, where it is used. Moreover, at the end of this paper there is an index, which contains all notions and symbols except ones defined in Chapter I and Chapter III.

We often consider several equations simultaneously. Then we denote them by E with different subscript or superscript. In such a case symbols denoting sets of boundaries, sets of dependencies and the other notions connected with a generalized equation will be created from letters given in the definition of a generalized equation by adding the same subscript or superscript which is added to E. For example, L_1 is the set of boundaries of E_1 and D_2^+ is the set of dependencies of E_2. A similar principle is applied to symbols and notions that will be introduced later.

Let $E = <L, D^+, D^-, F, W>$ be a generalized equation. A dependence $d = <<x_1, x_2, \ldots, x_n>, <y_1, y_2, \ldots, y_n>>$ is called **essential** if $[x_1, x_n) \neq [y_1, y_n)$. We say that d **applies to** $[u, w)$ if $[u, w) = [x_1, x_n)$ or $[u, w) = [y_1, y_n)$. A segment $[u, w)$ is **an object of** d if d applies to $[u, w)$. A segment $[u, w)$ is called **an object** if there exists $d \in D^+ \cup D^-$ such that $[u, w)$ is an object of d. An object $[x_1, x_n)$ is called the **first object** of d, $[y_1, y_n)$ is the **second** one. An object $[u, w)$ is called **essential** if there is an essential dependence d which applies to $[u, w)$. A segment $[u, w)$ is called **a coefficient** if $<u, w> \in dom\ W$. A dependence d **applies to** x if x belongs to an object of d. A coefficient $[u, w)$ **applies to** x if $x \in [u, w)$.

A boundary x is **inside** an object or a coefficient $[u, w)$ if $x \in (u, w)$. A boundary x is **inner** if there exists an object or a coefficient $[u, w)$ such that x is inside $[u, w)$. If a boundary x is not inner, then it is called a **closed**. A segment $[u, w)$ is **closed** if u and w are closed and there does not exist any closed boundary in (u, w). A segment $[u, w)$ **lies near** x if there exists a closed segment $[u_1, w_1)$ such that $[u, w) \subseteq [u_1, w_1)$ and $x \in [u_1, w_1)$.

We put

dep	is the number of dependencies,
essdep	is the number of essential dependencies,
bnd	is the number of boundaries,
inn	is the number of inner boundaries,
cls	is the number of closed boundaries less than ∞,
mcff	is the maximal length of a value of W,
essdep[u, w)	is the number of essential dependencies which apply to $[u, w)$,

essdep(x)   is $\sum\limits_{x \,\in\, [u, \,w)}$ essdep[u, w),

essdin[x, y) is $\sum\limits_{[u, \,w) \,\subseteq\, [x, \,y)}$ essdep[u, w),

essdpc(x)   is the sum of essdep(x) and the number of coefficients which apply to x.

near(x)   is the sum of numbers essdep[u, w), where [u, w) is an essential object lying near x,

rng(x)   is 0 if x is inner or near(x) = 0,

               is 1/2 if x is closed and near(x) = 1,

               is 1 if x is closed and near(x) > 1,

rng[u, w)   is $\sum\limits_{x \,\in\, [u, \,w)}$ rng(x),

rng   is rng[ø, ∞)

cpl   is essdep - rng.

The main boundary, denoted by **main**, is the greatest x such that [ø, x) is an essential object. [ø, **main**) is the **main object**. If the main boundary exists we define the main dependence **mndep** and the cut boundary **ctb**. The main dependence is chosen in a certain way from essential dependencies which apply to the main object. The cut boundary is the smallest boundary in the main object which belong to any segment [u, w) such that **main** is inside [u, w) and [u, w) is an essential object or a coefficient. The cut boundary is **main** if such segment does not exist.

A generalized equation E is **consistent** or **solvable** if there exists a solution of E. Let R be a solution of E. The length of R is lh(R[ø, ∞)). If E is consistent, then **lhms** is the minimal length of solutions of E. Notice that lhms depends only on E. The **periodicity exponent** of R is the greatest number p such that there exist an object [x, y), words v_1, v_2 and a non-empty word w for which

$$R[x, \,y) = v_1 w^p v_2.$$

We shall denote the periodicity exponent of R by **prd**. A generalized equation E is **evidently inconsistent** if there exists either

a) a dependence in D^+ with different coordinates whose objects are ordered by inclusion or,

b) a dependence in D^- with non-disjoint objects.

It is not difficult to see that

Lemma 3.1. If a generalized equation is evidently inconsistent, then it is not consistent.

Proof. Suppose that E is an evidently inconsistent generalized equation and R is a solution of E. If a) is satisfied, then there exists a boundary x such that $R(x) = \varepsilon$. If $[x, y)$ is an object of a dependence which satisfies b) or c), then $R[x, y)$ is not irreducible. ∎

Let R be a solution of a generalized equation E. The number

$$2 \cdot (\sum_{\substack{d \in D^+ \cup D^- \\ [x, y) \text{ is first object of d}}} lh(R[x, y)) \quad - \quad lh(R[\emptyset, \infty))).$$

is called an **excess** of an equation E with a solution R and is denoted by **exc(E, R)**.

IV. ELEMENTARY TRANSFORMATIONS

In this chapter we define some transformations of generalized equations. Definitions of these transformations will be given as descriptions of non deterministic algorithms which construct a quintuple E_{out}. If E_{out} can not be constructed, then these algorithms output "E is unsolvable". It may also happen that E_{out} is not a generalized equation, for example if $D^+_{out} \cap D^-_{out}$ is non empty. In this case elementary transformations should output "E is unsolvable". However, we shall not write down these parts of algorithms which check that the object obtained are generalized equations. The result of a transformation will be a finite set of generalized equations. This set contains all generalized equations which we can construct in accordance with the description of a transformation. Transformations (except ADB) will be applied also to pairs <E, R>, such that E is a generalized equation and R is a solution of E. In such case the result of transformation will be a pair $<E_1, R_1>$. We say about such transformations that they are consistently with R. We write $E_1 := TR(E, p1, p2, \ldots)$ if E_1 is in set which is obtained by applying the transformation TR with parameters p1, p2, ... to an equation E. Sometimes we omit some parameters in such notation if it does not lead to misunderstanding. If E is a generalized equation, then $E_1 := TR(E)$ implies E_1 is a generalized equation. The transformations which we will define now are called elementary.

Transformation ADB: addition of a boundary after x
Input: a generalized equation $E = \langle L, D^+, D^-, F, W \rangle$,
    $x \in L \setminus \{\infty\}$.

Let y be any element, which does not belong to L. Let $L_1 = L \cup \{y\}$ be ordered by the extension of the order of L such that the element y is between x and succ(x). We put $E_{out} = \langle L_1, D^+, D^-, F, W \rangle$.□

Transformation DVC: division of a coefficient
Input: a generalized equation $E = \langle L, D^+, D^-, F, W \rangle$,
    $\langle x, z \rangle \in$ dom W, $y \in (x, z)$.

Step 1. We choose two non-empty words w_1, w_2 such that
    $W(\langle x, z \rangle) = w_1 w_2$,
    if $\langle x, y \rangle \in$ dom W, then $W(\langle x, y \rangle) = w_1$,
    if $\langle y, z \rangle \in$ dom W, then $W(\langle y, z \rangle) = w_2$.
If such words do not exist, then we output "E is unsolvable".
Step 2. We define a function W_1 as follows
    dom $W_1 = (\text{dom } W \setminus \{\langle x, z \rangle\}) \cup \{\langle x, y \rangle, \langle y, z \rangle\}$,
    $W_1(\langle x, y \rangle) = w_1$, $W_1(\langle y, z \rangle) = w_2$.
    W_1 and W have the same values for the other arguments.
$E_{out} = \langle L, D^+, D^-, F, W_1 \rangle$.

If we transform E consistently with R, then in step 1 we choose
    $w_1 = R[x, y]$, $w_2 = R[y, z]$.
In this case, the result of DVC is $\langle E_{out}, R \rangle$.□

Transformation DVD: division of a dependence
Input: a generalized equation $E = \langle L, D^+, D^-, F, W \rangle$,
    $d = \langle\langle x_1, \ldots, x_n \rangle, \langle y_1, \ldots, y_n \rangle\rangle \in D^+ \cup D^-$,
    x_i for some i such that $1 < i < n$.

Let $F(d) = \langle a, b \rangle$. If $d \in D^+$, then we define
    $d_1 = \langle\langle x_1, \ldots, x_i \rangle, \langle y_1, \ldots, y_i \rangle\rangle$,
    $d_2 = \langle\langle x_i, \ldots, x_n \rangle, \langle y_i, \ldots, y_n \rangle\rangle$,
otherwise
    $d_1 = \langle\langle x_1, \ldots, x_i \rangle, \langle y_{n-i+1}, \ldots, y_n \rangle\rangle$,
    $d_2 = \langle\langle x_i, \ldots, x_n \rangle, \langle y_1, \ldots, y_{n-i+1} \rangle\rangle$.

Step 1: We choose a_1, $b_1 \in \mathcal{A}$ such that $a_1 b_1$ is irreducible. If $d_1 \in D^+ \cup D^-$ and $F(d_1) \neq \langle a, a_1 \rangle$ or $d_2 \in D^+ \cup D^-$ and $F(d_2) \neq \langle b_1, b \rangle$, then we output "E is unsolvable".

Step 2: If $d \in D^+$, then we define
$$D_1^+ = (D^+ \setminus \{d\}) \cup \{d_1, d_2\}, \quad D_1^- = D^-,$$
otherwise
$$D_1^- = (D^- \setminus \{d\}) \cup \{d_1, d_2\}, \quad D_1^+ = D^+.$$
Moreover, we put
$$\text{dom } F_1 = D_1^+ \cup D_1^-, \quad F_1(d_1) = \langle a, a_1 \rangle, \quad F_1(d_2) = \langle b_1, b \rangle,$$
$$F_1(d') = F(d') \text{ for the other } d' \in \text{dom } F_1.$$
$$E_{out} = \langle L, D_1^+, D_1^-, F_1, W \rangle.$$

If we transform E consistently with R, then we choose

the last letter of $R[x_1, x_i)$ as a_1 and

the first letter of $R[x_i, x_n)$ as b_1.

In this case the result of DVD is $\langle E_{out}, R \rangle$. $\square$

Transformation TRB: transfer of a boundary

Input: a generalized equation $E = \langle L, D^+, D^-, F, W \rangle$,
$$d = \langle\langle x_1, x_2, \ldots, x_n \rangle, \langle y_1, y_2, \ldots, y_n \rangle\rangle \in D^+ \cup D^-, \quad x \in [x_1, x_n).$$

Case 1: $x = x_i$. In this case $E_{out} = E$ and if we transform E consistently with R, then we obtain the pair $\langle E, R \rangle$.

Case 2: $x \in (x_i, x_{i+1})$.

Step 1. If $d \in D^+$, then we choose $z \in [y_i, y_{i+1})$, otherwise we choose $z \in [y_{n-i}, y_{n-i+1})$.

Step 2. We put $E_1 = E$ and $y = z$ or $E_1 := ADB(E, z)$ and y is the new boundary in E_1. The first possibility is used only if $z \neq y_i$ or $z \neq y_{n-i}$.

Step 3. If $d \in D^+$, then
$$d_1 = \langle\langle x_1, \ldots, x_i, x, \ldots, x_n \rangle, \langle y_1, \ldots, y_i, y, \ldots, y_n \rangle\rangle,$$
otherwise
$$d_1 = \langle\langle x_1, \ldots, x_i, x, \ldots, x_n \rangle, \langle y_1, \ldots, y_{n-i}, y, \ldots, y_n \rangle\rangle.$$
If $d_1 \in D^+ \cup D^-$ and $F(d) \neq F(d_1)$, then we output "E is unsolvable".

Step 4. We put

if $d \in D^+$, then $D_2^+ = (D^+ \setminus\{d\}) \cup \{d_1\}$ and $D_2^- = D^-$,

if $d \in D^-$, then $D_2^- = (D^- \setminus\{d\}) \cup \{d_1\}$ and $D_2^+ = D^+$,

$\text{dom } F_2 = D_2^+ \cup D_2^-$,

$F_2(d_1) = F(d)$, $F_2(d') = F(d')$ for the other $d' \in$ dom F_2.

$E_{out} = <L_1, D_2^+, D_2^-, F_2, W>$.

If we transform E consistently with R, then in step 1 we choose the greatest boundary z such that

$R[x_1, x) = R[y_1, z)A$ if $d \in D^+$,

$A(R[x_1, x))^{-1} = R[z, y_n)$ if $d \in D^-$

for some word A. Let B be the word such that $R(z) = AB$. We execute ADB in step 2 if A is non-empty. If A is empty, then $R_{out} = R$, otherwise we put

$R_{out}(z) = A$, $R_{out}(y) = B$,

R_{out} and R have the same values for the other arguments.

Now, the result of TRB is $<E_{out}, R_{out}>$.□

Transformation TRC: transfer of a coefficient $<x_i, x_j>$ over d

<u>Input</u>: a generalized equation $E = <L, D^+, D^-, F, W>$,

$d = <<x_1, x_2, \ldots, x_n>, <y_1, y_2, \ldots, y_n>> \in D^+ \cup D^-$,

$<x_i, x_j> \in$ dom W.

<u>Step 1.</u> If $d \in D^+$, $<y_i, y_j> \in$ dom W and $W(<y_i, y_j>) \neq W(<x_i, x_j>)$ or if $d \in D^-$, $<y_{n-i+1}, y_{n-j+1}> \in$ dom W and $W(<y_{n-i+1}, y_{n-j+1}>) \neq W(<x_i, x_j>)$, then we output "E is unsolvable".

<u>Step 2.</u> If $d \in D^+$, we put

dom $W_1 = ($dom $W \setminus \{<x_i, x_j>\}) \cup \{<y_i, y_j>\}$,

$W_1(<y_i, y_j>) = W(<x_i, x_j>)$,

if $d \in D^-$, then we put

dom $W_1 = ($dom $W \setminus \{<x_i, x_j>\}) \cup \{<y_{n-j+1}, y_{n-i+1}>\}$,

$W_1(<y_{n-j+1}, y_{n-i+1}>) = W(<x_i, x_j>)$.

Moreover, in both cases, W_1 and W have the same values for the other arguments.

$E_{out} = <L, D^+, D^-, F, W_1>$.

If we transform E consistently with R, then we get the pair $<E_{out}, R>$.□

Transformation TRD: transfer of a dependence d_1 over d

<u>Input</u>: a generalized equation $E = <L, D^+, D^-, F, W>$,

$d = <<x_1, x_2, \ldots, x_n>, <y_1, y_2, \ldots, y_n>> \in D^+ \cup D^-$,

$d_1 = <<u_1, \ldots, u_m>, <v_1, \ldots, v_m>> \in D^+ \cup D^-$ such that

$$\{v_1, \ldots, v_m\} \subseteq \{x_1, \ldots, x_n\}.$$

Assume that $v_i = x_{k_i}$. If $d \in D^+$, then we define

$$d_2 = \langle\langle u_1, \ldots, u_m \rangle, \langle y_{k_1}, \ldots, y_{k_m} \rangle\rangle,$$

and otherwise we define

$$d_2 = \langle\langle u_1, \ldots, u_m \rangle, \langle y_{n-k_m+1}, \ldots, y_{n-k_1+1} \rangle\rangle.$$

Step 1. If $d_2 \in D^+ \cup D^-$ and $F(d_2) \neq F(d_1)$, then we output "E is unsolvable".

Step 2. We define D_1^+, D_1^- and F_1 as follows

Case 1: $d, d_1 \in D^+$

$$D_1^+ = (D^+ \setminus \{d_1\}) \cup \{d_2\} \text{ and } D_1^- = D^-.$$

Case 2: $d \in D^+$ and $d_1 \in D^-$

$$D_1^- = (D^- \setminus \{d_1\}) \cup \{d_2\} \text{ and } D_1^+ = D^+.$$

Case 3: $d \in D^-$ and $d_1 \in D^+$

$$D_1^+ = (D^+ \setminus \{d_1\}) \text{ and } D_1^- = D^- \cup \{d_2\}.$$

Case 4: $d, d_1 \in D^-$

$$D_1^- = D^- \setminus \{d_1\} \text{ and } D_1^+ = D^+ \cup \{d_2\}.$$

$$\text{dom } F_1 = D_1^+ \cup D_1^-,$$

$$F_1(d_2) = F(d_1), \ F_1(d') = F(d') \text{ for the other } d' \in \text{dom } F_1.$$

$$E_{out} = \langle L, D_1^+, D_1^-, F_1, W \rangle.$$

If we transform E consistently with R, then we obtain the pair $\langle E_{out}, R \rangle$. □

Transformation ELB: elimination of a boundary

Input: a generalized equation $E = \langle L, D^+, D^-, F, W \rangle$,

$x \in L \setminus \{\infty\}$ such that x does not belong to any object and coefficient of E.

Let $L_1 = L \setminus \{x\}$. The order on L_1 is the restriction of the ordering of L. We obtain D_1^+, D_1^-, dom W_1 and dom F_1 by replacing all elements of the form $\langle y_1, \ldots, y_n, x \rangle$ by $\langle y_1, \ldots, y_n, \text{succ}(x) \rangle$. The change of arguments of W_1 and F_1 does not cause the change of values.

$$E_{out} = \langle L_1, D_1^+, D_1^-, F_1, W_1 \rangle.$$

If we transform E consistently with R we get the pair $\langle E_{out}, R_{out}\rangle$ such that R_{out} is the restriction of R to L_1.□

Transformation ELC: elimination of a coefficient
Input: a generalized equation $E = \langle L, D^+, D^-, F, W\rangle$,
  $\langle x, y\rangle \in$ dom W such that $[x, y)$ is disjoint with all objects and the other coefficients of E.

Step 1. We define W_1 as the restriction of W to dom W $\setminus \{\langle x, y\rangle\}$ and E_1 as $\langle L, D^+, D^-, F, W_1\rangle$. If we transform E consistently with R, then we obtain the pair $\langle E_1, R\rangle$.
Step 2. We transform E_1 using ELB until we eliminate all boundaries from $[x, y)$. E_{out} is an equation obtained in this way.□

Transformation ELDC: elimination of a dependence
Input: a generalized equation $E = \langle L, D^+, D^-, F, W\rangle$,
  $d = \langle\langle x_1, \ldots, x_n\rangle, \langle y_1, \ldots, y_n\rangle\rangle \in D^+ \cup D^-$ such that $\langle x_1, x_n\rangle \in$ dom W.

Step 1: If either
  1) $F(d) = \langle a, b\rangle$ and a is not the first letter of $W(\langle x_1, x_n\rangle)$
     or b is not the last letter of $W(\langle x_1, x_n\rangle)$,
  2) or $W(\langle x_1, x_n\rangle)$ is not irreducible,
then we output "E is unsolvable".
Step 2: We divide a coefficient $\langle x_1, x_n\rangle$ in $x_2, \ldots, x_{n-1}$ using DVC. Suppose that we get the equation $E_1 = \langle L, D^+, D^-, F, W_1\rangle$.
Step 3: If for some $i < n$
  $d \in D^+$, $\langle y_i, y_{i+1}\rangle \in$ dom W_1 and $W_1(\langle x_i, x_{i+1}\rangle) \neq W_1(\langle y_i, y_{i+1}\rangle)$
  or $d \in D^-$, $\langle y_{n-i}, y_{n-i+1}\rangle \in$ dom W_1 and $W_1(\langle x_i, x_{i+1}\rangle) \neq$
     $(W_1(\langle y_{n-i}, y_{n-i+1}\rangle))^{-1}$,
then we output "E is unsolvable".
Step 4: We define
  dom W_2 = dom $W_1 \cup \{\langle y_1, y_2\rangle, \ldots, \langle y_{n-1}, y_n\rangle\}$,
  if $d \in D^+$, then $W_2(\langle y_i, y_{i+1}\rangle) = W_1(\langle x_i, x_{i+1}\rangle)$, otherwise
  $W_2(\langle y_i, y_{i+1}\rangle) = (W_1(\langle x_{n-i}, x_{n-i+1}\rangle))^{-1}$ for $i < n$,
  the values of W_2 and W_1 are the same for the other arguments,
  if $d \in D^+$, then $D_2^+ = D^+ \setminus \{d\}$ and $D_2^- = D^-$, otherwise

$$D_2^+ = D^+ \text{ and } D_2^- = D^- \setminus \{d\},$$
$$F_2 \text{ is the restriction of F to } D_2^+ \cup D_2^-.$$
$$E_{out} = \langle L, D_2^+, D_2^-, F_2, W_2 \rangle.$$

If we transform E consistently with R, then we get the pair $\langle E_{out}, R \rangle$. □

Transformation ELED: elimination of an essential dependence

Input: a generalized equation E = $\langle L, D^+, D^-, F, W \rangle$,

$d = \langle\langle x_1, x_2, \ldots, x_n \rangle, \langle y_1, y_2, \ldots, y_n \rangle\rangle \in D^+ \cup D^-$ such that $[y_1, y_n)$ is disjoint with $[x_1, x_n)$ all coefficients and objects of the other dependencies,

$[y_1, y_n) = \{y_1, \ldots, y_{n-1}\}.$

Step 1. If $\langle\langle x_1, x_n \rangle, \langle x_1, x_n \rangle\rangle \in D^+$ and $F(\langle\langle x_1, x_n \rangle, \langle x_1, x_n \rangle\rangle) \neq F(d)$, then we output "E is unsolvable".

Step 2. If $d \in D^+$, then we define

$$D_1^+ = (D^+ \setminus \{d\}) \cup \{\langle\langle x_1, x_n \rangle, \langle x_1, x_n \rangle\rangle\} \text{ and } D_1^- = D^-$$

and in the other case we define

$$D_1^+ = D^+ \cup \{\langle\langle x_1, x_n \rangle, \langle x_1, x_n \rangle\rangle\} \text{ and } D_1^- = D^- \setminus \{d\}.$$

Moreover,

$$\text{dom } F_1 = D_1^+ \cup D_1^-,$$
$$F_1(\langle\langle x_1, x_n \rangle, \langle x_1, x_n \rangle\rangle) = F(d),$$
$$F_1(d') = F(d') \text{ for the other } d' \in \text{dom } F_1.$$

Let $E_1 = \langle L, D_1^+, D_1^-, F_1, W \rangle$. If we transform E consistently with R, then we obtain the pair $\langle E_1, R \rangle$.

Step 3. We execute the transformation ELB until we eliminate all boundaries from $[y_1, y_n)$. □

Transformation ELND1: elimination of a non-essential dependence

Input: a generalized equation E = $\langle L, D^+, D^-, F, W \rangle$,

a non-essential dependence $d = \langle\langle x, \text{succ}(x) \rangle, \langle x, \text{succ}(x) \rangle\rangle \in D^+$, which applies to an object which is disjoint with all coefficient and objects of the other dependencies.

Step 1. If d has only two letters and $F(d) = \langle a, a^{-1} \rangle$ or $F(d) = \langle a^{-1}, a \rangle$, then we output "E is unsolvable".

Step 2. We define
$$D_1^+ = D^+ \setminus \{d\},$$
F_1 is the restriction of F to $D_1^+ \cup D^-$.
Let $E_1 = \langle L, D_1^+, D^-, F_1, W \rangle$. If we transform E consistently with R, then we obtain the pair $\langle E_1, R \rangle$.
Step 3. We execute the transformation ELB until we eliminate all boundaries from the object of d. □

Transformation ELND2: elimination of a non-essential dependence
Input: a generalized equation $E = \langle L, D^+, D^-, F, W \rangle$,
a non-essential dependence d, which applies to [x, y) such that there
exists another dependence $d_1 \in D^+ \cup D^-$, which also applies to [x, y).

Step 1. If there exists $d_1 \in D^+ \cup D^-$, which applies to [x, y) and
$F(d) \neq F(d_1)$, then we output "E is unsolvable".
Step 2. We define
$$D_1^+ = D^+ \setminus \{d\}, \quad F_1 \text{ is the restriction } F \text{ to } D_1^+ \cup D^-.$$
$E_{out} = \langle L, D_1^+, D^-, F_1, W \rangle.$

If we transform E consistently with R, then we get the pair $\langle E_{out}, R \rangle$. □

Lemma 4.1. Assume that E is not evidently inconsistent. If d is an essential
dependence in E, $E_1 := DVD(E, d)$, d_1, d_2 are the parts of d in E_1, then d_1
and d_2 are essential. ■

Lemma 4.2. If for a given generalized equation E, an elementary
transformation always outputs "E is unsolvable", then E is unsolvable. ■

Lemma 4.3. The elementary transformations have the following properties
ADB	increases	bnd,
	does not decrease	inn,
	does not change	essdep, mcff,
DVC	does not increase	bnd, inn, essdep, mcff,
DVD	does not increase	bnd, inn, mcff,
	does not change	essdep if we divide a non-essential dependence,
TRB	does not decrease	bnd, inn,
	does not increase	essdep, mcff,
TRC	does not change	bnd, inn, essdep, mcff,
TRD	does not increase	bnd, inn, essdep, mcff,

ELB	decreases	bnd,
	does not change	inn, essdep, mcff,
ELC	decreases	bnd,
	does not increase	inn, mcff,
	does not change	essdep,
ELDC	does not increase	bnd, inn, essdep, mcff,
	decreases	essdep if we eliminate an essential dependence,
ELED	decreases	bnd, essdep,
	does not increase	inn, mcff,
ELND1	decreases	bnd,
	does not increase	inn,
	does not change	essdep, mcff,
ELND2	does not change	bnd, inn, essdep, mcff. ∎

Lemma 4.4. Suppose, that $<E_1, R_1>$ is a result of an elementary transformation of an equation E consistently with R. Then the following properties hold

1) if R is a solution of E, then R_1 is a solution of E_1,
2) if R is a minimal solution of E, then R_1 is a minimal solution of E_1
3) the length of R_1 is not greater than the length of R,
4) if $E_1 := ELB(E)$, then $lhms = lhms_1 + 1$,
5) if $E_1 := ELC(E)$, then $lhms_1 < lhms \leq lhms_1 + mcff$,
6) if $E_1 := ELND1(E)$, then $lhms_1 < lhms \leq lhms_1 + 3$,
7) if $E_1 := ELED(E)$, then $lhms_1 < lhms \leq 2 \cdot lhms_1$. ∎

Lemma 4.5. If a consistent generalized equation E_1 is a result of an elementary transformation of an equation E, then E is consistent. ∎

Lemma 4.6. The elementary transformations which are executed consistently with a solution do not increase prd. ∎

Lemma 4.7. If a generalized equation E_1 is a result of an elementary transformation of an equation E, rng W ⊆ C and C is a set of words closed with respect to subwords, then rng W_1 ⊆ C. ∎

V. THE REDUCING TRANSFORMATION

The reducing transformation have the following form

begin

        if condition_1 **then begin** transformation_1; exit **end**;

        if condition_2 **then begin** transformation_2; exit **end**;

        . . .

    end.

It means that execution of the reducing transformation has two parts. At first we find the first condition which is true and then we execute the corresponding transformation. Below we describe only conditions and corresponding transformations.

Reducing transformation: RTR

Input: a generalized equation E.

1). If E has only one boundary, then output "E is solvable".

2). If E is evidently inconsistent, then output "E is unsolvable".

3). If there exists an inner boundary x which is not inside an essential object, then we execute
the transformation RTR0:

    Input: a generalized equation E,

        an inner boundary x, which is not inside an essential object.

    We divide in x all coefficients and all non-essential dependencies using DVC and DVD.

4). If there exists a closed boundary $x < \infty$ such that $essdep(x) = 0$, then we execute
the transformation RTR1:

    Input: a generalized equation E,

        a closed boundary $x < \infty$ such that $succ(x)$ is also a closed boundary and $essdep(x) = 0$,

    We use ELDC and ELC, ELND2 and ELND1 or ELB to eliminate x.

5). If there exists an essential object disjoint with all others, then we execute

the transformation RTR2:

Input: a generalized equation E,

closed boundaries x, y such that [x, y) is an essential object disjoint with the others,

an essential dependence d, which applies to [x, y).

Step 1: Using TRB we transfer all boundaries from (x, y) over d.

Step 2: Using TRC and TRD we transfer over d all coefficients included in [x, y) and all dependencies whose objects are included in (x, y].

Step 3: Using ELED we eliminate d.

6). If there exists an essential object [x, y) such that essdep(x) = 1 or essdep(pred(y)) = 1, then we execute

the transformation RTR3:

Input: a generalized equation E,

an object [x, y) such that there exists exactly one essential dependence d, which applies to [x, y).

a boundary w ∈ [x, y) such that one of the following conditions is satisfied

a) [x, w) is disjoint with essential objects except [x, y),

b) [w, y) is disjoint with essential objects except [x, y).

Step 1: We divide all non-essential dependencies and coefficients in w using DVD and DVC.

Step 2: We transfer w over d using TRB.

Step 3: We divide d in w using DVD.

Step 4: We execute RTR2([x, w)) or RTR2([w, y)) respectively.

7). If there exists an essential object [x, y) and a boundary w ∈ [x, y) such that essdep(w) = 1, then we execute

the transformation RTR4:

Input: a generalized equation E,

an essential object [x, y) such that essdep(x) > 1 and essdep(pred(y)) > 1,

an essential dependence d, which applies to [x, y],

a boundary w ∈ [x, y) such that essdep(w) = 1.

Step 1: We divide in w all non-essential dependencies using DVD and all coefficients using DVC.

Step 2: We transfer w over d using TRB.

Step 3: We divide d in w using DVD.

Step 4: We execute RTR3([w, y], succ(w)).□

Lemma 5.1. The transformation RTR0 decreases the number of inner boundaries. The transformation RTR does not increase the number of inner boundaries.

Proof. It is obvious that RTR0 decreases the number of inner boundaries. By Lemma 4.3, only TRB and ADB increase the number of inner boundaries. ADB is not used in description of RTR. In RTR after application TRB we divide a dependence using DVD, which decreases the number of inner boundaries or we eliminate sufficiently many other inner boundaries using ELED. ∎

Lemma 5.2. Transformations RTR1, RTR2, RTR3 and RTR4 decrease lhms. The transformation RTR does not increase lhms if they are executed consistently with a minimal solution.

Proof. Lemma 5.2 follows easy from Lemma 4.4. ∎

Corollary 5.3. If $<E_1, R_1>$:= RTR(E, R) and $E_1 = E$, then $R_1 = R$.

Proof. If $R_1 \neq R$, then RTR consists in executing of RTR0, RTR1, RTR2, RTR3 or RTR4. In first case $E_1 \neq E$ is a consequence of Lemma 5.1. In the others cases it follows from Lemma 5.2. ∎

Lemma 5.4. Suppose that generalized equations E, E_1 have the same boundaries and closed boundaries of E are closed boundaries of E_1. If $[r, s)$ is closed in E, then

1) if $essdin[r,s) \leq essdin_1[r,s)$, then $rng[r, s) \leq rng_1[r, s)$,

2) if $essdin[r,s) - 1 = essdin_1[r,s)$, then $rng[r, s) - 1/2 \leq rng_1[r, s)$,

3) if $essdin[r,s) - 2 = essdin_1[r,s)$, then $rng[r, s) - 1 \leq rng_1[r, s)$.

Proof. Let $r_1, \ldots, r_n$ be all closed boundaries which belong to $[r, s)$ in E_1. It is obvious that

$essdin[r, s) = near(r)$,

$essdin_1[r, s) = near_1(r_1) + \ldots + near_1(r_n)$,

$rng[r, s) = rng(r)$,

$rng_1[r, s) = rng_1(r_1) + \ldots + rng_1(r_n)$.

1), 2) and 3) follow by an analysis of possible values of $near_1(r_i)$. ∎

Lemma 5.5. If E is a generalized equation, r, s are closed in E and essdin[r, s) ≥ 2, then rng[r, s) ≥ 1.
Proof. The method of proof of this lemma is the same as of Lemma 5.4. ∎

Lemma 5.6. 1) Transformations ADB, TRC, ELB, ELC, ELND1, ELND2, and transfer of a non-essential dependence using TRD do not change cpl.
2) Elimination of dependence, transformations DVC, TRB, ELDC, ELED and division of a non-essential dependence using DVD do not increase cpl.
Proof. By Lemma 4.3, transformations which are given in 1) do not change essdep. They do not change rng, because they do not change rng(x) and add or eliminate boundaries x such that rng(x) = 0. So, they also do not change cpl. The second part follows from Lemma 5.4 and the obvious equalities

$$rng = \sum_{[r, s) \text{ closed}} rng[r,s),$$
$$2 \cdot essdep = \sum_{[r,s) \text{ closed}} essdin[r, s). \blacksquare$$

Lemma 5.7. If E_1 := RTR2(E), then cpl_1 ≤ cpl.
Proof. Let E, x, y and d be correct data for RTR2. Notice that the lemma follows from Lemma 5.6 if any essential dependence is not transferred in step 2. A transfer of an essential dependence does not increase cpl in many cases. By Lemma 5.4, it is so if there are at least two essential dependencies different from d, which apply to a subset of [x, y). By the same argument, it is so also if TRD decreases essdep. So, we can assume that there are only two essential dependencies d and d_1 which apply to subset of [x, y) and transfer of d_1 does not decreases essdep. Then by Lemma 5.4, RTR2 does not decrease rng[u, w) for all closed [u, w) disjoint with [x, y). So, in this case, RTR2 decreases rng no more than by one and essdep by one. Consequently, RTR2 does not increase cpl. ∎

Lemma 5.8. If E_1 := RTR3(E), then cpl_1 ≤ cpl.
Proof. Let E and d be correct data for RTR3. It suffices to prove that the transformation which consists in dividing a dependence d and eliminating one part of d does not increase cpl since, by Lemma 5.6, the other parts of RTR3 do not increase cpl. Such a transformation does not change essdep or decreases it by one since there exists w in an object of d such that essdep(w) = 1. If essdep does not change, then essdin[r, s) also does not change for each closed segment [r, s) in E. So, by Lemma 5.4, rng does not decrease. If essdep decreases by one, then either essdin[r, s) decreases by two for one closed segment [r, s) or essdin[r, s) decreases by one for two

closed segments [r, s). In both cases, by Lemma 5.4, 2) and 3), rng decreases no more than by one. So, the complexity does not increase. ∎

Lemma 5.9. If E is not evidently inconsistent and E_1 := RTR4(E), then $cpl_1 \leq cpl$.
Proof. It suffices to prove that the first three steps of RTR4 do not increase cpl. Assume that in these steps we divide d, which applies to [x, y) in w ∈ (x, y) and d, x, y and w are the correct data for RTR4. Let [u, v) be a closed segment in E which includes (x, y]. After step 3 we have rng[u, w) ≥ 1/2, rng[w, v) ≥ 1. So, essdep may increase by 1, rng may increase no more than by 1/2 and consequently cpl may increase by 1/2. By part 1 of the Lemma 5.4, rng[r, s) does not decrease for [r, s) disjoint with [u, v). It is obvious, that if rng[u, w) = 1, then the complexity does not increase. Now, it suffices to notice that if rng[u, w) = 1/2, then the part of d which applies to [u, w) was in the set of dependencies. In this case the transformation which we consider does not change essdep and so decreases the complexity. ∎

Corollary 5.10. If E_1 := RTR(E), then $cpl_1 \leq cpl$. ∎

Lemma 5.11. If R is a minimal solution of E and $<E_1, R_1>$:= RTR(E, R), then either $lhms \leq lhms_1 + 3$, $lhms \leq lhms_1 + mcff$ or $lhms \leq 2 \cdot lhms_1$.
Proof. The lemma follows from Lemma 4.4. ∎

Lemma 5.12. If E_1 := RTR(E), then
$$bnd_1 \leq max(bnd, 2 \cdot cpl + inn + 2).$$
Proof. It follows easy from Lemma 4.3 that if E_1 := RTR1(E), then $bnd_1 < bnd$ and if E_1 := RTR0(E), then $bnd_1 = bnd$. If E_1 := RTR2(E), then $bnd_1 < bnd$, since in step 3 of RTR2, ELED eliminates one more boundary than RTR2 adds in step 1. As a consequence of the preceding fact we get that if E_1 := RTR3(E), then $bnd_1 \leq bnd$. Assume that E_1 := RTR4(E). It is obvious that $bnd_1 \leq bnd + 1$. Notice that if the equation E does not become transformed using RTR0, RTR1, RTR2 or RTR3, then for every closed boundary x holds near(x) ≥ 3. Therefore, 3·cls ≤ 2·essdep and also
$$cls \leq 2 \cdot essdep - 2 \cdot cls \leq 2 \cdot cpl.$$
Now, the lemma follows from the obvious equation bnd = inn + cls + 1. ∎

Theorem 5.13. Let $\langle E_0, R_0 \rangle, \ldots, \langle E_n, R_n \rangle$ be a sequence of generalized equations with solutions such that

    a) R_0 is a minimal solution of E_0,

    b) $\langle E_{i+1}, R_{i+1} \rangle := RTR(E_i, R_i)$ for all $i < n$,

    c) $E_{n-1} \neq E_n$,

    d) rng $W_i \subseteq C$ for $i \leq n$ and C has cf elements.

Then the following hold

    1) $E_i \neq E_j$ for $i < j \leq n$,

    2) $bnd_i \leq 2 \cdot cpl_0 + bnd_0 + 2$,

    3) $n \leq S_2(2 \cdot cpl_0 + bnd_0 + 2, cf)$ (S_2 is defined in Lemma 1.2),

    4) R_n is a minimal solution of E_n,

    5) $lhms_n \leq 3 \cdot 2^{S_2(2 \cdot cpl_0 + bnd_0 + 2, \, cf)} \cdot \max\{1, mcf\mathcal{S}_0\}$.

Proof. If $E_{i+1} = E_i$ for some $i < n$, then $E_{n-1} = E_n$ by Corollary 5.3. Hence, it follows from Lemma 5.1 and Lemma 5.2 that $inn_i < inn_j$ or $lhms_i < lhms_j$ for $i < j \leq n$ and consequently $E_i \neq E_j$ for $i < j \leq n$. Using Lemma 5.12, Corollary 5.10 and Lemma 5.1, it is easy to prove that if 2) holds for $i < n$, then it also holds for $i + 1$. So, 2) is true. The part 3) follows from 1) and Lemma 1.2. The part 4) is a consequence of Lemma 4.4. We can prove the last part by induction on n using Lemma 5.11. ∎

VI. THE MAIN AND THE BASIC TRANSFORMATIONS

In this chapter, we describe the main transformation and the basic transformation and we prove their simplest properties. The main transformation is an important element of the Makanin's algorithm and it is executed in the most difficult part of this algorithm. We shall study the main transformation in Chapter IX.

Main transformation MTR:

Input: a non evidently inconsistent generalized equation $E = \langle L, D^+, D^-, F, W \rangle$ for which the main boundary, the main dependence and the cut boundary are defined and are such that the main object is not a subset of any coefficient.

Step 1: We divide all non-essential dependencies in ctb using DVD.

Step 2: Using TRB, we transfer over mndep all boundaries x such that

a) $x \in [\emptyset, ctb)$,

b) $x = w$ such that $<u, w> \in$ dom W for some $u < ctb$,

c) $x = x_i$ for $i = 2, \ldots, n$, where $x_1 < ctb$ and $<x_1, \ldots, x_n>$ is a coordinate of a dependence different from the main one (the main dependence changes in this step).

Step 3: Using TRC and TRD, we transfer over mndep all coefficients which intersect $[\emptyset, ctb)$ and all dependencies whose objects intersect $[\emptyset, ctb)$.

Step 4: We divide the main dependence in ctb using DVD.

Step 5: Using ELED we eliminate the part of the main dependence which applies to $[\emptyset, ctb)$.□

The main transformation uses as an input data an equation which satisfies some conditions. The basic transformation tests if these conditions are satisfied and executes the main transformation if possible. The basic transformation will be described like RTR in Chapter V.

Basic transformation BTR:

Input: a generalized equation E.

1). If E has exactly one boundary, then we output "E is solvable".

2.) If E is evidently inconsistent, then we output "E is unsolvable".

3). If essdep($\emptyset$) = 0, then we execute

the transformation BTR0:

Input: a generalized equation E such that essdep($\emptyset$) = 0.

Step 1: We execute RTR0(E, succ($\emptyset$)).

Step 2: We execute RTR1(E, $\emptyset$).

4). If there exists a coefficient, which includes the main object, then we execute

the transformation BTR1:

Input: a generalized equation E such that there exists a main dependence in E and a coefficient $[\emptyset, x)$, which satisfies main $\leq x$.

<u>Step 1:</u> We divide all non-essential dependencies in **main** using DVD,

    <u>Step 2:</u> We divide a coefficient [ø, x) in **main** using DVC,

    <u>Step 3:</u> We eliminate the main dependence using ELDC.

5). We execute the main transformation MTR.□

Lemma 6.1. If $E_1 := BTR(E)$, then

    1) $essdep_1 \leq essdep$,

    2) $\dot{b}nd_1 < 2 \cdot bnd$.

Proof. It follows from Lemma 4.3 that BTR0 and BTR1 do not increase essdep and bnd. MTR increases essdep only by one in step 4 . Next, MTR eliminates one essential dependence in step 5. The elementary transformations except TRB do not increase bnd by Lemma 4.3. The inequality $bnd_1 < 2 \cdot bnd$ is obvious, because every execution of TRB increases bnd by at most one and TRB is executed in step 1 of MTR less than bnd times.∎

Lemma 6.2. If $E_1 := BTR1(E)$, then $essdep_1 < essdep$.∎

Lemma 6.3. Assume that R is a minimal solution of E. Then

    a) $<E_1, R_1> := BTR(E, R)$ implies $lhms_1 \leq lhms$,

    b) $<E_1, R_1> := BTR0(E,R)$ or $<E_1, R_1> := MTR(E,R)$ implies $lhms_1 < lhms$.∎

Lemma 6.4. If R_0 is a minimal solution of E_0 and $<E_{i+1}, R_{i+1}> := BTR(E_i, R_i)$ for $i < n$, then $E_i \neq E_j$ for $i < j \leq n$.

Proof. It follows immediately from Lemma 6.1, Lemma 6.2 and Lemma 6.3 that every two equations among $E_0, \ldots, E_n$ are different, since they have different parameters lhms or essdep.∎

Lemma 6.5. If $E_1 := MTR(E)$ and ctb = main, then $essdep_1 < essdep$ and $bnd_1 < bnd$.∎

Lemma 6.6. If essdep(main) > 0 and $E_1 := MTR(E)$ or $E_1 := BTR(E)$, then $cpl_1 \leq cpl$.

Proof. Observe, that an execution of the steps 4 and 5 of MTR can be replaced by an execution of RTR3. So, by Lemma 5.8, it does not increase the complexity. By Lemma 5.6, the steps 1) and 2) of MTR and TRC in step 3) do not increase cpl. If TRD decreases essdep, then by Lemma 5.6, it also does not increase the complexity. So, we can consider only such TRD, which does not change essdep.

First, assume that the main boundary is not closed. In this case, TRD in the step 2) does not change $rng(\emptyset)$, since there exist two essential objects lying near $\emptyset$ after the execution of TRD. Also it does not increase $rng[main, \infty)$ by Lemma 5.4. Thus, in this case TRD does not increase the complexity.

Now, assume that the main boundary is closed. In this case, TRD may increase cpl, but simultaneously it increases $essdin[u, v)$ for the closed segment $[u, v)$ which includes the second object of mndep and satisfies $u > 0$. In this situation, the step 4 of MTR will be omitted and the main dependence will be eliminated in the step 5. It follows from Lemma 5.4 that

$$rng_1 = rng_1[main, \infty) \geq rng[main, \infty) = rng - 1.$$

So, also in this case TRD does not increase the complexity. It suffice to prove now that BTR0 and BTR1 do not increase the complexity. It follows from Lemma 5.6 or Corollary 5.10. ∎

VII. REDUCED EQUATIONS

A generalized equation E is called reduced if

1) $essdep(x) \geq 2$ for all $x < \infty$

2) $near(x) \geq 3$ for all closed boundary $x < \infty$.

Notice that the main dependence, the main boundary and the cut boundary are defined for a reduced equation. It is not difficult to see that

Lemma 7.1. If $E = RTR(E)$, then E is reduced. ∎

Lemma 7.2. If $E = \langle L, D^+, D^-, F, W\rangle$ is a generalized equation such that $near(x) \geq 3$ for $x < \infty$ and $E_1 = \langle L, D_1^+, D_1^-, F_1, W\rangle$ is a generalized equation such that the set of essential dependencies in E_1 is a proper subset of the set of essential dependencies in E, then $cpl_1 < cpl$.
Proof. Lemma 7.2 follows from Lemma 5.5. ∎

Lemma 7.3. Assume that E is a reduced generalized equation, $E_1 := MTR(E)$ and $cpl_1 = cpl$. Then

1) $essdep_1(main) \geq 2$,

2) if $E_2 := BTR(E_1)$, then E_1 is reduced or $cpl_2 < cpl$.

Proof. First, we notice that when executing MTR, if some transformation TRD decreases essdep, then $cpl_1 < cpl$. In order to prove it we divide MTR into three parts: the first transformation TRD which decreases essdep, the transformation TR_1 which is executed before this TRD and the transformation TR_2 which is executed after this TRD. The transformation TR_1 does not increase cpl by Lemma 5.6 and since TRD does not increase cpl if there exist at least two the other essential dependencies with objects lying near the objects of a transferred dependence. TRD may decrease rng only if the second object of mndep does not lay near ø. It is obvious that after TR_1 near(x) ≥ 3 for each closed boundary $x > 0$. So, the next transformation TRD may decrease rng by no more than 1/2. It decreases cpl because we consider such a transformation TRD, which decreases essdep. Now, it suffices to prove that TR_2 does not increase cpl. It follows from Lemma 6.6 since if we replace TR_2 by MTR then as a result of transformation we get the same equation. The assumption essdep(**main**) > 0 of Lemma 6.6 holds because either **main** is in the second object of mndep or TR_1 and TRD do not change essdep(**main**). So, by the assumption, any transformation TRD, which is executed does not change essdep.

Assume that $x > ø$ and $[x, y)$ is a closed segment in E. It is easy to see that $essdin_1[x, y) ≥ essdin[x, y)$ since the number $essdin[x, y)$ may decrease if $[x, y)$ includes the object of the main dependence and the main dependence is eliminated. But in this case an object of a dependence applying to ø which is different from the main one is transferred to $[x, y)$. Notice also that similar arguments make it possible to prove that $essdep_1(x) ≥ 2$ for all x, **main** $≤ x < ∞$.

Now, we prove that all closed boundaries x in E_1 with **main** $≤ x$ are closed in E. It is obvious that closed boundaries x in E, $x > ø$ are closed in E_1. Suppose that there exists a closed boundary x in E_1 such that **main** $≤ x$ and x is inner in E. Notice that $rng_1(x) = 1$. We shall consider a few cases. If **main** is closed in E, then $ctb = $ **main** and $essdep_1 < essdep$. It is easy to see that also $rng_1 ≥ rng$. So, $cpl_1 < cpl$. If **main** is inside an essential object or a coefficient, then $ctb < $ **main** and $essdep_1 = essdep$. In this case $rng_1 > rng$ and so $cpl_1 < cpl$. If **main** is only inside non essential objects, then, because E is reduced, the first step of MTR increases rng and so decreases cpl. By Lemma 6.6, the next steps of MTR do not increase cpl and so $cpl_1 < cpl$. Hence, in all cases $cpl_1 < cpl$, which contradicts the assumption.

It is an immediate consequence of the preceding facts that $near_1(x) ≥ 3$ for all closed x in E_1 such that **main** $≤ x$. So, E_1 is reduced if $ctb = $ **main**.

If ctb < **main**, then the main boundary is inner in E_1. So, **main** lies near ø. Hence, $near_1(ø) \geq 3$ since an object of the part of mndep and objects of two essential dependencies which apply to **main** lay near ø. Also in this case $near_1(x) \geq 3$ for all closed boundaries in E_1.

Assume that ctb < **main** and E_1 is not reduced. Hence, there exists a boundary $x \in L_1$ such that $x < $ **main** and $essdep_1(x) = 1$. It means that there does not exist in E_1 any essential object which includes [ctb, **main**]. In this case [ctb, **main**) is the main object in E_1 and, by the definition of ctb, there exists a coefficient in E_1, which includes the **main** object. If we transform such E_1 using BTR, then we execute BTR1 and $cpl_2 < cpl_1$ by Lemma 7.2. ∎

VIII. LINEAR GENERALIZED EQUATIONS

A generalized equation E is called linear if $essdpc(x) \leq 2$ for all boundaries $x < \infty$. The transformation BTR has slightly simpler properties if it is applied to a linear equation. In particular, the set of boundaries described in step 1 of MTR is equal to [ø, ctb). Moreover, coefficients and objects which are transferred in step 2 of MTR are included in [ø, ctb) and are disjoint with each other. The above implies that

Lemma 8.1. If E is a linear generalized equation and $E_1 := BTR(E)$, then E_1 is also linear and $bnd_1 \leq bnd$. ∎

Lemma 8.2. If $E = \langle L, D^+, D^-, F, W \rangle$ is a linear generalized equation, $E_1 := BTR(E)$ and E_1 satisfies $bnd_1 = bnd$ and $essdep_1 = essdep$, then sets of values of functions W and W_1 coincide.
Proof. If $E_1 := BTR0(E)$, then $bnd_1 < bnd$ by Lemma 4.3. If $E_1 := BTR1(E)$, then $essdep_1 < essdep$, by Lemma 6.2. Thus $E_1 := MTR(E)$. But the transformation MTR does not change the set of values of W. ∎

Theorem 8.3. If E is a consistent linear generalized equation, then there exists a sequence $E_0, \ldots, E_n$ of generalized equations such that
   1) $E_0 = E$, E_n has only one boundary,
   2) $E_{i+1} = BTR(E_i)$ for $i < n$,
   3) $n \leq S_3(bnd)$, where $S_3(1) = 1 \cdot 2^{2 \cdot 1} \cdot S_1(1)$.

Proof. Let R be a minimal solution of E. Consider the maximal sequence of pairs $\langle E_i, R_i \rangle$ such that

    a) $\langle E_0, R_0 \rangle = \langle E, R \rangle$,

    b) $\langle E_{i+1}, R_{i+1} \rangle = BTR(E_i, R_i)$.

By Lemma 6.1 and Lemma 8.1, $essdep_i \le essdep$, $bnd_i \le bnd$ and all equations E_i are linear. The sequence $E_0, \ldots, E_n$ can be divided into subsequences such that every two equations in each of these subsequences $E_i, E_{i+1}, \ldots, E_j$ have the same numbers of boundaries and essential dependencies. There are at most $bnd \cdot essdep$ such subsequences. Lemma 6.4 says that every two elements of these subsequences are different. The length of such subsequences can be estimated by $S_1(bnd)$ using Lemma 1.1 and Lemma 8.2. So, the sequence $E_0, E_1, E_2, \ldots$ is finite and has at most $bnd \cdot essdep \cdot S_1(bnd)$ elements. Let E_n be the last element of that sequence. 3) follows from $essdep \le 2^{2 \cdot bnd}$. By Lemma 4.4, E_n is a consistent generalized equation. If E_n has more than one boundary, then we can define $E_{n+1} := BTR(E_n)$, which contradicts the definition of $E_1, \ldots, E_n$. $\blacksquare$

Lemma 8.4. If R is a minimal solution of E and $\langle E_1, R_1 \rangle := BTR(E, R)$, then $lhms \le lhms_1 + 3$, $lhms \le lhms_1 + mcff$ or $lhms \le 2 \cdot lhms_1$.
Proof. Lemma 8.4 follows from Lemma 4.4. $\blacksquare$

Theorem 8.5. (Makanin, [MA2]) The length of a minimal solution of a consistent linear generalized equation E is not greater than

$$3 \cdot 2^{S_3(bnd)} \cdot \max\{1, mcff\}.$$

Proof. We prove this theorem by induction on the length of the sequence defined in Theorem 8.3 using Lemma 8.4. $\blacksquare$

Lemma 8.6. Let E be a generalized equation such that $essdep(x) \ge 2$ for all $x < \infty$. Let R be a minimal solution of E. Then, there exist a generalized equation E_1 and a minimal solution R_1 of E_1 such that

    1) $bnd_1 \le 3 \cdot (bnd + 4 \cdot essdep^2)$,

    2) $mcff_1 \le \max\{mcff, exc(E, R)\}$,

    3) $lh(R_1[\emptyset, \infty)) = lh(R[\emptyset, \infty))$,

    4) $essdpc_1(x) = 2$ for all $x \in L_1 \setminus \{\infty\}$.

Proof. (first step) Let E and R satisfy the assumptions of the lemma. Let h be the greatest number of the form $essdep(x)$. We shall define three sequences: a sequence $E_h, \ldots, E_2$ of generalized equations, $R_h, \ldots, R_2$ a sequence of their solutions and $X_h, \ldots, X_2$ of sets such that X_1 is a subset of $C_i = (D_i^+ \cup D_i^-) \times \{1\} \cup (D_i^+ \cup D_i^-) \times \{2\}$. An element of C_i codes the first or the second coordinate of a dependence. We say that a segment $(u, w]$ corresponds to an element $<d, 1> \in C_i$ or $<d, 2> \in C_i$ if it is respectively the first or the second object of d. An element of C_i which belongs to X_i is called distinguished . The object corresponding to a distinguished element of C_i is called distinguished. If $X_i \subseteq C_i$, then $essdep_i^*(x)$ is the number of elements of X_i such that x belongs to the corresponding segment. A generalized equation E_i and a set X_i will be defined such that i should be the greatest number among $essdep_i^*(x)$.

Let $E_h = E$, $R_h = R$ and $X_h = \emptyset$. Suppose that E_i, R_i and X_i are defined. We shall define E_{i-1} and X_{i-1} by a repeated transformation as follows. We take the first boundary x such that $essdep_i^*(x) = i$. For such x there exists an object $[x, y)$, which corresponds to a non distinguished element of C_i. If every boundary w in $[x, y)$ satisfies $essdep_i^*(w) = i$, then we distinguish an element c. Otherwise, let z be the smallest boundary in $[x, y)$ such that $essdep_i^*(z) < i$. Consistently with R_i we divide in z the dependence from c using DVD. Simultaneously, we divide c into two parts c_1 and c_2 which correspond respectively to $[x, z)$ and $[z, y)$. Next, we distinguish c_1. The procedure is repeated until there does not exist z such that $essdep_i^*(z) = i$. By E_{i-1}, R_{i-1} and X_{i-1} we denote a generalized equation, a solution of it and a set which we get after transformations. So, E_2, R_2 and X_2 are defined.

Notice that

1) $bnd_2 \leq bnd + 4 \cdot essdep^2$,

2) $mcff_2 = mcff$, $exc(E_2, R_2) = exc(E, R)$, $lh(R_2[\emptyset, \infty)) = lh(R[\emptyset, \infty))$ and R_2 is minimal,

3) $essuep_2^*(x) = 2$ for all $x < \infty$.

1) follows from the following facts:

a) $h \leq 2 \cdot essdpc$,

b) z is the first element which belongs to some non distinguished object, so the number of steps in the construction of E_{i-1} is not greater than the number of non-distinguished objects and coefficients in E_i,

c) at most one boundary is added in each step,

d) the transformation reduces the number of non distinguished objects.

2) holds, since we have used only TRB and DVD and these transformations have similar properties. Notice that by 3)

$$\sum_{[x, y) \text{ is a non-distinguished}} lh(R_2[x, y)) = exc(E_2\ R_2).$$

(second step) We add new coefficients in E_2. If x belongs to a distinguished object or to a coefficient, then we add an element <x, succ(x)> to dom W_2 and we assign the value $R_2(x)$ to it. These coefficients will be called new. Then, if there exists a dependence d in E_2 such that exactly one coordinate of d is not distinguished and [x, y) is an object of this coordinate of d, then we create a coefficient <x, y> and we assign the value $R_2[x, y)$ to it. Finally, we eliminate all dependencies which have a distinguished coordinate.

Let E_2' be a generalized equation which is obtained by the method described above. It is obvious that

1) $bnd_2' = bnd_2$,

2) $mcff_2' \leq max\{mcff, exc(E, R)\}$,

3) R_2 is a minimal solution of E_2'

4) the number of essential dependencies and non-new coefficients which apply to x is equal 2 for all $x < \infty$.

(third step) We transform the equation E_2' into E_1 eliminating in some way all new coefficients. If <x, succ(x)> is a new coefficient, [u, w) is a coefficient and $x \in [u, w)$, then we divide (u, w] in x and succ(x) using DVC. In this case, DVÇ eliminates also a coefficient <x, succ(x)>). If <x, succ(x)> is a new coefficient, [u, w) is an object, d is an essential dependence, which applies to [u, w) and $x \in [u, w)$, then

1) we transfer x and succ(x) over d using TRB consistently with R_2,

2) we divide d in x and succ(x) using DVD consistently with R_2,

3) using ELDC consistently with R_2, we eliminate the part of d which applies to <x, succ(x)>.

After elimination of all new coefficients we get an equation E_1, which satisfies

a) $bnd_1 \leq 3 \cdot bnd_2'$,

b) $mcff_1 \leq mcff_2'$,

c) $essdpc_1(x) = 2$ for $x < \infty$.

As the result of these transformations we get also the solution R_1 of E_1. It is easy to see that R_1 is a minimal solution of E_1 and the lengths of R_1 and R_2 are equal. ■

Theorem 8.7. (Makanin, [MA2]) If E is a generalized equation such that essdep(x) $\geq$ 2 for x $<$ ∞ and R is a minimal solution of E, then

$$\text{lhms} \leq S_4(\text{bnd}) \cdot \max \{1, \text{ mcff, exc}(E, R)\},$$

where $S_4(1) = 3 \cdot 2^{S_3}(2^{4 \cdot 1 + 4})$ and S_3 is defined in Theorem 7.3.

Proof. It follows from Lemma 8.6, Theorem 8.5 and the obvious inequality essdep $< 2^{2 \cdot \text{bnd}}$. ∎

IX. EQUATIONS WITH NUMBERED DEPENDENCIES

In this chapter, we shall consider a triple $\langle E, R, N \rangle$ such that

1) E is a generalized equation and R is a solution of E,

2) essdep(x) $\geq$ 2 for all x $<$ ∞, so the main dependence is defined,

3) if $E_1 := \text{MTR}(E)$, then $\text{essdep}_1 = \text{essdep}$,

4) the main object of E is not a subset of a coefficient,

5) N is a function on $\{1, \ldots, \text{essdep}\}$ onto the set of all essential dependencies of E.

Such a triple is called an equation with a solution and numbered dependencies or briefly an equation with numbered dependencies. N(i) is called the ith dependence of E or a dependence number i.

Equations with numbered dependencies will also be transformed using MTR. If we transform $\langle E, R, N \rangle$ using MTR, then we get $\langle E_1, R_1, N_1 \rangle$ such that

1) $\langle E_1, R_1 \rangle := \text{MTR}(E, R)$,

2) if N(i) = mndep, then $N_1(i)$ is the part of mndep which exists in E_1,

3) if an object of N(i) is included in the main object, then $N_1(i)$ is the result of the transfer of N(i) over the main dependence,

4) $N_1(i) = N(i)$ in the other cases.

If $\langle E_1, R_1, N_1 \rangle$ is the result of such a transformation of $\langle E, R, N \rangle$, then we write $\langle E_1, R_1, N_1 \rangle := \text{MTR}(E, R, N)$. If N(i) = mndep, then the transformation MTR of $\langle E, R, N \rangle$ is called the ith transformation.

In this chapter, we shall consider a sequence

(1) $\qquad \langle E_i, R_i, N_i \rangle$ for i = 1, $\ldots$, n

of equations with numbered dependencies such that $\langle E_{i+1}, R_{i+1}, N_{i+1} \rangle :=$ $\text{MTR}(E_i, R_i, N_i)$ for i $<$ n. We fix the following notations

δ_i is the number such that $N_i(\delta_i) = \text{mndep}_i$,

q_i is the number of integers j, such that $1 \leq j < n$ and $\delta_j = i$,

$$\alpha_i = \frac{1}{(prd_i + 1)},$$

$$\beta_i = \frac{1}{essdep_i * (prd_i + 2)},$$

$$\gamma_i = \frac{1}{2 * essdep_i^2 * (prd_i + 2)},$$

$$l_i(j) = lh(R_i[x, y)),$$

$$f_i(j) = lh(R_i[\emptyset, x)),$$

$$s_i(j) = lh(R_i[\emptyset, u)),$$

where x, y and u satisfy $N_i(j) = \langle\langle x, \ldots, y\rangle, \langle u, \ldots, w\rangle\rangle$.
In addition, we assume that the main dependencies $mndep_1$ and $mndep_2$ satisfy $f_1(\delta_1) = 0$ and $f_2(\delta_2) = 0$.

Lemma 9.1. If $m \le essdep_1$, then $(1 - \beta_1)^{-m} - 1 \le \alpha_1$.
Proof. This lemma follows from Bernoulli inequality. ∎

Lemma 9.2. For a given sequence (1) of equations with numbered dependencies, we have $l_n(\delta_j) \le l_1(\delta_{j+1})$.
Proof. The transformation MTR changes the length of $R[x, y)$ for objects $[x, y)$ involved in only one dependence. So, $l_1(\delta_j) \ge \ldots \ge l_j(\delta_j) > l_{j+1}(\delta_j)$ $\ge \ldots \ge l_n(\delta_j)$ for $j \le n$. Lemma 9.2 follows from the properties of $mndep_{j+1}$, since $l_n(\delta_j) \le l_{j+1}(\delta_j) \le l_{j+1}(\delta_{j+1}) \le l_1(\delta_{j+1})$. ∎

Lemma 9.3. (Makanin, [MA2]) For a given sequence (1) of equations with numbered dependencies, if $\delta_1 = \delta_n$ and $\delta_i \ne \delta_j$ for $i < j < n$, then there exists $j < n$ such that

$$(1 - \beta_1) \cdot l_1(\delta_j) > l_n(\delta_j).$$

Proof. Assume that $(1 - \beta_1) \cdot l_1(\delta_j) \le l_n(\delta_j)$ for all $j < n$. Now, we shall prove that

(2) $\qquad \alpha_1 \cdot l_1(\delta_1) \le \sum_{j=1}^{n-1} (l_1(\delta_j) - l_n(\delta_j)).$

(2) will follow from the following facts.
Fact 1. For $1 \le j < n$ and $1 \le i \le essdep_1$, we have

$$f_j(i) \le f_{j+1}(i) + l_1(\delta_j) - l_n(\delta_j),$$
$$s_j(i) \le s_{j+1}(i) + l_1(\delta_j) - l_n(\delta_j).$$

Proof. We begin by proving the first inequality. If the position of the first object of the ith dependence has not been changed during the execution of $MTR(E_j)$, then the left and the rigth hand side of the inequality are equal. In the other case, $f_j(i) \le l_1(\delta_j) - l_n(\delta_j)$. The second inequality has a similar proof. □

Fact 2. The smaller of the numbers $f_n(\delta_1)$ and $s_n(\delta_1)$ is 0.

Proof. It is so, since $\delta_1 = \delta_n$ and $N_n(\delta_n)$ is the main dependence. □

Fact 3. If $n > 3$, then

$$\min \{f_3(\delta_1), s_3(\delta_1)\} \leq \sum_{j=3}^{n-1} (1_1(\delta_j) - 1_n(\delta_j)).$$

If $n = 3$, then $\min \{f_3(\delta_1), s_3(\delta_1)\} = 0$.

Proof. Fact 3 follows from Fact 1 and Fact 2. □

Fact 4.

$$\alpha_1 \cdot 1_1(\delta_1) < s_1(\delta_1),$$

if $n > 2$, then $\alpha_1 \cdot 1_1(\delta_2) < s_2(\delta_2)$.

Proof. Let $mndep_1 = \langle\langle\varnothing, \ldots, x\rangle, \langle y, \ldots, z\rangle\rangle$, $u = R_1[\varnothing, x)$, $w = R_1[\varnothing, y)$.
It is obvious that $1_1(\delta_1) = lh(u)$ and $s_1(\delta_1) = lh(w)$. Assume that $y < x$. In

this case, $mndep_1 \in D^+$ and there exists a word v such that $wu = uv = R_1[\varnothing, z)$. Therefore, $u = w^p v'$ for a subword v' of w and some $p > 0$. By the definition of the periodicity exponent, we have $p \leq prd_1$. So, $lh(u) < (prd_1 + 1) \cdot lh(w)$. If $x \leq y$, then the first part of the Fact 4 is obvious. The second part of Fact 4 follows by a similar analysis of E_2, since $1_1(\delta_2) = 1_2(\delta_2)$ and $prd_2 \leq prd_1$. □

Fact 5. If $n = 2$, then $s_1(\delta_1) = 1_1(\delta_1) - 1_n(\delta_1)$. If $n > 2$, then $s_2(\delta_2) \leq 1_1(\delta_2) - 1_n(\delta_2) + f_3(\delta_1)$.

Proof. Observe that if $n = 2$, $\delta_1 = \delta_2$ and $\langle E_2, R_2, N_2\rangle$ is an equation with numbered depediencies, then $mndep_1$ has the form

$$\langle\langle\varnothing, main_1\rangle, \langle ctb_1, y\rangle\rangle.$$

It is obvious that $mndep_1 = \langle\langle\varnothing, main_1\rangle, \langle x, y\rangle\rangle$. If $y \leq main_1$, then E_1 is inconsistent. If $y > main_1$, then $ctb_1 \leq x$ by the definition of ctb. If $ctb < x$ and $\delta_1 = \delta_2$, then $[ctb_1, main_1)$ is the main object of E_2 and there exists a coefficient or an essential object in E_2 which includes the main one. It contradicts the definition of an equation with numbered depedencies or the definition of the main dependence. If $mndep_1$ has a given form, then the first part of Fact 5 is obvious.

Assume that $n > 2$. It is obvious that the first objects of $N_2(\delta_1)$ and $N_2(\delta_2) = mndep_2$ have the form $[\varnothing, x)$. So, $N_2(\delta_1)$ is transferred over $N_2(\delta_2)$. Suppose that $[y, z) \subseteq L_3$ is the first object of $N_3(\delta_1)$ and $[u, w) \subseteq L_2$ is the second object of $N_2(\delta_2)$. It is easy to see that $y \in [u, w)$ in E_2 and both sides of the inequality from the Fact 5 are equal respectively to $lh(R_2[\varnothing, u))$ and to $lh(R_2[\varnothing, y))$. □

Now, we shall prove the inequality (2). If $n = 2$, then (2) is a consequence of Fact 3 and Fact 4. If $n > 2$, then

$$\alpha_1 \cdot l_1(\delta_2) < s_2(\delta_2) \le (l_1(\delta_2) - l_n(\delta_2)) + s_3(\delta_2).$$

Now,

$$\alpha_1 \cdot l_1(\delta_1) \le \alpha_1 \cdot l_1(\delta_1) + \alpha_1 \cdot (l_1(\delta_2) - l_n(\delta_1)) \le$$
$$\le \alpha_1 \cdot (l_1(\delta_1) - l_n(\delta_1)) + \alpha_1 \cdot l_1(\delta_2) <$$
$$< (l_1(\delta_1) - l_n(\delta_1)) + s_2(\delta_2) \le$$
$$\le (l_1(\delta_1) - l_n(\delta_1)) + (l_1(\delta_2) - l_n(\delta_2)) + f_3(\delta_1)$$

and

$$\alpha_1 \cdot l_1(\delta_1) < s_1(\delta_1) \le (l_1(\delta_1) - l_n(\delta_1)) + (l_1(\delta_2) - l_n(\delta_2)) + s_3(\delta_1).$$

Finally, the inequality (2) follows from Fact 3 and Fact 5.

In the order to get a contradiction it suffices to prove that

$$(3) \qquad \alpha_1 \cdot l_1(\delta_1) > \sum_{j=1}^{n-1} (l_1(\delta_j) - l_n(\delta_j)).$$

Observe that Lemma 9.2 and the assumption imply that $l_n(\delta_{n-1}) \le l_1(\delta_1)$ and

$$(1 - \beta_1) \cdot l_1(\delta_j) \le l_n(\delta_j) \le l_1(\delta_{j+1})$$

by the assumption of a proof by contradiction and Lemma 9.2. So, by Lemma 9.1 and $n - 1 \le \text{essdep}_1$ we have

$$\sum_{j=1}^{n-1} (l_1(\delta_j) - l_n(\delta_j)) \le \beta_1 \cdot \sum_{j=1}^{n-1} l_1(\delta_j) \le$$

$$\beta_1 \cdot l_n(\delta_{n-1}) \cdot \sum_{j=1}^{n-1} (1 - \beta_1)^{-(n-1-j)} =$$

$$[\beta_1 \cdot (1 - \beta_1)^{-1} \cdot ((1 - \beta_1)^{-1} - 1)^{-1}] \cdot ((1 - \beta_1)^{-n-1} - 1) \cdot l_n(\delta_{n-1}) \le$$

$$((1 - \beta_1)^{-n-1} - 1) \cdot l_n(\delta_{n-1}) \le$$

$$\alpha_1 \cdot l_n(\delta_{n-1}) \le \alpha_1 \cdot l_1(\delta_1). \blacksquare$$

Lemma 9.4. For a given sequence (1) of equations with numbered dependencies, if

$$(1 - \beta_1) \cdot l_1(\delta_j) \le l_n(\delta_j) \text{ for } j < n \text{ and}$$

$$(1 - \beta_1) \cdot l_1(\delta_n) > l_n(\delta_n),$$

then $q_j \le 1$ for any j such that $1 \le j \le \text{essdep}_1$ and so $n \le \text{essdep}_1$.

Proof. Otherwise, there exist i, j, $1 \le i < j \le n$ such that $\delta_i = \delta_j$ and every two elements among δ_i, δ_{i+1}, ..., δ_{j-1} are different. So, the sequence $\langle E_i, R_i, N_i \rangle$, ..., $\langle E_j, R_j, N_j \rangle$ satisfies the assumptions of Lemma 9.3 and consequently there exists $k < j$ such that

$$(1 - \beta_k) \cdot l_1(\delta_k) > l_j(\delta_k).$$

Thus,

$$(1 - \beta_1) \cdot 1_1(\delta_k) \geq (1 - \beta_i) \cdot 1_i(\delta_k) > 1_j(\delta_k) \geq 1_n(\delta_k),$$

which contradicts the assumption. The second part of Lemma 9.4 follows immediately from the first one. ∎

Lemma 9.5. For a given sequence (1) of equations with numbered dependencies, if

$$(1 - \beta_1) \cdot 1_1(\delta_j) \leq 1_n(\delta_j) \text{ for } j < n \text{ and}$$

$$(1 - \beta_1) \cdot 1_1(\delta_n) > 1_n(\delta_n),$$

then

$$\gamma_1 \cdot \sum_{j=1}^{n} 1_n(\delta_j) \leq \sum_{j=1}^{n} (1_1(\delta_j) - 1_n(\delta_j)).$$

Proof. We have

$$\gamma_1 \cdot \sum_{j=1}^{n} 1_n(\delta_j) \leq \gamma_1 \cdot \sum_{j=1}^{n} 1_1(\delta_j) \leq \gamma_1 \cdot 1_1(\delta_n) \cdot \sum_{j=1}^{n} (1 - \beta_1)^{-(n - j)} \leq$$

$$\leq \gamma_1 \cdot \beta_1^{-1} \cdot (1 - \beta_1) \cdot ((1 - \beta_1)^{-n} - 1) \cdot 1_1(\delta_n) \leq$$

$$\leq [\gamma_1 \cdot \beta_1^{-2} \cdot (1 - \beta_1) \cdot \alpha_1] \cdot [\beta_1 \cdot 1_1(\delta_n)] <$$

$$< 1/2 \cdot (\alpha_1 / (\alpha_1 + 1)) \cdot (1_1(\delta_n) - 1_n(\delta_n)) \leq$$

$$\leq 1_1(\delta_n) - 1_n(\delta_n)) \leq \sum_{j=1}^{n} (1_1(\delta_j) - 1_n(\delta_j)).$$

The successive lines of the proof follow from the assumption and Lemma 9.2, a simple computation, Lemma 9.1 and Lemma 9.4. ∎

Lemma 9.6. (Makanin, [MA2]) For a given sequence (1) of equations with numbered dependencies, if

$$(1 - \beta_1) \cdot 1_1(\delta_n) > 1_n(\delta_n),$$

then

(4)
$$\gamma_1 \cdot \sum_{j=1}^{essdep_1} q_j \cdot 1_n(j) \leq \sum_{j=1}^{essdep_1} (1_1(j) - 1_n(j)).$$

Proof. We prove Lemma 9.6 by induction on the number m of integers j such that $0 < j < n$ and $(1 - \beta_1) \cdot 1_1(\delta_j) > 1_n(\delta_j)$. Observe that if m = 1, then $q_j \leq 1$ for $j \leq n$ by Lemma 9.4. So, the inequality (4) follows from Lemma 9.5, since

$$\gamma_1 \cdot \sum_{j=1}^{essdep_1} q_j \cdot 1_n(j) \leq \gamma_1 \cdot \sum_{j=1}^{n} 1_n(\delta_j) \leq$$

$$\leq \sum_{j=1}^{n} (1_1(\delta_j) - 1_n(\delta_j)) \leq \sum_{j=1}^{essdep_1} (1_1(j) - 1_n(j)).$$

If $m > 1$, then there exists $k < n$ such that

$$(1 - \beta_1) \cdot l_1(\delta_k) > l_n(\delta_k).$$

Now, the inequality (4) follows from $prd_{k+1} \leq prd_1$ and inequalities (4) which follow from induction hypotesis for $\langle E_1, R_1, N_1 \rangle, \ldots, \langle E_k, R_k, N_k \rangle$ and $\langle E_{k+1}, R_{k+1}, N_{k+1} \rangle, \ldots, \langle E_n, R_n, N_n \rangle$. ∎

Theorem 9.7. (Makanin, [MA2]) For a given sequence (1) of equations with numbered depedencies, if R_1 is a minimal solution of E_1 and $exc(E_1, R_1) > mcff_1$, then there exists j such that

(5) $q_j \leq S_5(bnd_1) \cdot (prd_1 + 2),$

where

$$S_5(1) = 8 \cdot 2^{8 \cdot 1} \cdot S_4(1) \text{ and } S_4 \text{ is defined in Theorem 8.7.}$$

Proof. It is easy to see, that (5) holds if $n \leq essdep_1$. So, we assume that $essdep_1 > n$. Assume also that for $j \leq essdep_1$ we have $q_j > S_5(bnd_1) \cdot (prd_1 + 2)$. By Lemma 9.3, there exists $k \leq n$ such that

1) a sequence $\langle E_1, R_1, N_1 \rangle, \ldots, \langle E_k, R_k, N_k \rangle$ satisfies the assumptions of Lemma 9.6,

2) every two elements among $\delta_{k+1}, \ldots, \delta_n$ are different.

It follows from Theorem 8.7 that

$$lhms_1 \leq S_4(bnd_1) \cdot exc(E_1, R_1).$$

Observe that the condition 2 from the definition of an equation with numbered dependencies imply that

$$lhms_n \leq \sum_{j=1}^{essdep_1} l_n(j).$$

From Lemma 9.6 we get that

$$(1 + \gamma_1 \cdot S_5(bnd_1) \cdot (prd_1+2)) \cdot \sum_{j=1}^{essdep_1} l_n(j) \leq$$

$$\leq \sum_{j=1}^{essdep_1} (1 + \gamma_1 \cdot q_j) \cdot l_n(j) \leq \sum_{j=1}^{essdep_1} l_1(j) \leq$$

$$\leq essdep_1 \cdot lhms_1 \leq essdep_1 \cdot S_4(bnd_1) \cdot exc(E_1, R_1) =$$

$$essdep_1 \cdot S_4(bnd_1) \cdot exc(E_n, R_n) \leq essdep_1 \cdot S_4(bnd_1) \cdot (2 \cdot essdep_1 - 2) \cdot lhms_n \leq$$

$$\leq 2 \cdot essdep_1^2 \cdot S_4(bnd_1) \cdot \sum_{j=1}^{essdep_1} l_n(j).$$

Thus

$$\frac{1}{2 \cdot essdep_1^2} \cdot S_5(bnd_1) < 2 \cdot essdep_1^2 \cdot S_4(bnd_1) \leq$$

$$\leq 2 \cdot 2^{4 \cdot \text{bnd}_1} \cdot S_4(\text{bnd}_1) \leq \frac{1}{2 \cdot \text{essdep}_1^2} \cdot S_5(\text{bnd}_1),$$

which is not possible. ∎

X. THE MAKANIN'S THEOREM

Let $<E, R, N>$ be a generalized equation with numbered dependencies. Given a set $X \subseteq \{1, \ldots, \text{essdep}\}$ we define two sets A and B,

    $A = \{x \in L$: there exists $j \in X$ such that x is the first or the last
            element of any coordinate of $N(j)$ or there exists $j \notin X$
            such that x is an element of any coordinate of $N(j)$ or
            there exists $<y, z> \in \text{dom } W$ such that $x = y$ or $x = z\}$,

    $B = \{x \in L$: x does not belong to an object of any dependence whose
            number is in $X\}$.

Lemma 10.1. Assume that $<E_1, R_1, N_1>$ is a generalized equation with numbered dependencies, $E_2 := \text{MTR}(E_1)$, $X \subseteq \{1, \ldots, \text{essdep}_1\}$, the number of mndep_1 belongs to X and A_1, A_2, B_1 and B_2 are defined above. Then

    1) $B_1 \subseteq B_2$ and $A_1 \cap B_1 \subseteq A_2 \cap B_2$,

    2) if B_1 is a proper subset of B_2, then $A_1 \cap B_1$ is a proper subset of $A_2 \cap B_2$.

Proof. 1) follows from the fact that MTR does not change anything outside objects of the main dependence. To prove 2) we consider several cases. Assume that mndep_1 is in D_1^+. Let $x \in B_2 \setminus B_1$ and let y be the greatest element in A_2 not greater than x. Observe that $x \in B_2$ if and only if $y \in B_2$. So, $y \in A_2 \cap B_2$ and we can assume that $y \in A_1 \cap B_1$. Suppose also that $x \in L_1$. It can happen only if x is in the first part of the second object of mndep_1. The first element of this object is in A_2 since $\text{essdep}_1(\emptyset) \geq 2$. So, y is in the object of mndep_1, which is not possible because it means that $y \notin B_1$. If $x \notin L_1$, then it is also in the first part of the second object of mndep_1 and y is also in the second object of mndep_1. So, $y \notin B_1$. If $\text{mndep}_1 \in D_1^-$, then we use similar arguments. ∎

Given a generalized equation with numbered dependencies $<E, R, N>$ and a set $X \subseteq \{1, \ldots, \text{essdep}\}$ we define equations E' and E". To obtain E' we add to E coefficients $<x, y>$ for all x, y such that $[x, y)$ is an object of a dependence $N(j)$, $j \in X$. We assign the value $R[x, y)$ to $<x, y>$. Next, using ELDC consistently with R we eliminate all dependencies $N(j)$ for $j \in X$. The equation E" has the same sets of dependencies and the function F as E', but L" is equal to $A \cup B$. We define also the function R" such that $R"(x) = R[x, succ"(x))$ for $x \in L"$, $x < \infty$ ($[x, succ"(x))$ denotes a subset of L). Let $mx\{R, X\}$ be the greatest of $lh(R[x, y))$ such that $[x, y)$ is an object of $N(j)$, $j \in X$. The equation E" and the solution R" have the following properties.

Lemma 10.2. If $<E, R, N>$ is a generalized equation with numbered dependencies and $X \subseteq \{1, \ldots, essdep\}$, then

    1) R" is a solution of E" and $lh(R[\emptyset, \infty)) = lh(R"([\emptyset, \infty))$,

    2) if R is minimal, then R" is also minimal,

    3) $prd" \leq prd$ and $mcff" \leq max \{mcff, mx\{R, X\}\}$,

    4) if E is reduced and X is non-empty, then $cpl" < cpl$.

Proof. The condition 1) is obvious. On the other hand, if R" is a solution of E", then the function R' such that $R'(x) = R"(x)$ for $x \in B$ and $R'(x) = R(x)$ for the other elements of L is a solution of E and the length of it is equal to the length of R". The third condition is an immediate consequence of the definition of E". The forth condition follows from Lemma 7.2. ∎

The sequence (1) of generalized equations with numbered dependencies is X-invariant if

    1) $\{\delta_1, \ldots, \delta_{n-1}\} \subseteq X$,

    2) there exists a sequence $E_1", \ldots, E_n"$ of a generalized equation with solutions $R_1", \ldots, R_n"$ such that for all $j \leq n$

        2.1) $lh(R_j"[\emptyset, \infty)) = lh(R_j[\emptyset, \infty))$,

        2.2) $prd_j" \leq prd_j$ and $mcff_j" \leq max \{mcff_j, mx\{R_j, X\}\}$,

        2.3) $cpl_j" < cpl_j$,

        2.4) $bnd_j" \leq 2 \cdot bnd_i$ for all $i < j \leq n$.

Lemma 10.3. Every sequence (1) of an equation with numbered dependencies such that E_1 is reduced and $cpl_1 = cpl_n$, for any X, $\{\delta_1, \ldots, \delta_{n-1}\} \subseteq X \subseteq \{1, \ldots, essdep_1\}$ is a concatenation of at most $S_6(bnd_1)$ X-invariant subsequences, where

$$S_6(1) = 3 \cdot 1 \cdot 2^{2 \cdot 1}.$$

Proof. First we shall prove that if $B_1 = B_n$, then the sequence (1) of an equation with numbered dependencies is X-invariant. The definition of equations $E_j"$ is given in the paragraph preceding the formulation of Lemma

10.2. The conditions 2.1 and 2.2 follow from Lemma 10.2. It follows from Lemma 6.6 and Lemma 7.3 that E_i is reduced for $i \leq n$. So, the condition 2.3 is also a consequence of Lemma 10.2. The condition 2.4 follows from the fact that L_j'' is the union of B_j, which is equal to B_i, and A_j, which is the image of A_i by some function.

Every sequence of equations with numbered dependencies can be divided into subsequences such that $B_k = B_{k+1}$ holds for elements of these subsequences. By Lemma 10.1, the number of these subsequences is estimated by the cardinality of $A_n \cap B_n$ or by the cardinality of A_n. By definition there are three kinds of elements of A_n. There exists at most $4 \cdot x$ elements of the first kind, where x is the cardinality of X, $2 \cdot bnd_1 \cdot (essdep_1 - x)$ elements of the second kind and at most bnd_1^2 other elements. So A_n has no more than $3 \cdot bnd_1 \cdot 2^{2 \cdot bnd_1}$ elements. ∎

In the next lemmas we shall consider a sequence (1) of generalized equations with numbered dependencies, which satisfies

(6)

$\quad$ 7.1) R_1 is a minimal solution of E_1,

$\quad$ 7.2) E_1 is reduced,

$\quad$ 7.3) $cpl_1 = cpl_n$,

$\quad$ 7.4) if $[x, y)$ is an essential object in E_n, then
$$lh(R_n[x, y)) > mcff_1.$$

We shall also assume that there exists an increasing function S such that

(7) $\quad$ $lhms \leq S(bnd, prd) \cdot mcff$

for any generalized equation E for which $cpl < cpl_1$, and there exists a minimal solution of E, whose the periodicity exponent is prd.

Lemma 10.4. (Makanin, [MA3]) Suppose, that a X-invariant sequence (1) of equations with numbered dependencies satisfies (6) and

$\quad$ 5) X is a proper subset of $\{1, \ldots, essdep_1\}$.

Let S be a function which satisfies (7). Then there exists $j \in \{\delta_1, \ldots, \delta_{n-1}\}$ such that
$$q_j \leq S_7(bnd_1, prd_1),$$
where

(A) $\quad$ $S_7(1, p) = 2^{6 \cdot 1} \cdot (p + 2) \cdot S(1, p) \cdot S(2 \cdot 1, p)$.

Proof. Assume that $q_j > S_7(bnd_1, prd_1)$ for all $j \in X$. First notice that

(8) $\quad$ $mx \{R_n, X\} \cdot (1 + \gamma_1 \cdot S_7(bnd_1, prd_1)) \leq essdep_1 \cdot mx \{R_1, X\}$.

In order to use Lemma 9.6 we have to shorten given sequence of equations like in the proof of Theorem 9.7. So, we get the inequality (4), which we can represent in the form

(9) $\qquad \sum_{j=1}^{essdep_1} (1 + \gamma_1 \cdot q'_j) \cdot l_n(j) \leq \sum_{j=1}^{essdep_1} l_1(j),$

where q'_j is such that $q_j - 1 \leq q'_j \leq q_j$. By the condition 1) of the definition X-invariance it follows that for $j \notin X$ we have $q_j = q'_j = 0$ and $l_n(j) = l_1(j)$. So, in the both sides of the inequality (9) we can omit all components which correspond to j such that $j \notin X$. Now, the inequality (8) easy follows from (9).

Let E_1''' be a generalized equation which is obtained from E_1 as described in the paragraph preceding Lemma 10.2 for the set $X^c = \{1, \ldots, essdep_1\} \setminus X$. By Lemma 10.2 3) and (6) it is clear, that $mcff_1''' \leq \max \{mcff_1, mx\{R_1, X^c\}\} = mx \{R_1, X^c\}$. So, by (6), (7) and Lemma 10.2,

(10) $\qquad mx \{R_1, X\} \leq lhms_1 = lhms_1''' \leq S(bnd_1, prd_1) \cdot mx \{R_1, X^c\} =$

$\qquad = S(bnd_1, prd_1) \cdot mx \{R_n, X^c\} \leq S(bnd_1, prd_1) \cdot lhms_n.$

Finally, it follows from X-invariance of a given sequence of equations that

(11) $\qquad lhms_n = lhms_n''' \leq S(2 \cdot bnd_1, prd_1) \cdot mx \{R_n, X\}.$

From the inequalities (8), (10) and (11) we can easily derive a contradiction. ∎

Lemma 10.5. If (1) is a sequence of equations with numbered dependencies such that $\delta_1 = \delta_2 = \ldots = \delta_{n-1}$, then $n \leq prd_1 + 1$.
Proof. We shall use the notation from Lemma 9.3. It is easy to see that for $i < n$

$\qquad s_i(\delta_i) = l_{i+1}(\delta_i) - l_i(\delta_i),$
$\qquad s_{i+1}(\delta_{i+1}) = s_{i+1}(\delta_i) = s_i(\delta_i).$

So, by Fact 4 from Lemma 9.3,

$\qquad (n - 1) \cdot s_1(\delta_1) = l_1(\delta_1) - l_n(\delta_1) \leq l_1(\delta_1) < (prd_1 + 1) \cdot s_1(\delta_1). ∎$

Lemma 10.6. Suppose that a sequence (1) of equations with numbered dependencies satisfies (6) and

5) $\{\delta_1, \ldots, \delta_{n-1}\}$ is a proper subset of $\{1, \ldots, essdep_1\}$.
Let S be a function which satisfies (7). Then

$\qquad n \leq S_8(bnd_1, prd_1),$

where the functions S_8 and f, g are defined by induction as follows

$$S_8(1, p) = f(2^{2 \cdot 1}, 1, p),$$
$$f(1, 1, p) = p + 2,$$
$$f(d + 1, 1, p) = g(d, S_6(1), 0, 1, p),$$

(B)
$$g(d, 0, 0, 1, p) = 0,$$
$$g(d, i + 1, 0, 1, p) = g(d, i, S_7(1 \cdot 2^{g(d, i, 0, 1, p)}, p), 1, p)$$
$$g(d, i, j + 1, 1, p) =$$
$$= g(d, i, j, 1, p) + 1 + f(d, 1 \cdot 2^{g(d, i, j, 1, p)}, p)$$

(S_6 is defined in Lemma 10.3, S_7 is defined in Lemma 10.4).

Proof. First, we observe that $bnd_i \le 2^{i-1} \cdot bnd_1$ by Lemma 6.1 2) and $prd_i \le prd_1$ by Lemma 4.6. It is easy to see that the lemma follows from the fact that for all sequences of equations with numbered dependencies satisfying the assumptions of the lemma we have

(12) if $\{\delta_1, \ldots, \delta_{n-1}\}$ has d elements, then $n \le f(d, bnd_1, prd_1)$.

We prove (12) by induction on d. If $d = 1$, then (12) an immediate consequence of Lemma 10.5. To prove that (12) holds for $d + 1$ we use Lemma 10.3 and the following property of g:

for every sequence (1) satisfying the assumption of Lemma 10.6, if
$X = \{\delta_1, \ldots, \delta_{n-1}\}$ has $d + 1$ elements and there exists an increasing sequence of numbers $m_0, \ldots, m_{i+j}$ such that

$$m_0 = 0, \quad m_{i+j} = n + 1,$$
$$(E_{m_k + 1}, R_{m_k + 1}, N_{m_k + 1}), \ldots, (E_{m_{k+1}}, R_{m_{k+1}}, N_{m_{k+1}}) \text{ is}$$
X-invariant for $k < i$,
$$\{\delta_{m_k + 1}, \ldots, \delta_{m_{k+1} - 1}\} \text{ is a proper subset of X for k such}$$
that $i \le k < i + j$,

then $n \le g(d, i, j, bnd_1, prd_1)$.

We prove the above by induction on i and j using Lemma 10.4 and (12) as the induction hypothesis. ∎

Lemma 10.7. Suppose that a sequence (1) of equations with numbered dependencies satisfies (6) and let S be a function which satisfies (7). Then
$$n \le S_9(bnd_1, prd_1),$$
where the functions S_9 and h are defined by induction as follows

$$S_9(1, p) = h(S_5(1) \cdot (p + 2) + 1, 1, p),$$

(C)
$$h(0, 1, p) = 0,$$
$$h(i + 1, 1, p) = h(i, 1, p) + 1 + S_8(1 \cdot 2^{h(i, 1, p)}, p).$$

Proof. It follows from Theorem 9.7 that there exist δ, $k \le S_5(bnd_1)(prd_1 + 2)$ and an increasing sequence $m_1, \ldots, m_k$, such that

$\delta = \delta_{m_1} = \delta_{m_2} = \ldots = \delta_{m_k}$ and $\delta \neq \delta_j$ for the other $j \leq n$. We put $m_{k+1} = n$. By easy induction on i we prove using Lemma 10.6 that $m_i \leq h(i, 1, p)$ for $i \leq k + 1$. $\blacksquare$

Lemma 10.8. Suppose that E_1 is a reduced generalized equation and R_1 is a minimal solution of E_1. Let $(E_1, R_1), \ldots, (E_n, R_n)$ be a sequence of a generalized equations with solutions such that

    1) $(E_{i+1}, R_{i+1}) := BTR(E_i, R_i)$ for $i < n$,

    2) $cpl_i = cpl_1$ for $i \leq n$,

    3) $exc(E_i, R_i) > mcff_i$ for $i \leq n$,

    4) if $[x, y)$ is an essential object in E_n, then $lh(R_n[x, y)) > mcff_n$.

If S is a function which satisfies (7), then

$$n \leq S_{10}(bnd_1, prd_1),$$

where S_{10} and k are defined by induction as follows

$$S_{10}(1, p) = k(2^{2 \cdot 1}, 1, p) + 1,$$

(D) $k(0, 1, p) = 0,$

$$k(i + 1, 1, p) = k(i, 1, p) + S_9(1 \cdot 2^{k(i,1,p)}, p).$$

Proof. It is easy to see that $S_{10}(1, p) \geq 2$. So, we can assume that $n \geq 3$. If $(E_2, R_2) := BTR1(E_1, R_1)$, then $cpl_2 < cpl_1$ by Lemma 7.2. So, $(E_2, R_2) := MTR(E_1, R_1)$. Now, Lemma 7.3 implies that E_2 is reduced. In this way, by induction we can prove that $(E_{i+1}, R_{i+1}) := MTR(E_i, R_i)$ and E_{i+1} is a reduced for all $i < n - 1$. By Lemma 6.1, there exists an increasing sequence $m_0, m_1, \ldots, m_{j+1}$ such that

$$m_0 = 1, \; m_{j+1} = n - 1 \text{ and } j \leq essdep_1,$$

$$essdep_i = essdep_{i+1} \text{ if } m_u \leq i < m_{u+1}, \; u \leq j.$$

Now, we can define N_i such that for $u \leq j$

$$(E_{m_u}, R_{m_u}, N_{m_u}), \ldots, (E_{m_{u+1}} - 1, R_{m_{u+1}} - 1, N_{m_{u+1}} - 1)$$

are sequences of equations with numbered dependencies. It is easy to see that these sequences satisfy the assumptions of Lemma 10.7. We can prove by induction on i using Lemma 6.1 that $m_i \leq k(i, 1, p)$ for $i \leq j + 1$. $\blacksquare$

Decidability of solvability problem for generalized equations follows from the above lemmas, in particularly from Lemma 10.8. To make sure we shall study the Makanin's algorithm which is described below. This algorithm consists of repeated transformations of a generalized equation into a reduced generalized equation and next on decreasing of complexity of this reduced equation using the basic transformation, more precisely:

THE MAKANIN'S ALGORITHM

Input: a generalized equation E.

repeat
  repeat
    E_1 := E;
    E := RTR(E)
  until E_1 = E;
  repeat
    c := cpl;
    choose either 1), 2) or 3)
      1) test if E has a solution of the length not greater than S_4(bnd)·mcff, if it so, then we output "E is solvable",
      2) for any essential object [x, y) we choose an essential dependence d which applies to d, we create a coefficient <x, y> and we assign a word of the length not greater than mcff to it and then we eliminate d using ELDC,
      3) E := MTR(E)
  until cpl < c
until cpl = 0.

If we transform E consistently with R, then we choose 1) if exc(E, R) ≤ mcff and we choose 2) if there exists an essential object [x, y) such that lh(R[x, y)) ≤ mcff. In the other case we choose 3). □

The Makanin's algorithm is non-deterministic. It means that the Makanin's algorithm is not rigorously defined: if we execute this algorithm, then in many places we should choose one of finitely many possibilities. But we can prove that the Makanin's algorithm has the following properties:

if we transform a generalized equation E according to the Makanin's algorithm, then and we get an answer "E is solvable", then E is solvable,

if a generalized equation E is consistent, then we can transform E according to the Makanin's algorithm in such a way that we get a positive answer making at most certain well defined number of transformations.
It is not difficult to see that we can define also a usual deterministic algorithm which decides the solvability problem for generalized equations. Now, we shall prove the above properties of the Makanin's algorithm.

Theorem 10.9. If we get an output "E is solvable" during a transformation of a generalized equation E according to the Makanin's algorithm, then E is solvable.

Proof. Only RTR and MTR give an output "E is solvable". So, Theorem 10.9 follows easy from Lemma 4.5 and from simple observation (we do not execute elementary transformations only). ∎

Lemma 10.10. For every a generalized equation E
   1) $cpl \geq 0$,
   2) if E is reduced and $cpl = 0$, then $bnd = 1$,
   3) if $essdep = 1$, then $cpl = 0$.

Proof. Let l_0, ..., l_b be an increasing sequence of all closed boundaries in E. It is obvious that $rng(l_i) \leq 1/2 \cdot near(l_i) = 1/2 \cdot essdin[l_i, l_{i+1})$. So, $rng = \sum_{i<b} rng(l_i) \leq 1/2 \cdot \sum_{<b} essdin[l_i, l_{i+1}) = essdep$ and as a consequence of it, $cpl \geq 0$. Now, assume that $\emptyset \neq \infty$. If E is reduced, then $2 \cdot cpl \geq cls$ (see proof of Lemma 5.12). So, $cpl > 0$. It means that if E is reduced and $cpl = 0$, then $\emptyset = \infty$ and $bnd = 1$. If $essdep = 1$, then either there exists exactly one closed boundary x, $rng(x) > 0$ and $rng(x) = 1$ for such x or there exist exactly two closed boundaries x, y, $rng(x) > 0$, $rng(y) > 0$ and these boundaries satisfy $rnd(x) = rnd(y) = 1/2$. In the both cases $rng = 1$ and so, $cpl = 0$. ∎

Makanin's Theorem 10.11. (Makanin, [MA3]) Let R be a minimal solution of a generalized equation E. If we transform E using the Makanin's algorithm consistently with R, then we execute the transformations RTR and BTR at most $S_{11}(cpl, bnd, prd, cf)$ times, where cf is the number of words in $\mathcal{A}^*$ of the length not greater than mcff and S_{11} is defined by induction as follows:
$$S_{11}(0, 1, p, x) = S_2(1 + 2, x),$$
$$S_{11}(c + 1, 1, p, x) = S_2(2 \cdot c + 1 + 4, x) + S_{10}(2 \cdot c + 1 + 4, p)$$
$$+ S_{11}(c, (2 \cdot c + 1 + 4) \cdot 2^{S_{10}(2 \cdot c + 1 + 4, p)}, 1, x),$$
$$S(1, p) = S(c, 1, p, x) = 3 \cdot 2^{S_{11}(c, 1, p, x)}$$

S_{10} is defined by systems of equalities (A), (B), (C) and (D), (the function S and functions in (A), (B), (C), (D) have two additional arguments c and x; S_2 is defined in Lemma 1.2.).

Proof. Let R be a minimal solution of a generalized equation E. Let C be the set of all words in $\mathcal{A}^*$ of the length not greater than mcff. It follows from Lemma 4.7 and from a simple observation that all generalized equations E_1 which are given during the execution of the Makanin's algorithm satisfy rng $W_1 \subseteq C$. Let cf be the number of elements of C.

We shall prove this theorem by induction on cpl. If cpl = 0, then by Lemma 7.1, after several executions of RTR we get a reduced generalized equation E_1. By Corollary 5.10 and Lemma 10.10, $cpl_1 = 0$ and so $bnd_1 = 1$. The next execution of RTR outputs "E is solvable" and finishes the execution of the Makanin's algorithm. By Theorem 5.13, the transformation RTR is executed no more than $S_2(bnd + 2, cf) = S_{11}(0, bnd, prd, cf)$ times.

Assume that cpl > 0. We shall use Lemma 10.8. First, we notice that the function S, $S(1, p) = S(cpl - 1, 1, p, cf)$, which is defined above satisfies the condition (7). It is a consequence of the induction hypothesis, Lemma 5.11 and Lemma 8.4. We can prove it in a similar way as Theorem 8.5 or Theorem 5.13 5).

The equation E is first transformed using RTR. By Lemma 7.1, Lemma 4.4, Lemma 4.6, Theorem 5.13 and Corollary 5.10 we eventually get a reduced generalized equation E_1 and a minimal solution R_1 of E_1 such that $prd_1 \leq prd$, $bnd_1 \leq 2 \cdot cpl + bnd + 2$ and $cpl_1 \leq cpl$. By Theorem 5.13, it takes no more than

(13) $S_2(2 \cdot cpl + bnd + 2, cf)$

of executions of RTR. If $exc(E_1, R_1) \leq mcff_1$, then the Makanin's algorithm does not use the transformation RTR or BTR subsequently. So, in this case Theorem 10.11 follows from the obvious inequality $S_2(2 \cdot cpl + bnd + 2, cf) \leq S_{11}(cpl, bnd, prd, cf)$.

Now, we can assume that $exc(E_1, R_1) > mcff_1$. It follows from Lemma 10.8 and Lemma 10.10 that after execution of BTR no more than

(14) $S_{10}(bnd_1, prd_1) \leq S_{10}(2 \cdot cpl + bnd + 2, prd)$

times we get a generalized equation E_2 and a minimal solution R_2 of E_2 such that $cpl_2 < cpl_1$ or there exists an essential object $[x, y]$ in E_2 such that $lh(R_2[x, y)) < mcff_2$. By Lemma 6.1, $bnd_2 \leq 2^{S_{10}(bnd_1, prd_1)} \cdot bnd_1$. In the second case the Makanin's algorithm replaces some dependence by a coefficient. So, by Lemma 7.2 we get in this way an equation E_3 such that $cpl_3 < cpl_1$ and $bnd_3 = bnd_2$. By induction hypothesis, the Makanin's algorithm transforms E_2 or E_3 applying RTR or BTR no more than

(15) $S_{11}(cpl - 1, (2 \cdot cpl + bnd + 2) \cdot 2^{S_{10}(2 \cdot cpl + bnd + 2, prd)}, prd, cf)$

times. So, if we transform E using the Makanin's algorithm consistently with R, then we execute the transformations RTR or BTR no more times than the sum of numbers (13), (14) and (15). This sum is equal to $S_{11}(cpl, bnd, prd, cf)$. ■

XI. THE NUMBER OF STEPS OF MAKANIN'S ALGORITHM

We conclude this paper with the proof that the functions S_8, S_9, S_{10}, S_{11} and S are not primitive recursive. These functions have definitions, which contain a part similar to the definition of S_9 and therefore, they grow faster than the Ackermann's function Ac. Ac is defined by recursion as follows:

$$Ac(0, n) = n + 1,$$
$$Ac(m + 1, 0) = Ac(m, 1),$$
$$Ac(m + 1, n + 1) = Ac(m, Ac(m + 1, n)).$$

It is well known that Ac has the following properties.

Theorem 11.1. The function Ac is increasing with respect to both arguments. If p is a primitive recursive function, then there exists a natural number m such that

$$p(x_1, \ldots, x_n) \leq Ac(m, \max \{x_1, \ldots, x_n\})$$

for all $x_1, \ldots, x_n$. ∎

It follows easy by Theorem 11.1 that functions which grow faster then Ac are not primitive recursive.

Notice that functions S, S_7, g, f, S_8, h, S_9, k, S_{10} and S_{11} defined in the Chapter X are non-decreasing and they satisfy

$$S(c - 1, 1, p, x) = S(1, p) \leq S_7(1, p) \leq g(d, 1, 0, 1, p) \leq$$
(16) $$\leq f(d + 1, 1, p) = S_8(1,p) \leq h(1, 1, p) \leq S_9(1, p) \leq$$
$$\leq k(1, 1, p) \leq S_{10}(1, p) \leq S_{11}(c, 1, p, x) \leq S(c, 1, p, x)$$

for $d = 2^{2 \cdot 1} - 1$ and for all the other arguments (as in Chapter X, arguments c and x are omitted).

We can prove that

Lemma 11.2. Let S, T, u, W be non negative integer functions and let $U(c, 1) = u(c, 1, W(c, 1))$. Assume that $u(c, 1, 0)$, $T(c, 1)$ are non decreasing with respect to 1 and $W(c, 1)$ is increasing with respect to 1. If

1) $S(c, 1 + 1) \leq T(c, 1 + 1)$,
2) $u(c, 1, i + 1) = u(c, 1, i) + 1 + T(c, 1 \cdot 2^{u(c,1,i)})$,
3) $U(c, 1 + 1) \leq S(c + 1, 1 + 1)$

for any c, 1, i, then

(17) $$Ac(c, 1) < U(c, 1 + 2)$$

for any c, 1.

Proof. It is obvious that the function u is increasing with respect to i. By induction on i we prove that u is non-decreasing with respect to l. The inequality (17) holds for c = 0 since

$$U(0, l + 2) = u(0, l + 2, W(0, l + 2)) \geq l + 2 > Ac(0, l).$$

Assume that (17) holds for some c. By induction on l we prove that it holds for c + 1. Notice that if $U(c + 1, l + 2) > Ac(c + 1, l)$, then

$$U(c + 1, l + 3) = u(c + 1, l + 3, W(c + 1, l + 3)) \geq$$
$$\geq u(c + 1, l + 2, W(c + 1, l + 2) + 1) \geq$$
$$\geq T(c + 1, (l + 2) \cdot 2^{u(c + 1, l + 2, W(c + 1, l + 2))}) \geq$$
$$\geq T(c + 1, U(c + 1, l + 2) + 2) \geq S(c + 1, U(c + 1, l + 2) + 2) \geq$$
$$\geq U(c, U(c + 1, l + 2) + 2) > Ac(c, U(c + 1, l + 2)) >$$
$$> Ac(c, Ac(c + 1, l)) = Ac(c + 1, l + 1).$$

Similarly we can prove that $U(c + 1, 2) > Ac(c + 1, 0).$ ∎

Theorem 11.3. The functions S_8, S_9, S_{10}, S_{11} and S are not primitive recursive.

Proof. We put

$$S(c, l) = T(c, l) = f(c, l, p),$$
$$u(c, l, j) = g(c, 0, j, l, p),$$
$$W(c, l) = S_7(l, p) \ (= S_7(l \cdot 2^{g(c, 0, 0, l, p)}, p)).$$

These functions satisfy the assumptions of Lemma 11.2. So,

$$Ac(c, l) < U(c, l) = g(c, l, 0, l, p)$$

and $g(c, l, 0, l, p)$ is not primitive recursive. If $c = 2^l - 1$, then

$$Ac(c, l) < g(c, l, 0, l, p) \leq f(c + 1, l, p) = S_8(l, p).$$

Hence $Ac(2^l - 1, l) < S_8(l, p)$ and by Theorem 11.1, S_8 is not primitive recursive. The inequalities (16) imply that S_9, S_{10}, S_{11} and S are not primitive recursive. We can use also Lemma 11.2. In order to prove in this way that S_9 is not primitive recursive we should put

$$S(c, l) = S(c - 1, l, p, x),$$
$$T(c, l) = S_8(l, p) = S_8(c, l, p, x),$$
$$u(c, l, i) = h(i, l, p) = h(c, i, l, p, x),$$
$$W(c, l) = S_5(l) \cdot (p + 2) + 1.$$

In this case, $U(c, l) = S_9(c, l).$ ∎

INDEX OF NOTIONS AND SYMBOLS EXCEPT GIVEN IN CHAPTER I AND CHAPTER III

1. Non-elementary transformations
   RTR0, RTR1, RTR2, RTR3, RTR4, RTR, Reducing Transformation
                                           - begin of Chapter V,
   MTR, Main Transformation, BTR0, BTR1, BTR, Basic transformation
                                           - begin of Chapter VI.
2. Kinds of generalized equations
   reduced generalized equations   - begin of Chapter VII,
   linear generalized equations   - begin of Chapter VIII,
   generalized equations with (a solution and) numbered dependencies
                                            - begin of Chapter IX.
3. Notions connected with generalized equations with numbered dependencies
   i^{th} dependencies, δ_i, q_i, α_i, β_i, γ_i $l_i(j)$, $f_i(j)$, $s_i(j)$
                                           - begin of Chapter IX,
   A, A_1, B, B_i, E', E_i', E", E_i'', R", R_i''
   X - invariant sequence of a generalized equation
                                           - first three pages of Chapter X.
4. Functions

S_1	- Lemma 1.1,		S_7	- Theorem 10.4,
S_2	- Lemma 1.2,		S_8, f, g	- Theorem 10.6,
S_3	- Theorem 7.3,		S_9, h	- Theorem 10.7,
S_4	- Theorem 8.7,		S_{10}, k	- Theorem 10.8,
S_5	- Theorem 9.7,		S_{11}	- Theorem 10.11,
S_6	- Theorem 10.3,		Ac	- Chapter XI,
S	- (7) in Chapter X, Theorem 10.11.			

REFERENCES

[JAF] J. **Jaffar**, *Minimal and Complete Word Unification*, Journal of the Association for Computing Machinery, 37, 1(1990), pp. 47 - 85.
[KP1] A. **Kościelski**, L. Pacholski, *Complexity of Unification in Free Groups and Free Semi-Groups*, Proceedings 31st Annual Symposium on Foundations of Computer Science, v. II, (1990), pp. 824 - 829.
[KP2] A. **Kościelski**, L. Pacholski, *Is Makanin's Algorithm Deciding Solvability of Equations in Free Groups Primitive Recursive?*.
[KP3] A. **Kościelski**, L. Pacholski, *On the Exponent of Periodicity of a Minimal Solution of a Word Equation*.
[MA1] G. S. **Makanin**, *The problem of solvability of equations in a free semigroup*, Matematiceskii Sbornik, 103, 2(1977), pp. 147 - 236 (in Russian), English translation in Math. USSR Sbornik, 32, 2(1977), pp. 129 - 198.
[MA2] G. S. **Makanin**, *Equations in a free group*, Izviestiya AN SSSR, 46(1982), pp. 1199 - 1273 (in Russian), English translation in Math. USSR Izv. 21(1983).
[MA3] G. S. **Makanin**, *Decidability of the universal and the positive theories of a free group*, Izviestiya AN SSSR, 48, 4(1984), pp. 735 - 749 (in Russian), English translation in Math. USSR Izv., 25(1985).
[RAZ] A. A. **Razborov**, *On systems of equations in a free group*, Izviestiya AN SSSR, 48(1984), pp 779 - 832 (in Russian), English translation in Math. USSR Izv., 25(1985).

Implementation of Makanin's Algorithm

Habib Abdulrab
LIR/LITP
Faculté des Sciences, B.P. 118, 76134 Mont-Saint-Aignan Cedex †
and
LMI/INSA de Rouen
B.P. 08, 76131 Mont-Saint-Aignan Cedex

E.m.: abdulrab@geocub.greco-prog.fr

ABSTRACT : *This paper presents an introduction to Makanin's Algorithm, and discusses its potential role in the core of a programming system. The basic notions and steps are described via some examples. Some simplifications and improvements to this algorithm leading to an effective implementation are described. Particular attention is paid to the description of the elimination of equations with schemes which have no solution, the strategy of the construction of the algorithm's tree, the representation of position equations, and the resolution of systems of linear diophantine equations (SLDE). This last problem is presented here as a direct application of Gomory's algorithm finding integer solutions to SLDE. Our implementation of Makanin's algorithm presents an interactive system written in LISP and running on LISP Machine, and on VAX.*

1. Introduction

The problem of solving word equations arises in many areas of theoretical computer science, but especially in the unification of formal systems ([6], [7]). The string unification in PROLOG-3 [5] illustrates the potential application of solving word equations in programming languages. The attempts to fuse LISP and logic programming [10], by importing basic mechanisms used in logic programming, especially the unification concept, provides another example. The present implementations of word unification in programming languages impose a very important restriction to the resolution of word equations. On the other hand, Makanin's algorithm provides a general method for solving word equations. But the implementation as described in this paper, is not efficient enough to be used as a unification module in a programming language. We discuss the algorithmic reasons for this and some perspectives for the development of the implementation.

The study of word equations has been tackled by Markov who gave an algorithm to decide whether a word equation in two variables has a solution or not. Hmelevskii [9] solved equations in three variables. Makanin [11] showed that solving arbitrary equations is decidable. He gave an algorithm to decide whether a word equation with constants has a solution or not. His labour-consuming algorithm is described in 70

† This work was also supported by the Greco de Programmation du CNRS and the PRC Programmation Avancée et Outils pour l'Intelligence Artificielle.

pages. Pécuchet [13] gave a new description of Makanin's algorithm. We have provided [1] some simplifications and improvements to this algorithm leading to an effective implementation.

Let us start with some formal definitions.

Let X be a finite set (alphabet); we denote by X^* the set of all finite sequences (words) over X. The empty word is denoted by the symbol 1. The length of a word w (the number of letters composing it) is denoted by $|w|$.

Given two disjoint alphabets $\mathbf{V}$ and $\mathbf{C}$, a _word equation_ e is an ordered pair (e_1, e_2) of elements of $\mathbf{L}^* = \{\mathbf{V} \bigcup \mathbf{C}\}^*$.

The alphabet $\mathbf{V}$ is called the alphabet of variables (denoted by $x, y, z \dots$). $\mathbf{C}$ is called the alphabet of constants (denoted by $A, B, C \dots$).

A _solution_ of the equation e is a mapping $\alpha : \mathbf{V} \longrightarrow \mathbf{L}^*$ such that the images of e_1 and e_2 obtained by substituting $\alpha(v)$ for each variable v in e.

the substitution of each variable v by $\alpha(v)$ in e, are identical.

EXAMPLE 1 :

The equation $e = (AyB , xxz)$ has the solution α:
$$\alpha(x) = AB, \; \alpha(y) = BA, \; \alpha(z) = 1.$$

EXAMPLE 2 :

The equation $e = (Axxy , yyB)$ has no solution.

A solution α in which no variable $v \in \mathbf{V}$ has the image $\alpha(v) = 1$ is said to be **continuous**. The **projections** of an equation f are given by deleting each subset of $\mathbf{V}$ from f.

EXAMPLE 1 (continued) :

The projections of the previously considered equation $e = (AyB , xxz)$ are given by:

$(AB , 1), (AyB , 1), (AB , xx), (AB , z), (AyB , z), (AB , xxz), (AyB , xx), (AyB , xxz$

The projection (AyB , xx) of e has the continuous solution α, given by $\alpha(x) = AB, \alpha(y) = BA$.

It is obvious that an equation has a solution if and only if one of its projections has a continuous solution.

2. The unification in PROLOG-3

It is well-known that the algorithm of unification [14] is the heart of the PROLOG language. PROLOG-3 [5] proposes a major modification to this algorithm. In this section the modification is discussed and the potential role of Makanin's algorithm in this area is demonstrated.

Here is an example, given in [5], of a program written in PROLOG-3:

$$\{z : 10, \; < A, B, C > .z \; = \; z. < B, C, A >\}?$$

The result of running of this program is

$$\{z \ = \ < A, B, C, A, B, C, A, B, C, A >\}$$

More precisely, this program computes the list z, which produces the same list if it is appended at the right hand side of the list $< A, B, C >$ and at the left hand side of the list $< B, C, A >$, such that the length of z is equal to 10.

In other terms, this program computes a solution α of the equation $ABCz = zBCA$ such that $|\alpha(z)| = 10$. A solution satisfying this condition is $z = ABCABCABCA$

It must be observed from this example that PROLOG-3 takes into account the associativity of the operation of concatenation (denoted by .). In equational terms, this implies that the value of a variable given by the solution α may be a sequence of terms, not simply one term as in the case of a classic PROLOG. This major difference provides a very powerful tool of formal computation. On the other hand, it is costly as the size of the unification program is multiplied by 50 in PROLOG-3 [5].

However, there is an important restriction: it is necessary to specify the length l of each variable x, used in the equation to solve, by the condition $x : l$. In the program given above, $z : 10$ is an example of a constraint which must be given explicitly. Note that the classic unification algorithm is replaced by an algorithm solving a **system of constraints** which becomes the heart of PROLOG-3. Of course, this must be efficient, which is why Colmeraur [5] justifies, the necessary condition on the length of variables mentioned above. Note that this restriction avoids solving word equations which have no solution.

3. Informal description of Makanin's algorithm

We describe here the basic notions of Makanin's algorithm and its general behavior via an example.

Consider the above-mentioned equation $e = (AyB , xxz)$.

The first step of the algorithm consists in the computation of all the projections of e in order to find a continuous solution to one of these projections.

The second step consists in associating, for each projection $p = (p_1, p_2)$ of e, all the possible ways of choosing the positions of the symbols of p_1 according to those of p_2. The following diagram illustrates one possibility for the projection (AyB, xx). (see Figure 1).

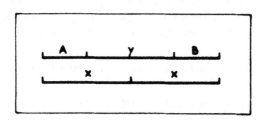

Now, this **scheme applicable** to p will be transformed into a so-called **position equation**. This new object inherits the seven boundaries of the scheme and of all occurrences of constants (these occurrences are called **constant bases**), but variables will be treated in a special manner.

Single occurrence variables, such as y, will disappear.

The n occurrences $(n > 1)$ of other variables are replaced by $2n - 2$ new variables associated via a symmetrical binary relation (called duality relation). These new variables are called **variable bases**.

The position equation E_0 computed from the previous scheme applicable to e is: (see Figure 2)

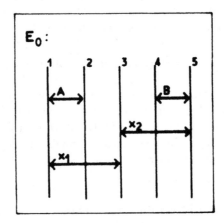

Here, x_1 is called the **dual** of x_2 and conversely.

After the second step of the algorithm (i.e. the computation of all the schemes applicable to all the projections of e), the algorithm develops a tree level by level. The tree is denoted by $\mathcal{A}$ and its levels by $L_i (i \geq 0)$.

The first level of $\mathcal{A}$ contains the position equations computed from the schemes applicable to the projections. In our example, L_0 contains the previous position equation E_0. It is important to observe that it is the only step where new variable bases will be generated. Their number will thereafter remain bounded.

The step from level $L_i (i >= 0)$ to level L_{i+1} is based on the **transformation** of position equations. There are five distinct types of position equations. According to its type, each position equation E existing in $\mathcal{A}$ is transformed into a set $T(E)$ of position equations.

We will describe here how to transform the position equation E_0 of L_0 in order to generate L_1.

The largest leftmost variable base of E_0 (i.e. x_1) is called the **carrier**. The first occurrence of A with a left boundary equal to 1 is called **leading base**. Having a carrier and another leading base characterizes one of the five types of position equations. The transformation of a position equation of this type consists in transferring the leading base A, in all the possible ways, under the dual of the carrier. There are two distinct ways to do the transfer. Either A takes all the space between the boundaries 3 and 4, or a part of this space. So $T(E_0)$ has the following two position equations (denoted respectively by E_1, E_2): (see Figure 3).

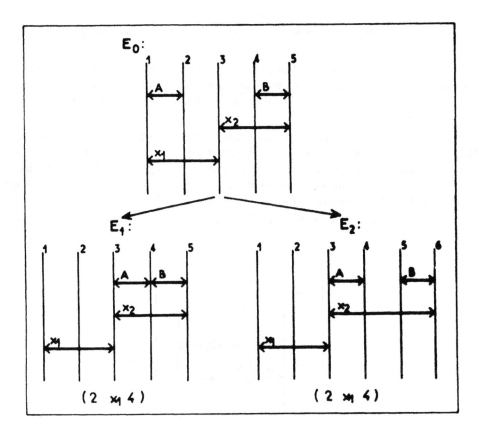

The list $(2\ x1\ 4)$ of the last two position equations is called a **connection**. Such an object is created in order to avoid any loss of information during this move. It plays the role of a link between old and new positions of A. This connection $(2\ x1\ 4)$ indicates that the prefix of $x1$ ending at boundary 2 is equal to the prefix of its dual (i.e. $x2$) ending at boundary 4.

Here, we transform only the first position equation E_1. The transformation of the second one is realized in the same way.

E_1 has a carrier with a right boundary greater than 2, and no leading base. This situation characterizes another type of position equations. The transformation of such a type consists in the transfer of all the boundaries, existing between the left and right boundaries of the carrier, into the dual of the carrier. In our example, we transfer the boundary 2, in all the possible ways, between the boundaries 3 and 5. This move can be realized in three ways:

1) The boundary 2 will be located between the boundaries 3 and 4.

2) The boundary 2 will be located between the boundaries 4 and 5.

3) The boundary 2 will be identified with the boundary 4.

Note here that the first two possibilities contradict the information, given by the connection- i.e: the segment between the boundaries 1 and 2 is equal to the segment between the boundaries 3 and 4. These two possibilities are not **admissible**, and must be eliminated.

More precisely, a system of linear diophantine equations (called the system of **length equations**) $AX = B$, A and B with integer entries, is associated with each position equation. A position equation E is called _admissible_ when this system has a non-negative integer solution. Fundamentally, this system has a non-negative integer solution whenever the lengths of the bases of E are consistent.

The transformation of E_1 gives rise to the following position equation E_3: (see Figure 4).

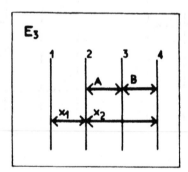

Note that the connection is deleted.

The type of this position equation features each position equation having a carrier with right boundary equal to 2, and no other leading base.

The transformation of such a position equation consists in deleting the carrier and its dual, leading to the following position equation: (see Figure 5).

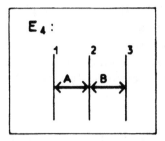

This last position equation has no carrier. The transformation of each position equation of this type consists in deleting the first boundary, and the leading base (if one exists). So we obtain: (see Figure 6)

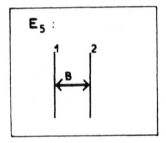

This last position equation is **simple** (that is, it has only one constant letter), and so the initial equation e has a solution.

Note that, the algorithm develops a tree level by level until we obtain an empty level or a level **equivalent** to a previous level, (this notion will be described in 4.2) in which case the initial equation e has no solution, or a level containing a simple position equation, in which case e has a solution.

The concept of position equation [13] is a constrainted version of the notion of generalized equation introduced by Makanin [11]. The only difference between this notion and that of Makanin's generalized equation is that the boundaries are *totally* ordered and that the right boundary of a constant is the successor of its left boundary. This concept provides a "geometrical" interpretation to the concept of generalized equation used by Makanin, and allows the processes of transformation to be described graphically. The transformation of position equations in the previous example differs from that of the original algorithm [11] by the reduction from 7 to 5 of the number of types of position equation [13]. The formal description of the notions and the operations used by the algorithm is given in [4].

The purpose of the algorithm being to decide whether an equation admits a solution or not, we provide an algorithm [2] which, by taking advantage of the tree $\mathcal{A}$, computes effectively a solution to the initial equation e whenever e is solvable. The idea is to compute a solution of e, from the scheme which generates the root of the subtree containing a simple position equation

In our example, the following solution $x = AB, y = BA, z = 1$ can be deduced from the scheme applicable to the projection (AyB, xx) given previously.

4. Implementation

The purpose of this section is to describe the main characteristics of the implementation. These provide some simplifications and improvements to this algorithm thus allowing an effective implementation solving non trivial equations.

4.1. Schemes applicable to an equation

Essentially, we [1] introduce the formal notion of a scheme applicable to an equation e to formalize the concept, used by Makanin's algorithm, of "mixing" the positions of the symbols of an equation, in all the possible ways.

Obviously, there are many possible ways of choosing the positions of the symbols of e_1 according to those of the symbols of e_2.

EXAMPLE 3 :

the following diagrams illustrate some possibilities for the equation $e = (xAz , AzB)$: (see Figure 7)

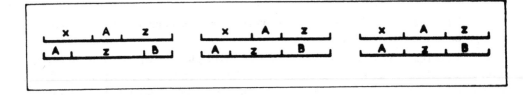

Informally, a scheme applicable to an equation $e = (e_1, e_2)$ indicates how to locate the positions of the symbols of e_1 according to those of e_2 in a possible solution of e.

Formally, a _scheme_ is any word $s \in = \{=, <, >\}^* =$, that is a word over the alphabet $\{=, <, >\}$ beginning and ending with the letter $=$.

A scheme s is called _applicable_ to an equation $e = (e_1, e_2), |e_1| \neq 0$, and $|e_1| \neq 0$, if the following conditions are satisfied [1]:

1) $S_< + S_= = |e_1| + 1$.
2) $S_> + S_= = |e_2| + 1$.

where S_ϕ, $\phi \in \{=, <, >\}$ is the number of occurrences of ϕ in s.

The left and right boundaries of a symbol t in a scheme s applicable to e are denoted by $lb(t)$ and $rb(t)$.

An equivalent definition is given by the function **schemes** shown below:

```
(defun schemes (l1 l2)
    (product '= (S l1 l2)))
(defun product (x l)
    (mapcar (lambda (y)
                    (cons x y))
            l))
(defun S (l1 l2)
    (if (and l1 l2)
        (if (not (or (cdr l1) (cdr l2)))
            '((=))
            (append
                    (product '= (S (cdr l1) (cdr l2)))
                    (product '< (S (cdr l1) l2))
                    (product '> (S l1 (cdr l2)))))))
```

Here is, for example, the value of $T_{(AxB,zzz)}$:

```
? (schemas '(A x B) '(z x z))
 ((= = = =) (= = < > =) (= = > < =) (= < = > =) (= < < > > =)
  (= < > = =) (= < > < > =) (= < > > < =) (= > = < =)
  (= > < = =) (= > < < > =) (= > < > < =) (= > > < < =))
```

One can observe [1] that the size of all the schemes applicable to an equation e, denoted by, T_e grows exponentially with the length of e. More precisely:

$$Card(T_e) = \sum_{i=0}^{n_1} C_{n_1}^i C_{n_1+n_2-i}^{n_1}$$

where $n_1 = |e_1| - 1$, and $n_2 = |e_2| - 1$.

In addition, T_e contains generally a very important number of schemes that can be eliminated because they imply some contradictions on the lengths or the values of the letters of e.

EXAMPLE :

- The length of z in the second scheme of Example 3 is both greater than and equal to the length of a constant base.

The value of z in the third scheme of Example 3 is equal to both A and B.

These observations lead to a definition of the concept of a **solution** of a scheme applicable to an equation. The formal definition is given in [1].

We prove that an equation e admits a continuous solution if and only if one of its applicable schemes admits a solution. The concept of a solution of a scheme does not provide a procedure enabling a solution to the initial equation to be computed. On the other hand, it enables us to state some necessary conditions [1] satisfied by each scheme $s \in T_e$ which has a solution, and so, the application of the algorithm can be restricted to the only subset S_e of applicable schemes which satisfy these conditions. This approach plays a principal role in our construction of $\mathcal{A}$, the size of which can be greatly reduced by eliminating some types of schemes which have no solution.

EXAMPLE :

Consider the equation $e = (xyxyA, yxyByy)$. T_e contains 3653 schemes applicable to e, whereas S_e contains only 2 schemes applicable to e.

Here are two examples of the necessary conditions used in the definition of S_e:

1) Consider the smallest equivalence relation among the letters of e, defined as follows: two symbols of a scheme s applicable to e are equivalent when they have the same left and right boundaries. The concept of a solution of a scheme applicable to an equation enables us to show that in every scheme which has a solution, every class of symbols has at most one constant letter. Consequently, one of the characteristics of S_e is that the classes of symbols of every one of its schemes separate the constants.

2) Consider the smallest transitive relation (denoted by $<$), over the classes of symbols of the previous equivalence relation, satisfying the following two conditions:

a) let $class_1$ and $class_2$ be two distinct classes, then, by definition:
$$class_1 < class_2 \text{ if } lb(x_1') \geq lb(x_2') \text{ and } rb(x_1') \leq rb(x_2'),$$
where x_1' and x_2' are respectively two symbols of $class_1$ and $class_2$.

b) let $class$ be a class of symbols and $class_0$ be any class of symbols containing a constant symbol, then, by definition:
$class_0 < class$ if $rb(class) - lb(class) > 1$. ($class$ is said to be longer than a constant class).

The concept of a solution of a scheme makes it possible to prove that in every scheme which has a solution, the relation $<$ is a strict order relation, and so, all the schemes of S_e verify this condition.

We show in [1] how S_e can be constructed effectively. Of course, we do not proceed by computing first T_e and then removing all the schemes that do not satisfy the necessary conditions, but we compute S_e directly.

Note that the computation of S_e arises in several steps of the algorithm: the position equations of the first level of $\mathcal{A}$ are directly computed from S_e. The transformation of every position equation uses this computation to realize all the possibilities of the transfer. Finally, the computation of S_e arises in our algorithm which computes a solution to the equation e whenever one exists.

Note also that the computation of S_e can solve some types of equations and thus avoid calling Makanin's algorithm.

EXAMPLE :

-If for each projection p of an equation e, S_p is empty, then e has no solution.

-Let p be a projection of an equation e and l_p be the list of all the classes of symbols which are longer than a constant class. If l_p is empty, then e has a solution.

4.2. Representation of a position equation

The representation of a position equation, in our implementation, is designed to achieve conveniently the basic operations of the algorithm. One of these operations is to test whether, given a position equation, there exists an **equivalent** equation in $\mathcal{A}$.

We discuss here this notion and show how it is implemented.

In fact, the equality of two position equations is based on the following equivalence relation among position equations: two position equations E_1 and E_2 are called equivalent when they differ only by renaming of variables or constants.

Note that the correspondence of the names of the bases of E_1 and E_2 must conserve the definition of the relation between two dual variables. More precisely, if x_1 and x_2 are two dual bases of E_1, and x'_1 and x'_2 are the bases of E_2 corresponding to x_1 and x_2, then x'_1 and x'_2 must be duals in E_2.

Consider the three following position equations: (see Figure 8)

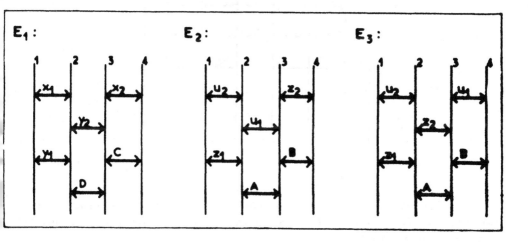

It is clear that E_1 and E_3 are equivalent, but E_1 and E_2 are not.

The equivalence test among the levels of $\mathcal{A}$, used in the halt test of the original algorithm of Makanin, consists in testing, whenever a level L of $\mathcal{A}$ is constructed, if there exists a level L', already constructed, such that the position equations of L are equivalent to those of L'. In which case the initial equation to be solved has no solution. This test can be deleted advantageously [1] from the algorithm by replacing it in the following manner: a position equation is deleted whenever it has an equivalent equation already existing in $\mathcal{A}$.

Two relations [1] are defined to realize efficiently the equivalence test between position equations. The first one is an equivalence relation among the bases: two bases are equivalent if and only if they have a common right and left boundary and if their duals have a common left and right boundary. The second one is a total order relation, denoted by $<$, among the classes (defined by the last equivalence relation) of bases having the same left boundary. It is defined as follows:

Let $class_1$ and $class_2$ be two classes of bases having the same left boundary. $class_1 < class_2$ if and only if the right boundary of a base of $class_1$ is less than the right boundary of a base of $class_2$ or if the left boundary of the dual of a base of $class_1$ is less than the left boundary of the dual of a base of $class_2$.

For example, all the bases of the following position equation are ordered according to $<$: (see Figure 9)

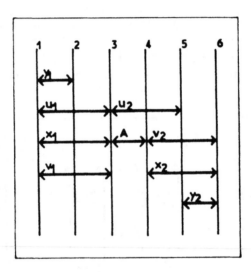

One can note that x_1 and v_1 belong to the same class, y_1 and u_1 are single in their classes. $class(u_1) < class(x_1)$ because $lb(u_2) < lb(x_2)$, and $class(y_1) < class(u_1)$ because $rb(y_1) < rb(u_1)$.

The list of bases having the same left boundary, in our representation of a position equation, is always ordered by the relation $<$. When a base is transferred, it is merged in the list of bases having the new left boundary. The relation $<$ plays a LISP's predicate role for the merging. The reason for this order is to facilitate the search of a bijection among the bases of E_1 and E_2. In fact, such a bijection is built on the correspondence, from left to right, between the elements of the ordered lists.

We now give an example of the internal representation of a position equation and discuss it briefly.

EXAMPLE :

Let E be the following position equation: (see Figure 10)

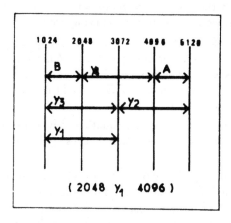

The representation of E is given by:

```
#[(#[1024 () (y1 y3 B)]    ; E is represented by an array of 7 elements.
   #[2048 (B) (y4)]        ; The first element is the list of the
   #[3072 (y1) (y2)]       ; boundaries. Each boundary b is represented by
   #[4096 (y4) (A)]        ; an array of three elements : 1) number of b.
   #[5120 (y2 A) ()])      ; 2) The list of the bases with right boundary
                           ; equal to the number of b.
                           ; 3) The ordered list of bases with left
                           ; boundary equal to the number of b.
   ((2048 y1 4096))        ; The second one is the list of connections.
   ((y1 . 1024) ... )      ; The third one is an A-list composed of the
                           ; variable bases and their right boundaries.
   ((y1 . 3027) ... )      ; The fourth one is an A-list composed of the
                           ; variable bases and their left boundaries.
   (y1 y3)                 ; The fifth one is a list composed from half of
                           ; the variables. Each variable has its dual
                           ; variable, in its P-list, under the
                           ; indicator dual.
   ((B . 1)(A . 1))        ; The sixth one is an A-list composed of each
                           ; constant and its number of occurrences.
   5]                      ; The last one is the number of boundaries.
```

Here are some observations concerning this representation:

1) If the boundaries of E are numbered by $1, 2 \ldots$, it is necessary to renumber these boundaries after each insertion or deletion of a boundary, and so, it is necessary to regenerate the data structures of the representation of E. In order to avoid this systematic renumbering, we proceed in the following way:

The boundaries of the position equations of the first level of $\mathcal{A}$ are numbered by $p, 2p, 3p \ldots$ (p is a power of 2, for example: 1024). When a new boundary is inserted between two boundaries k, k', it is numbered by $(k + k')/2$. Only one array of three elements is created for each new boundary of E, and all the areas of the other boundaries are those of the predecessor of E in $\mathcal{A}$.

2) The operation on each position equation E which consists in testing if E is simple, is realized easily in this representation, because (cdr (aref E 5)) = () if and only if E is simple.

3) All the functions operating on a position equation such as defining or reading its components, are defined as operations on a data type. Each computation over position equation in any module of our program uses these functions. This provides a great degree of freedom if we want, for example, to experiment other data structures for E. In fact, we have only to change these functions without altering the code of the other modules. Here are some simple examples of these functions:

```
(defun bases-with-left-boundary (boundary)
   (aref boundary 2))

(defun dual (variable)
   (get variable 'dual))

(defun create-boundary (number left-boundary right-boundary)
   (vector number left-boundary right-boundary))
```

. . .

4.3. Construction of $\mathcal{A}$

As we have mentioned above, the algorithm constructs $\mathcal{A}$ according to the "breadth first" strategy. After the construction of each level L_i, $i = 1, 2 \ldots$ it tests if L_i is equivalent to an already existing level.

We prove [1] that the equivalence test among the levels of $\mathcal{A}$, used in the original algorithm of Makanin, can be deleted advantageously from the algorithm by introducing the fact that a position equation must be deleted when it has an equivalent equation already existing in $\mathcal{A}$. Thus the complexity of the algorithm can be reduced by one exponential, and other strategies for constructing $\mathcal{A}$ can be introduced. The one described in the following paragraph [1] allows a faster halting of the algorithm when the initial equation has a solution.

Since the size of each subtree of $\mathcal{A}$ grows exponentially with the number of variable letters of the projection associated with its root, the first idea of this strategy is to construct the small subtrees before constructing the big subtrees. Intuitively, that is because:

1) If the simple position equation, which provides the halt of the algorithm, exists in a small subtree then, this idea generally provides a faster halting, because no level of the big subtrees is constructed.

2) Otherwise, when the simple position equation E exists in a big subtree, then the halt of the algorithm produced by this idea is not, in general, later than "breadth first" strategy. In fact, this strategy consists in starting at the first level, and then going down one level at a time, in order to construct all the nodes of each level. Consequently, the small subtrees are often already constructed by "breadth first" strategy, before finding the simple position equation E.

So, the projections of e are ordered according to the following relation:

$$(*) \; P_i < P_j \iff P_j \text{ has more variable letters than } P_i.$$

For example, the projections given in Example 1 are ordered according to (*).

The following function computes all the projections of an equation, ordered according to (*).

```
(defun projections (equation)
; computes all the projections of a word according to the relation (*).
   (if (null *alphabet-of-variables*)
       (list equation)
       (nconc
         (spprojections
           (set-difference equation *alphabet-of-variables*)
           (cdr *alphabet-of-variables*))
         (list equation)))))
(defun spprojections (projections alphabet)
  (cond
    ((null alphabet) projections)
    ((null (cdr alphabet))
     (append (set-difference (car projections) alphabet)
             projections))
    (t (nconc
         (append
           (spprojections
             (set-difference (car projections) alphabet)
```

```
        (cdr alphabet))
      (spprojections (cdr projections) (cdr alphabet)))
  (list (car projections))))))
```

Let $P_1 \dots P_{2\,card(v)}$ be the projections of e, ordered according to (*), and let $E(P_{k\,s(1)}), E(P_{k\,s(2)}) \dots$ be the position equations computed from the schemes $P_{k\,s(1)}, P_{k\,s(2)} \dots$ applicable to the projection P_k. The construction of $\mathcal{A}$ begins with the construction of the subtree whose root is $E(P_{1\,s(1)})$, followed by $E(P_{1\,s(2)}) \dots E(P_{j\,s(i)})$ ..., as shown by the following diagram: (see Figure 11)

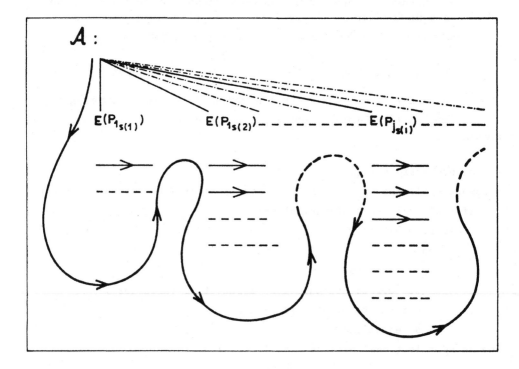

It can be observed that the two grand strategies "breadth first" and "depth first", are conjugated here.

The previous order (*) is based on the relationship between the size of a subtree and the number of variables n in the projection connected to its root. But, it is sometimes possible to find a subtree with a small number n, which is big enough to retard the construction of a further subtree containing a simple position equation. This is illustrated by the following diagram: (see Figure 12)

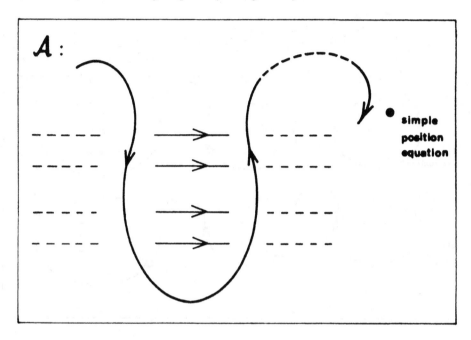

In order to avoid this situation, and to allow to our strategy to be more efficient, in all cases, the second idea is to add the following mechanism of **suspension**: the construction of each subtree is suspended after a number of levels proportional to the length of the projection connected with this subtree. When a subtree is suspended, its last level is saved in a queue. The next subtree is then constructed. As long as there is no simple position equation, the continuation of the construction of A from the queue is done with the same mechanism of suspension, i.e. the transformation of each level existing in the queue is suspended after the same number of levels indicated above. This mechanism is illustrated by the following diagrams: (see Figure 13)

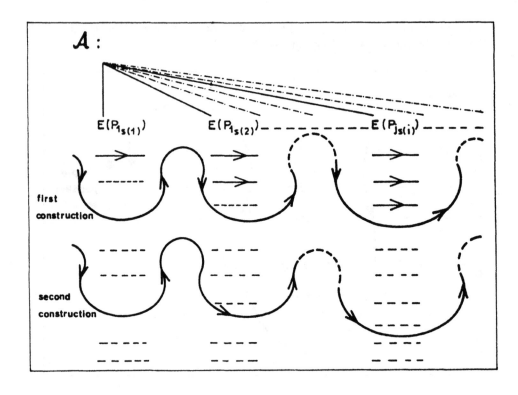

Note that, in order to solve word equations, in the sense of associative unification [4] (i.e. if the notion of the solution of word equation is defined as a mapping $\alpha : V \longrightarrow L^* - \{1\}$), it is sufficient to redefine the function projections as follows :

```
(defun projections (equation)
    (list equation))
```

Note also that when the initial equation has no solution, the two strategies construct the same number of position equations.

4.4. *Solving Systems of Linear Diophantine Equations*

Testing if a position equations E is admissible, *i.e* if the system of linear diophantine equations $Ax = B$ associated with E has a non-negative integer solution, is one of the important operations used in Makanin's algorithm. It is used to eliminate the non-admissible position equations from the tree $\mathcal{A}$. We show in [1] how to use Gomory's algorithm, called "cutting planes" method [8], to test whether a position equation is admissible. The idea of Gomory's algorithm is to use, as a first step, the simplex method in order to test if the system $Ax = B$ has a non-negative real solution. If such a solution is integer, then, it is an integer solution of the system $Ax = B$. Otherwise, a new constraint is imposed. This new constraint is a new equation computed from the solution resulted from the first step, and the entries of the system to be solved. This equation states an hyperplan whose intersection with the solution space S of $Ax = B$ gives a new solution space S', such that the following three conditions are satisfied:

a) Every integer solution in S remains in S'.

b) A portion of S, containing the solution of $Ax = B$ falls out S'.

c) The number of constraites is finite. That is, after a finite number of steps, either there will be a solution with integer values, or no solution.

In order to realize an efficient implementation of the admissibility test, we have experimented two algorithms of the "cutting plane" family. The first one is that outlined above. The second is the "all integer" algorithm, which differs from the first one by the fact that the simplex method is not used. Our expermimentation of both algorithmes consisted in comparing the resolutions of systems of linear diophantine equations given in literature, and especially be generating aleatorly, during many dayes, systems to be solved by both algorithms. We have observed that the first algorithm provides faster results. This may be explained by:

1) It is well-known that the simplex method is an efficient algorithm whose computational results are particularly promissing. Thus, it can be used to advantage for the integer problem.

2) If the system to be solved has no non-negative integer solution, because simply there is no positive real solution (this is rather frequent in this case), the halt of the resolution is obtained fast at the first step of the first algorithm.

3) Especially for grand systems, the convergence is, in general, faster in the first algorithm, because of both the number of iterations and the computational time of each iteration. On the other hand, the second algorithm can be implemented, with less difficulty, and gives interesting results in the case of small systems. Note that, new versions combining several methos [12] for solving SLDE, and new families of interesting cuts are given since Gromory's algorithm, but in order to conserve a reasonable volume for our system solving word equations, we have only used in our implementation the primitive cuts given by Gromory. (The module of solving SLDE constitutes about 7 per cent of the whole code of our system).

5. Why LISP?

The implementation of solving word equations is motivated by their importance for several applications, and in order to be used as a unification module, in a programming system based on symbolic computation. The choice of LISP is, of course, particularly interesting.

There are several other reasons justifying the choice of LISP to implement the algorithm. First, the objects used by the algorithm such as connections, boundaries ... etc, naturally have the list structure. The insertion or the deletion of a letter or a boundary is one of the primitive operations used frequently on these objects. It is well-known that LISP provides a very powerful tool for list computations.

Fundamentally, the algorithm explicitly constructs the tree $\mathcal{A}$ by allocating memory dynamically for all its position equations. If a position equation E_1, resulting from the transformation of another position equation, is not admissible or has another equivalent position equation already existing in $\mathcal{A}$, then E_1 is eliminated, its areas are no longer in use and must be collected. LISP provides several strategies for automatic "garbage collection" and frees the programmer from the management of memory.

Furthermore, LISP allows the programmer to allocate the data structures used in his program and manipulate their references easily. This allows us to let the position equations of $\mathcal{A}$ share the maximum of areas of lists and arrays.

LISP provides a variety of structures for flow of control. Consider the example of non-local exit. We have mentioned above that the computation of S_e is used for several applications in our implementation. But each application has its own interpretation of the results of the computation of S_e, and its own continuation after the computation. Both the structure of non-local exit, and the possibility to pass as argument a function which interprets the result and specifies its continuation means that it is possible to use the same text of LISP's functions for several similar applications.

More generally, all the basic tools given by LISP provide a very powerful programming environment, which is especially useful in implementing a very hard algorithm such as Makanin's algorithm. Note that this is the first implementation since Makanin's paper. Our implementation has also been coded in CAML by Rouaix [15].

6. Practical realization

The complete text of the LISP program is available in [1]. It runs on VAX780 under UNIX and LISP Machine. It is divided into the following modules:

Given two non empty words e_1 and e_2, the first module computes the sufficient set $S_{(e_1,e_2)}$ (cf. 4.1) of applicable schemes.

The second module computes the position equation created from a scheme s applicable to an equation e.

Given a position equation E, the third module determines whether E is admissible or not. We give a method [1] for the effective construction of the diophantine linear system associated with every position equation, and show how to adapt the "cutting plan" method [8] for solving such a system. The first component of this module is the computation of such a system. The second is divided into two parts calling one another: the first is the simplex program computing a non-negative real solution to systems of linear diophantine equations and the second is the program of the "cutting plan" method designed for the computation of a non-negative integer solution.

The fourth module contains the operations on the internal representation of a position equation.

The fifth module enables the equivalence test between two position equations to be realized.

The sixth module implements the core of the algorithm; it first determines the type of a position equation and then transforms it into a set of admissible and *normalized* position equations. The notion of normalization of position equations is described in [1]. This module is based on some technical results, described in [3], providing an efficient implementation of the algorithm from algorithmic and programming points of view. In fact, we show [3] that the position equations resulting from the transformation of four types of position equations (among five) are always normalized. We also show that one of the three conditions of normalization is always satisfied. These theoretical results play an interesting role, reducing the cost of testing the normalization. The computation of the position equations resulting from the transformation of a position equation is realized with minimal allocations of data structures.

We give an algorithm which computes a solution to the initial equation. The idea is to compute a solution to the scheme which generates the root of the subtree containing a simple position equation (we prove that such a scheme has a solution). So, given a scheme s applicable to an equation, the seventh module gives a solution to the initial equation e computed from s.

The eighth module contains graphical tools for drawing schemes and position equations.

The ninth module contains the program which constructs $\mathcal{A}$ according to the strategy described above. It calls all the other modules.

The user can choose interactively, either a "talkative" session allowing the tree $\mathcal{A}$, the applicable schemes and some information about the construction of $\mathcal{A}$ to be visualized, or a "silent" session providing only the result of the computation.

7. Perspectives

Our program was designed with several modules to enable us to extend our program to a general system providing a rich tool of formal computation on equations, and later, to extend this new system to a system of formal computations on words.

The size of the tree $\mathcal{A}$ grows exponentially with the length of the initial equation e, and its elements must be explicitly memorized. Some steps of the algorithm, such as determining the admissibility of a position equation, have a high degree of algorithmic complexity. These important obstacles prevent the efficient use of this implementation in the core of a programming system. However, we think that the extension of the approach which eliminates schemes which have no solution, the generalization of this approach for the position equations, a direct implementation of the module containing the operations on position equations in the core of our system, and a particular treatment of some families of equations will give some efficient tools. These can provide a new version of our implementation which may be a component of a programming system.

****

Some examples of execution of the LISP Machine's version and some computer outputs are given below:

Acknowledgements: I thank Jean-Pierre Pécuchet and Klaus Schulz for the improvements they suggested and for their helpful comments.

References:

[1] H. Abdulrab : Résolution d'Équations sur les Mots : Étude et Implémentation LISP de l'Algorithme de Makanin. (Thèse), University of Rouen (1987). And Rapport LITP 87-25, University of Paris-7 (1987).

[2] H. Abdulrab : Equations in Words. RAIRO d'Informatique Théorique, vol. 24, n. 2, p. 109-130 (1990) .

[3] H. Abdulrab : On the normalization lemma of Makanin's Algorithm. Research Rapport, LIR (1990) .

[4] H. Abdulrab, J-P. Pécuchet : Solving Word Equations, Journal of Symbolic Computation,(1989) 8, 499-521.

[5] A. Colmeraur : Une Introduction à PROLOG-3. Faculté des sciences de Luminy, Marseille. Juillet (1987).

[6] F. Fages, G. Huet : Complete Sets of Unifiers and Matchers in Equational Theories. Theoretical Computer Science, 43, (1986), p. 189-200.

[7] W. M. Farmer : A Unification Algorithm for Second-Order Monadic Terms, The Mitre Corporation, Bedford, Massachusettes, 10, (1986).

[8] R.E. Gomory : An Algorithm for Integer Solutions to Linear Programs. Recent advances in mathematical programming, Eds R.L Graves et p. Wolfe. p. 269-302 (1963).

[9] Yu. I. Hmelevskii : Equations in Free Semigroups. Trudy Mat. In st. Steklov, 107, (1971).

[10] T. Ito, T. Yuasa : Some Non-Standard Issues on LISP Standardization. International workshop on LISP evolution and standardization. AFCET, Paris (1988).

[11] G.S. Makanin : The Problem of Solvability of Equations in a Free Semigroup. Mat. Sb. 103(145) (1977) p. 147-236 English transl. in Math. USSR Sb. 32 (1977).

[12] M. Minoux : Programmation Mathmatique: Thorie et Algorithmes, Dunod, (1983). Mat. Sb. 103(145) (1977) p. 147-236 English transl. in Math. USSR Sb. 32 (1977).

[13] J.P. Pécuchet : Équations Avec Constantes et Algorithme de Makanin. (Thèse), University of Rouen, (1981).

[14]   A. Robinson :   A Machine-Oriented Logic Based on The Resolution Principle. Journal of the ACM, 12 decembre (1965).

[15]   F. Rouaix : Une Implémentation de l'Algorithme de Makanin en CAML. Mémoire de DEA d'Informatique, University of Paris-7, (1987).

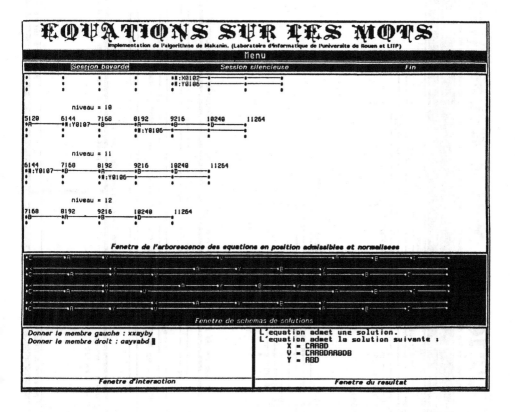

ÉQUATIONS SUR LES MOTS

Implementation de l'algorithme de Makanin. (Laboratoire d'informatique de l'université de Rouen et LITP)

Menu   Session silencieuse   fin

```
7552   7616   7632   7648   7664   7680   8192   9216   10240  11264
       •#:X0103  •#:Y0106                          •B     •R
•#:Y0109                                           •#:Y0107
•#:X0104
       •#:X0105
              •B
              •#:Y0106

(7632 X0105 Y0109 7664)
(7600 Y0109 10240)
(7616 Y0109 7664)


7552   7616   7680   8192   9216   10240  11264
       •#:X0103  •R     •#:X0102  •B     •R
•#:Y0109         •#:Y0106         •#:Y0107
•#:X0104
       •#:X0105
              •B
              •#:X0102
              •#:Y0106

(7600 X0105 Y0109 9216)
(7600 Y0109 10240 9216)
(7616 Y0109 9216)
```

Fenetre de l'arborescence des equations en position admissibles et normalisees

```
•B ——————•Y ——————•X ——————•Y ——————•R ——————•Y ——————•R
•X ——————•Y ——————•X ——————•Y ——————•B ——————•R ——————•R

•B ——————•Y ——————•X ——————•Y ——————•R ——————•B ——————•R
•X                                                      •R
Je continue a explorer le sous arbre dont la racine est engendree par le schema suivant :

•B ——————•Y ——————•X ——————•Y ——————•R ——————•R ——————•R
•X ——————•Y ——————•X ——————•Y ——————•R ——————•R ——————•R
```

Fenetre de schemas de solutions

L'equation n'admet pas de solution.

Fenetre du resultat

Donner le membre gauche : byxay
Donner le membre droit : xxyaba

Fenetre d'interaction

Makanin's Algorithm for Word Equations-
Two Improvements and a Generalization

Klaus U. Schulz

Centrum für Informations- und Sprachverarbeitung (CIS)
Universität München
Leopoldstr. 139, D-8000 München, Germany

Abstract: In 1977 G.S. Makanin [Mak] proved that it is decidable whether a word equation has a solution or not. Here we describe two improvements of Makanin's algorithm which bring it nearer to the area of practical applicability: a simple pre-algorithm is suggested which decides the solvability of word equations with not more than two occurrences of each variable and which partially solves and simplifies the decision procedure for all other equations. A new transformation procedure is given which applies to arbitrary position equations and has several advantages. In a separate part we generalize Makanin's result and show that the solvability of word equations with variables x1,...,xn remains decidable when we specify regular languages L1,...,Ln over the coefficient alphabet and ask for solutions where the i-th components belongs to Li.

Introduction

Makanin's algorithm [Mak] shows that it is decidable whether a word equation has a solution or not. The original decision procedure is extremely complex and was not designed for a direct implementation. Improvements were given by Pécuchet [Péc], Jaffar [Jaf], Kościelski-Pacholski [KoP] and a first implementation is available now (Abdulrab [Abd]).

Still, the procedure in its actual version is not efficient enough to be useful as a component of a practically interesting programming system. Here we want to give two improvements which may be regarded as further steps towards such a practical applicability. First we describe a pre-algorithm which allows a very simple partial analysis of word equations (section 2). For all equations where variables do not have more than two occurrences this analysis is even complete. For the remaining equations at least all variables with (one or) two occurrences are resolved, in a sense which will become clear. Moreover, the pre-algorithm influences and simplifies the remaining analysis which follows then Makanin's strategy.

Our second contribution concerns the transformation of position equations. The transformation procedure is the central subprocedure and may be seen as the motor of Makanin's algorithm. In the existing formulations of the algorithm, various types of position equations are distinguished and accordingly transformed in distinct manner. In section 4 we give a single transformation which applies to arbitrary position equations and has in most cases a better behaviour than case-dependent transformations. In particular, a maximal number of bases is transported, a non-trivial left part of the position equations is always erased and the introduction of new boundary connections is often completely avoided.

In combination our results imply that a solution S of a solvable equation E is found after at most ($nl(E)$ trivial and) I complex (non-deterministic) transformation steps, where $nl(E)$ is the notational length of E and I is the sum of the lengths of the components X_i of S.

The paper is structured as follows: in the first section we describe the basic algorithm for word equations by Lentin [Len], Plotkin [Plo] and Siekman[Si1]. In the second section we show how the transformation steps of this algorithm may be used for the simplifying pre-algorithm. In the third section we provide the theoretical background which is necessary for a self-contained representation of the complete algorithm. The fourth section describes the new transformation strategy and the complete search tree. The technical proof of the transformation theorem is omitted here. It may be found in [Sc3]. In section five we discuss examples and add some remarks. Eventually we give the following generalization of Makanin's result in section 6: for every word equation E with variables $x_1, \ldots, x_n$ and coefficients in the alphabet $\mathcal{C}$ and for arbitrary regular languages $\mathcal{L}_1, \ldots, \mathcal{L}_n$ over $\mathcal{C}$ it is decidable whether E has a solution $(X_1, \ldots, X_n)$ where $X_i \in \mathcal{L}_i$ ($i = 1, \ldots, n$). From an algebraic point of view this implies, for example, that it is decidable whether the word equation has a solution where the components may be found within given boolean combinations of finitely generated submonoids.

1 The Basic Algorithm

The concept of a word equation and its solutions is introduced. We describe the basic algorithm for word unification given by Lentin[Len], Plotkin [Plo] and Siekmann [Si1]. In contrast to Makanin's [Mak] algorithm, the basic algorithm does not always terminate. The transformation steps which are used are, however, much simpler than the corresponding steps of Makanin's procedure. We include this section since our reformulation of Makanin's algorithm is based on the idea of using this simple transformation steps as long as possible.

Let

$$C = \{a_1, \ldots, a_r\} \qquad (r \geq 1)$$
$$\mathcal{X} = \{x_1, \ldots, x_n\} \qquad (n \geq 0)$$

be two disjoint finite alphabets of *coefficients* and *variables*. A *word equation* is an equation E of the form

$$\sigma_1 \ldots \sigma_k == \sigma_{k+1} \ldots \sigma_{k+l}, \qquad (1.1)$$

where $k, l > 1, \sigma_i \in \mathcal{C} \cup \mathcal{X}$ $(1 \leq i \leq k + l)$. The number $k + l$ is the *notational length* $nl(E)$ of E. A *solution* of E is a sequence

$$S = (X_1, \ldots, X_n) \qquad (1.2)$$

of non-empty words over $\mathcal{C}$ such that both sides of E become graphically identical when we replace all occurrences of x_i by X_i $(1 \leq i \leq n)$. The word $S(\sigma_1 \ldots \sigma_k)$ is defined in the obvious way, regarding $S : (\mathcal{C} \cup \mathcal{X})^+ \to \mathcal{C}^+$ as the unique morphism mapping a_i to a_i $(1 \leq i \leq r)$ and x_j to X_j $(1 \leq j \leq n)$. The standard terminology of unification theory may be simplified in the case of word equations. A *unifier* of E is a sequence

$$U = (U^{(1)}, \ldots, U^{(n)}) \qquad (1.3)$$

of non-empty words over $\mathcal{C} \cup \mathcal{X}$ such that both sides of E become graphically identical when we replace all occurrences of x_i by $U^{(i)}$ $(1 \leq i \leq n)$. Thus solutions are special unifiers. The unifier $V = (V^{(1)}, \ldots, V^{(n)})$ is an instance of (1.3) if there exists a morphism $M : (\mathcal{C} \cup \mathcal{X})^+ \to (\mathcal{C} \cup \mathcal{X})^+$ mapping a_i to a_i $(1 \leq i \leq r)$ such that $V^{(i)} = M(U^{(i)})$ $(1 \leq i \leq n)$. A set $\mathcal{U}$ of unifiers for E is *complete* if every solution is an instance of a unifier $U \in \mathcal{U}$. $\mathcal{U}$ is *minimal* if no element $V \in \mathcal{U}$ is an instance of another element $U \in \mathcal{U}$.

We shall now describe the basic algorithm for word unification. To stress geometrical intuitions and the relationship to position equations (which will be introduced later), we shall represent the word equation (1.1) also in the following form:

$$\left| \begin{array}{c} \sigma_1 \ldots \sigma_k \\ \sigma_{k+1} \ldots \sigma_{k+l} \end{array} \right| \qquad (1.4)$$

We say (σ_1, σ_{k+1}) is the *head* and $\sigma_2 \ldots \sigma_k == \sigma_{k+2} \ldots \sigma_{k+l}$ is the *tail* of (1.4). We distinguish the following types of heads:

Type 1: all heads (σ, σ) with two indentical entries.
Type 2: all heads (x_i, x_j), where $x_i \neq x_j$ are variables.
Type 3: all heads (x_j, a_i) or (a_i, x_j), where $a_i \in \mathcal{C}, x_j \in \mathcal{X}$.
Type 4: all heads (a_i, a_j), where $a_i \neq a_j$ are coefficients.

The algorithm starts with equation (1.4) which is augmented by the *substitution list* $(x_1, \ldots, x_n)$ which is trivial at the beginning. Let us treat an example before we formally define the search tree $\mathcal{T}_{basic}(E)$.

Example 1.1: Suppose we want to find the solutions (unifiers) of the equation $E : axbzx == zczyyy$ with variables x, y and z and coefficients a, b and c. We take

$$\left(\left|\begin{array}{c} axbzx \\ zczyyy \end{array}\right|, (x, y, z)\right)$$

as top node of $T_{basic}(E)$. For any solution (X, Y, Z) of $axbzx == zczyyy$, either $a = Z$ or $Z = aZ_1$ for a non-empty word Z_1. Accordingly we treat two subcases: we try to solve the equation $axbax == acayyy$ (here z has been replaced by a) and the equation $axbaz_1x == az_1caz_1yyy$ (here z has been replaced by az_1). But now we may cut off the left a on both sides of these equations and try to solve the equations $xbax == cayyy$ and $xbazx == zcazyyy$ (for solvability it is meaningless whether we use z_1 or z as a variable name). In the substitution list we store that z has been replaced by a (respectively az). Thus we get the following two immediate successor nodes:

$$\left(\left|\begin{array}{c} xbax \\ cayyy \end{array}\right|, (x, y, a)\right), \qquad \left(\left|\begin{array}{c} xbazx \\ zcazyyy \end{array}\right|, (x, y, az)\right)$$

Let us continue with node $(xbax == cayyy, (x, y, a))$ whose equation has head of type 3. We may replace x by c or by cx. Here we choose the second possibility. The new augmented equation is $(xbacx == ayyy, (cx, y, a))$ which then leads (for example) to $(xbacax == yyy, (cax, y, a))$. This node of type 2 has three successors which correspond to the possibilities $X = Y, X = YX_1$ and $Y = XY_1$ (where now (X, Y) denotes a solution of the actual equation). Replacing y by x and erasing the trivial left part we get $(bacax == xx, (cax, x, a))$. Finally, after the further replacements of x first by bx, then by ax, cx and ax we reach node

$$\left(\left|\begin{array}{c} baca \\ baca \end{array}\right|, (cabaca, baca, a)\right).$$

Since we only erased identical left parts of the respective equations it is clear that $X = cabaca$, $Y = baca$ and $Z = a$ is a unifier of E.

Definition 1.2: The pair

$$\left(\left|\begin{array}{c} \sigma_1 \dots \sigma_k \\ \sigma_{k+1} \dots \sigma_{k+l} \end{array}\right|, (x_1, \dots, x_n)\right)$$

is the label of the top node of the unordered, finitely branching search tree $T_{basic}(E)$ for the equation E of the form (1.4). Suppose now that η is any node of $T_{basic}(E)$ with label (E', U), where the equation E' may have empty sides, generalizing for a moment the concept, and U is a substitution list of the form (1.3). In the following cases η is a leaf:

- if both sides of E' are graphically identical (successful leaf),
- if only one side of E' is empty (blind leaf),
- if E' has head of type 4 (blind leaf).

In the other cases, the immediate successors of η depend on the transformation of E' defined below. For every element (E_i', U_i) of $Trans(E', U)$, the node η has one immediate successor η_i labelled with (E_i', U_i).

- **Transformation** (of the word equation E with substitution list U)

Let $E^{\sigma \to v}$ ($U^{\sigma \to v}$) denote the result of simultaneously replacing all occurrences of σ in E (in U) by v.

(T$_1$) if E has head of type 1, then $(TAIL(E), U)$ is the only element of $Trans(E, U)$,

(T$_2$) if E has head (x_i, x_j) of type 2, then $Trans(E, U)$ has three elements:
$(TAIL(E^{x_j \to x_i}), U^{x_j \to x_i})$, $(TAIL(E^{x_j \to x_i x_j}), U^{x_j \to x_i x_j})$
and $(TAIL(E^{x_i \to x_j x_i}), U^{x_i \to x_j x_i})$.

(T$_3$) If E has head (a_i, x_j) or (x_j, a_i) of type 3, then $Trans(E, U)$ has two elements:
$(TAIL(E^{x_j \to a_i}), U^{x_j \to a_i})$ and $(TAIL(E^{x_j \to a_i x_j}), U^{x_j \to a_i x_j})$.

The set
$$\mathcal{U}_T = \{U;\ T_{basic}(E) \text{ has a successful leaf } (E', U)\}$$
is a complete and minimal set of unifiers for E. A formal proof is given in [Si2].

As a matter of fact, $T_{basic}(E)$ may have infinite paths. Thus, even if we organize the search in a breadth first manner we only have a semi-decision procedure for solvability. There are, however, cases where we get a decision procedure:

Lemma 1.3: *If no variable occurs twice in the equation E, then $T_{basic}(E)$ is finite and the length of the paths of $T_{basic}(E)$ does not exceed $nl(E)$.*

Proof: The transformation of E may be described by means of a replacement step (which may be trivial) followed by a step where we take the tail and cut off the head of the new equation. The replacement step introduces at most one additional symbol. By the following step two symbols are erased. Thus the notational length decreases. As a matter of fact, no variable occurs twice in a successor equation and the argument may be repeated. If eventually an equation with notational length 2 is reached, then the immediate successor is a leaf. ∎

Definition 1.4: The equations E_1 and E_2 are *isomorphic* if E_1 and E_2 become graphically identical when we replace all occurrences of variables x in E_1 by $\Psi(x)$ and all occurrences of coefficients a_i by $\Phi(a_j)$, for permutations Ψ of $\mathcal{X}$ and Φ of $\mathcal{C}$.

Lemma 1.5: *If no variable occurs more than twice in the equation E, then solvability may be decided by means of a finite subtree of $T_{basic}(E)$ the length of whose paths does not exceed $nl(E)^2 (nl(E)!)$.*

Proof: The replacement step of the transformation introduces at most two new symbols. By the second step, two symbols are erased afterwards. Thus the notational length of any (not necessarily immediate) successor equation does not exceed the notational length of E. Since the set of all word equations over $\mathcal{C}, \mathcal{X}$ with notational length not exceeding $nl(E)$ is finite any infinite path π of $\mathcal{T}$ must have two nodes η_1 and η_2 which are labelled with two isomorphic equations E_1 and E_2. Suppose, for example, that η_2 is a successor of η_1 (not necessarily immediate, of course). Suppose the path π leads from η_2 to the successful leaf η_4. Let η_3 be the last node on π labelled with an equation which is isomorphic to E_1 before η_4 is reached. Using the isomorphism and simulating the transformations of the final subpath from η_3 to η_4 we may now find a corresponding successful leaf on a path starting directly from η_1. On this path, no node will be labelled with an equation isomorphic to E_1, by construction. If this path contains two isomorphic equations of another type, then the argument may be repeated. Thus eventually a solution is found in a path where all word equations are non-isomorphic. If we only want to decide solvability, then we may therefore cut off every branch as soon as two isomorphic equations have occurred (solutions, respectively unifiers, may be lost by this strategy). If we consider the equations E' which may occur in such a reduced path, there are not more than $nl(E)$ possibilities for the notational length of E'. For a given length l, there are at most $nl(E)!$ possibilities to arrange a sequence of l symbols of the respective alphabets. For every arrangement, there are at most $nl(E)$ possibilities to distinguish two sides of the equation. Thus the total number of these equations cannot exceed $nl(E)^2 (nl(E)!)$. ∎

Various refinements of the basic algorithm are possible. See Livesey and Siekmann [LiS] for a discussion of these issues.

2 The Flat Part of the Search Tree

The concept of a special multi-equation system (sme-system) and its solutions is introduced. Sme-systems allow a reformulation of Makanin's algorithm. The simplicity of the basic algorithm becomes now available for the first part of the search tree, called the *flat* part. If no variable occurs more than twice in the word equation, then the whole search tree is flat. In any case, the simple transformation steps of the basic algorithm are applied as long as possible.

Definition 2.1: An l-sided multi-equation ME over $\mathcal{C}, \mathcal{X}$ has the form

$$\sigma_{1,1} \ldots \sigma_{1,k_1} == \sigma_{2,1} \ldots \sigma_{2,k_2} == \ldots == \sigma_{l,1} \ldots \sigma_{l,k_l}, \tag{2.1}$$

where $l \geq 2$, $k_i \geq 1$ $(1 \leq i \leq l)$ and $\sigma_{i,j} \in \mathcal{C} \cup \mathcal{X}$ $(1 \leq j \leq k_i, 1 \leq i \leq l)$. The number $k_1 + \ldots + k_l$ is the *notational length* $nl(ME)$ of ME. A *solution* of ME is a sequence

$$S = (X_1, \ldots, X_n) \tag{2.2}$$

of non-empty words over $\mathcal{C}$ such that all sides of (2.1) become graphically identical when we replace every occurrence of x_i by X_i $(1 \leq i \leq n)$. For $k \leq k_j$, the word $S(\sigma_{j,1} \ldots \sigma_{j,k})$ is defined in the obvious way, regarding S as a morphism similar as in section 1 $(1 \leq j \leq l)$.

Definition 2.2: A *multi-equation system* (me-system) over $\mathcal{C}, \mathcal{X}$ is a finite set of multi-equations over $\mathcal{C}, \mathcal{X}$. An me-system is a *special* multi-equation system (sme-system) if no variable x_i occurs more than twice in the system. The notational length $nl(SME)$ of the sme-system SME is the sum of the notational lengths of its elements. A sequence (2.2) is a solution of the (special) multi-equation system $SME = \{ME_1, \ldots, ME_r\}$ if S is a simultaneous solution of all multi-equations ME_i $(1 \leq i \leq r)$. An sme-system SME may be empty, in this case every sequence of the form (2.2) is a solution. For convenience we shall sometimes omit the prefix "sme" and speak about a system when we mean a sme-system.

Definition 2.3: Let SME be a system. The variable $x_i \in \mathcal{X}$ is called a *single* (with respect to SME) if it has only one occurrence and *paired* if it has two occurrences in SME.

Definition 2.4: Let E be an equation of the form (1.1). The sme-system $SME(E)$ *canonically associated with* E is defined by the following algorithm:

• **Algorithm**

Begin
    $SME(E) := \emptyset$.
  - For $i = 1$ until $k + l$ do
      if σ_i is the hth occurrence of the variable x_j in E, then replace σ_i by the new
      variable $\sigma_i' = x_{j,h}$.
  - Add the equation

$$\sigma_1' \ldots \sigma_k' == \sigma_{k+1}' \ldots \sigma_{k+l}'$$

    to $SME(E)$.
  - For $i = 1$ until n do
      if x_i has $n_i \geq 2$ occurrences in E, then add the multi-equation

$$x_{i,1} == x_{i,2} == \ldots == x_{i,n_i}$$

    to $SME(E)$.
End of Algorithm.

The equation $\sigma'_1 \ldots \sigma'_k == \sigma'_{k+1} \ldots \sigma'_{k+l}$ is called the *principal equation* of $SME(E)$, the equation $x_{i,1} == x_{i,2} == \ldots == x_{i,n_i}$ is the *multi-equation associated with* x_i.

Example/Definition 2.5: The best graphical way to represent the translation of E into $SME(E)$ is the matrix representation: the equation $axbzx == zczyyy$ of example 1.1 is transformed into

$$\begin{vmatrix} ax_1bz_1x_2 \\ z_2cz_3y_1y_2y_3 \end{vmatrix} \begin{vmatrix} x_1 \\ x_2 \end{vmatrix} \begin{vmatrix} z_1 \\ z_2 \\ z_3 \end{vmatrix} \begin{vmatrix} y_1 \\ y_2 \\ y_3 \end{vmatrix}$$

with *principal column* $(ax_1bz_1x_2, z_2cz_3y_1y_2y_3)^t$ and the *columns associated with* x, y and z.

Lemma 2.6: (a) *The number of variables of* $SME(E)$ *is the number of variable occurrences in* E, $nl(SME(E)) \leq 2nl(E)$.
(b) *If* E *has a solution* $S = (X_1, \ldots, X_n)$, *then* $SME(E)$ *has a corresponding solution* S' *which assigns* X_i *to every variable* $x_{i,j}$.
(c) *If* $SME(E)$ *has a solution* S', *then the words* $S'(x_{i,j})$ *coincide* $(1 \leq j \leq n_i)$. *The assignment* $S(x_i) = S'(x_{i,1})$ $(1 \leq i \leq n)$ *defines a solution of* E.

Proof: trivial. ∎

Definition 2.7: An sme-system SME is *flat* if it is empty or contains at least one two-sided equation.

As a matter of fact, all sme-systems of the form $SME(E)$ are flat.

Definition 2.8: The sme-systems SME_1 and SME_2 over the alphabets $C, \mathcal{X}$ are called *isomorphic* if SME_2 and SME_1 become graphically identical when we replace all occurrences of variables x in SME_2 by $\Psi(x)$ and all occurrences of coefficients a_j by $\Phi(a_j)$, for permutations Ψ of $\mathcal{X}$ and Φ of C.

Definition 2.9: An sme-system SME is *trivially unsolvable* in two cases:
(a) if SME contains a multi-equation ME which has two sides with prefixes (suffixes) v and w over the alphabet C such that neither v is a prefix (suffix) of w nor vice versa,
(b) if the sme-system $SME^{(0)}$ which we get when we replace all occurrences of coefficients in SME by the single coefficient a_1 does not have a solution.
An sme-system which is not trivially unsolvable is called *admissible*.

Lemma 2.10: *There exists an algorithm which decides whether a given sme-system* SME *is admissible or trivially unsolvable.*

Proof: It is trivial to check whether SME satisfies condition (a) of definition 2.9. The solvability of $SME^{(0)}$ may be expressed as an existential first order formula of arithmetic without multiplication (Presburger arithmetic). It is well known that this fragment is decidable (see, for example, Cooper [Coo]). ∎

We are now ready to start the description of the search tree $T_{Mak}(E)$ of the word equation E in our reformulation of Makanin's algorithm. If the sme-system $SME(E)$ associated with the word equation E is not admissible, then it is clear that E does not have a solution. If $SME(E)$ is admissible, the description of $T_{Mak}(E)$ will employ (at most) three steps: first, we shall describe the initial part $T_{ini}(E)$ of $T_{Mak}(E)$. The generation of $T_{ini}(E)$ describes the different possibilities to resolve the principal column and corresponds (in a sense which will be described in section 6) to the generation of the first set of position equations associated with E in the standard formulation of Makanin's algorithm. Second, we describe the tree $T_{flat}(E)$ which extends $T_{ini}(E)$. In simple situations, the tree $T_{flat}(E)$ (or even $T_{ini}(E)$) is already the complete search tree. In the other cases, $T_{flat}(E)$ is extended to the tree $T_{Mak}(E)$ which will always show whether E is solvable or not.

The Initial Tree

Definition 2.11: For every word equation E, the *initial search tree* $T_{ini}(E)$ is defined as follows:
- The top node of $T_{ini}(E)$ is labelled with $SME(E)$. (We assume, of course, that $SME(E)$ is admissible.) We introduce in addition marks which indicate the type (principal equation, multi-equation associated with $x_1,...$) of any multi-equation in $SME(E)$.

Suppose η is any node of $T_{ini}(E)$, labelled with the admissible system SME. In the following cases, η is a leaf of $T_{ini}(E)$:
- If SME is empty, then η is a *successful leaf*.
- If the principal column has been completely resolved, i.e. if all multi-equations of the non-empty system SME are associated with variables, then η is an *open leaf* of $T_{ini}(E)$ (the term "open" indicates that η is not a leaf of $T_{Mak}(E)$).

In the other case, if $SME \neq \emptyset$ is admissible and has non-empty principal equation, the successors of η are defined by means of two procedures which are described below: first, SME is *transformed*. The result of the transformation is the set $Trans(SME)$. Second, all elements of $Trans(SME)$ are *simplified* by means of a simplification procedure. As a result, some sme-systems may be erased, others may become empty (this is a significant difference here). We get the set $Simpl(Trans(SME))$ of at most three sme-systems. For every admissible sme-system $SME_i \in Simpl(Trans(SME))$ the node η has one successor η_i labelled with SME_i. If, however, $Simpl(Trans(SME))$ does not contain an admissible system, then η is a *blind leaf*, i.e. SME is unsolvable.

The following transformation steps follow exactly the corresponding transformation steps of the basic algorithm. They are formulated more generally as it would be necessary for the initial tree. For the initial tree, the distinguished equation is always the principal equation.

- **Transformation** (of the admissible sme-system SME with distinguished equation ME)

(T$_1$) Suppose that ME has head (σ, σ) with two identical entries. Then take the tail of ME and leave the other multi-equations unmodified. The resulting system is admissible and the label of the unique successor of η.

(T$_2$) Suppose that ME has head $(x_{i_1,j_1}, x_{i_2,j_2})$ with two distinct variables. $Trans(SME)$ has three elements SME_i $(1 \leq i \leq 3)$:

(1) To get SME_1, replace all occurrences of x_{i_2,j_2} in SME by x_{i_1,j_1}. Take then the tail of the distinguished equation and leave the other multi-equations unmodified.

(2) To get SME_2, replace all occurrences of x_{i_2,j_2} in SME by $x_{i_1,j_1} x_{i_2,j_2}$. Then take the tail of the distinguished equation and leave the other multi-equations unmodified.

(3) To get SME_3, replace all occurrences of x_{i_1,j_1} in SME by $x_{i_2,j_2} x_{i_1,j_1}$. Then take the tail of the distinguished equation and leave the other multi-equations unmodified.

(T$_3$) Suppose that ME has head $(x_{i,j}, a_h)$ or $(a_h, x_{i,j})$, where a_h is a coefficient symbol. $Trans(SME)$ has two elements SME_1, SME_2:

(1) To get SME_1, replace all occurrences of $x_{i,j}$ in SME by a_h. Then take the tail of the distinguished equation and leave the other multi-equations unmodified.

(2) To get SME_2, replace all occurrences of $x_{i,j}$ in SME by $a_h x_{i,j}$. Then take the tail of the distinguished equation and leave the other multi-equations unmodified.

After the transformation it might happen that one side of the distinguished equation of a structure SME_i is empty while the other is not. Then SME_i is not an sme-system in the sense of definitions 2.1 and 2.2. We erase it. If both sides of the principal equation are empty after the transformation, then the principal equation is erased (resolved).

Lemma 2.12: *Let SME be an admissible sme-system with distinguished equation ME. Then SME is solvable if and only if an sme-system in $Trans(SME)$ is solvable.*

Proof: Trivial if ME has head (a_i, a_i) or $(x_{i,j}, x_{i,j})$. If ME has head $(x_{i_1,j_1}, x_{i_2,j_2})$ and SME has the solution S which assigns the words X_{i_1,j_1}, X_{i_2,j_2} to x_{i_1,j_1}, x_{i_2,j_2}, then there are three possibilities: either (i) X_{i_1,j_1} and X_{i_2,j_2} are identical or (ii) X_{i_1,j_1} is a proper prefix of $X_{i_2,j_2} = X_{i_1,j_1} X'_{i_2,j_2}$ or (iii) X_{i_2,j_2} is a proper prefix of $X_{i_1,j_1} = X_{i_2,j_2} X'_{i_1,j_1}$. Now consider the systems SME_i $(1 \leq i \leq 3)$ as defined in the transformation rule (T$_2$). In case (i), SME_1 has the solution S' which we get from S by omitting the component X_{i_2,j_2}. In case (ii), SME_2 has the solution $S'(x_{i_2,j_2}) = X'_{i_2,j_2}$, $S'(x_{i,j}) = S(x_{i,j})$ for $(i,j) \neq (i_2,j_2)$. In case (iii), SME_3 has the solution

$S'(x_{i_1,j_1}) = X'_{i_1,j_1}$, $S'(x_{i,j}) = S(x_{i,j})$ for $(i,j) \neq (i_1,j_1)$. Conversely, if SME_1 has a solution S', then $S(x_{i_2,j_2}) = S'(x_{i_1,j_1})$, $S(x_{i,j}) = S'(x_{i,j})$ for $(i,j) \neq (i_2,j_2)$ defines a solution for SME. If SME_2 has a solution S', then $S(x_{i_2,j_2}) = S'(x_{i_1,j_1})S'(x_{i_2,j_2})$, $S(x_{i,j}) = S'(x_{i,j})$ for $(i,j) \neq (i_2,j_2)$ defines a solution for SME. If SME_3 has a solution S', then $S(x_{i_1,j_1}) = S'(x_{i_2,j_2})S'(x_{i_1,j_1})$, $S(x_{i,j}) = S'(x_{i,j})$ for $(i,j) \neq (i_1,j_1)$ defines a solution for SME. The proof for head $(x_{i,j}, a_h)$ or $(a_h, x_{i,j})$ is similar, compare (T$_3$). ∎

Remark 2.13: The previous proof shows that the solution S of an sme-system SME determines a unique successor system SME' of $Trans(SME)$ which has a corresponding solution S'. We say that S leads to SME'. We shall see immediately that simplification of a solvable system leads again to a unique solvable successor system. Thus the solution S of an sme-system SME occurring in $T_{ini}(E)$ determines a unique path of $T_{ini}(E)$ where all nodes are labelled with solvable sme-systems.

• **Simplification** (of the sme-system SME)

The following *simplification rules* are applied until the system is erased or a system SME' is reached which cannot be further simplified by the rules.

(S$_1$) If the multi-equation ME in SME has two identical sides of the form $x_{i,j}$ (where $x_{i,j}$ is a variable), then erase both sides. If now ME has only one side, then erase ME.

(S$_2$) If SME contains a multi-equation ME which has a side of the form a_i (where a_i is a coefficient symbol) and if all other sides of ME have length 1 and are variables, then replace all occurrences of these variables in SME by a_i. Erase ME.

Lemma 2.14: *Let SME be an sme-system. Suppose we get SME' by applying one of the rules (S$_1$), (S$_2$). Then SME is solvable if and only if SME' is solvable.*

Proof: Trivial for (S$_2$). The only-if part of the proof is trivial in any case. Consider (S$_1$). If we erase two occurrences of the same variable $x_{i,j}$ in ME and get ME' which is not erased afterwards, then the solution S' of SME' assigns via composition a unique word w to all sides of ME'. To get a solution of SME, we assign w to $x_{i,j}$ and use the components of S' to get values for the remaining variables. If ME' is erased, then we assign any non-empty word w to $x_{i,j}$ and proceed as above. In both cases it is simple to see that we get in fact a solution of SME. ∎

Theorem 2.15: (a) *The maximal length of a path in $T_{ini}(E)$ does not exceed the notational length $nl(E)$ of E.*
(b) *If E has a solution, then the corresponding solution of $SME(E)$ (compare lemma 2.6 (b)) leads to a successful leaf of $T_{ini}(E)$ or to an open leaf of $T_{ini}(E)$ which is labelled with a solvable sme-system. If $T_{ini}(E)$ has a successful leaf or an open leaf which is labelled with a solvable sme-system, then E is solvable.*

(c) *If all variables of E occur only once, then $T_{ini}(E)$ does not have an open leaf. E has a solution if and only if $T_{ini}(E)$ has a successful leaf.*

Proof: (a) Since the principal equation of $SME(E)$ does not have two occurrences of the same variable the notational length of the principal equation decreases at every transformation step. Thus the number of such steps cannot exceed $nl(E)$. If the principal equation has two sides of one symbol only, then both symbols disappear in the following transformation step, by (S_0).
(b) A trivial inductive argument based on remark 2.13 and lemmata 2.12, 2.14.
(c) In this case the set of multi-equations of $SME(E)$ which are associated with variables is empty. Thus $T_{ini}(E)$ cannot have any open leaf and (c) follows from (b). ■

The Flat Tree

We may now describe the second part of the search tree. Recall that all leaves of $T_{ini}(E)$ are successful leaves (labelled with an empty sme-system) or blind leaves (labelled with an unsolvable sme-system) or open leaves (labelled with a non-empty sme-system where the principal equation has been completely resolved).

Definition 2.16: The unordered, finitely branching tree $T_{flat}(E)$ is an extension of $T_{ini}(E)$. Suppose η is any node of $T_{flat}(E)$ labelled with the admissible sme-system SME. In the following cases, η is a leaf of $T_{flat}(E)$:
  - If η is a successful leaf of $T_{ini}(E)$ or if SME is empty, then η is a successful leaf.
  - If SME is isomorphic to a system SME' which has occurred earlier in the path, then η is a blind leaf.
  - If SME is non-flat, then η is an open leaf of $T_{flat}(E)$.
In the other case, if η is labelled with the flat (admissible, non-empty) sme-system SME, then we assume that a two-sided multi-equation ME of SME is distinguished (if not, then we choose one). Then η has one successor η_i for every admissible system SME_i in $Simpl(Trans(SME))$ where simplification and transformation procedure are defined as earlier.

Theorem 2.17: (a) *The maximal length of a path in $T_{flat}(E)$ does not exceed the number $(d!)^3$, where $d = 2nl(E)$.*
(b) *E has a solution if and only if $T_{flat}(E)$ has a successful leaf or $T_{flat}(E)$ has an open leaf which is labelled with a solvable admissible sme-system.*
(c) *If no variable occurs more than twice in E, then $T_{flat}(E)$ does not have an open leaf. E is solvable if and only if $T_{flat}(E)$ has a successful leaf.*

Proof: (a) $SME(E)$ contains at most $2nl(E)$ symbols. There are at most $d!$ possibilities to write these symbols in some arbitrary order. For every possibility, there are at most $d!$ possibilities to distinguish by means of marks arbitrary non-empty subsequences (which represent sides of multi-equations). For every marked sequence, there are at most $d!$ possibilities to distinguish principal marks (which separate distinct multi-equations).

(b) similar as in theorem 2.15. The exclusion of nodes which are labelled with an sme-system which is isomorphic to the label of a predecessor node may be justified exactly as in the proof of lemma 1.5.

(c) If no variable has more than two occurrences in E, then all multi-equations in $SME(E)$ are two-sided. It is trivial that the number of the sides of an equation may only decrease by transformation and simplification. Thus a non-flat sme-system of the tree is always empty. Now (c) follows from (b). ∎

Remark 2.18: When we resolve a non-principal equation in a path of $T_{flat}(E)$, then it is not difficult to detect whether an sme-system has occurred which is isomorphic to an earlier system. First, note that in any case neither the number of symbols of the actual sme-system nor the number of symbols in the distinguished equation which we try to resolve increases by a transformation step. This follows immediately from the fact that all variables have at most two occurrences. Second, if the number of symbols of the distinguished equation does *not* decrease at a transformation step, then necessarily a variable has been replaced for which the second occurrence falls into the same equation. Let us call this phenomenon a local replacement, since in this case all other parts of the actual sme-system remain unchanged. If the resolution of the distinguished equation does not terminate, there must finally occur a situation where only local replacements are made. But in this case, two identical distinguished equation will occur within a finite number of steps. Thus we only have to store the distinguished equation and we may erase all stored equations whenever the notational length of the distinguished equation decreases.

Remark 2.19: It is simple to keep track of the solutions, as we did in the first section for the basic algorithm. We augment $SME(E)$ by the additional substitution list

$$(x_{1,1}, x_{2,1}, \ldots, x_{n,1}).$$

All replacements which occur during transformation or simplification are also applied to this sequence. If the rule (S_1) is applied with the variable $x_{i,1}$ and ME is not empty, then we replace $x_{i,1}$ in the substitution list by one of the remaining sides, etc. If we reach a leaf labelled with an empty sme-system, then the substitution list defines a unifier of the equation E. The proof is not difficult and omitted. The technique may even be optimized since the structure of the columns associated with the variables reflects the actual substitution list. An example will be discussed below.

The Tree of a System of Equations

We may generalize the procedure to systems of equations $E_1, \ldots, E_k$. There is only one essential difference: in the algorithm of definition 2.4 we count now the number of occurrences within the whole system and introduce a principal equation E_i' for every equation E_i. Thus the sme-system $SME(E_1,\ldots,E_k)$ associated with $E_1,\ldots,E_k$ has now k principal equations which do not have two occurrences of the same variable. When we resolve the principal equations, the two sides of every principal equation must become empty at the same time, otherwise the system is erased.

As an example, consider the system $xsyxbxb == rzycycy, ycyxb == azazy, zazaz == asyxs$ with variables r,s,x,y and z. The canonically associated sme-system has the following representation:

$$\left|\begin{matrix} x_1s_1y_1x_2bx_3b \\ r_1z_1y_2cy_3cy_4 \end{matrix}\right| \left|\begin{matrix} y_5cy_6x_4b \\ az_2az_3y_7 \end{matrix}\right| \left|\begin{matrix} z_4az_5az_6 \\ as_2y_8x_5s_3 \end{matrix}\right| \left|\begin{matrix} s_1 \\ s_2 \\ s_3 \end{matrix}\right| \left|\begin{matrix} x_1 \\ \cdots \\ x_5 \end{matrix}\right| \left|\begin{matrix} y_1 \\ \cdots \\ y_8 \end{matrix}\right| \left|\begin{matrix} z_1 \\ \cdots \\ z_6 \end{matrix}\right|.$$

Example 1.1 revised

Let us show how example 1.1 is treated now. $SME(axbzx == zczyyy)$ has the following matrix representation:

$$\left|\begin{matrix} ax_1bz_1x_2 \\ z_2cz_3y_1y_2y_3 \end{matrix}\right| \left|\begin{matrix} x_1 \\ x_2 \end{matrix}\right| \left|\begin{matrix} y_1 \\ y_2 \\ y_3 \end{matrix}\right| \left|\begin{matrix} z_1 \\ z_2 \\ z_3 \end{matrix}\right|.$$

Replacing z_2 by a and using (S_2) we get

$$\left|\begin{matrix} x_1bax_2 \\ cay_1y_2y_3 \end{matrix}\right| \left|\begin{matrix} x_1 \\ x_2 \end{matrix}\right| \left|\begin{matrix} y_1 \\ y_2 \\ y_3 \end{matrix}\right| \qquad (-,-,a).$$

Here a is the z-entry of the substitution list. It is not necessary to store x- and y-entries since we might use any line of the corresponding columns as substitution value. Combining now two steps, we replace x_1 by cax_1:

$$\left|\begin{matrix} x_1bax_2 \\ y_1y_2y_3 \end{matrix}\right| \left|\begin{matrix} cax_1 \\ x_2 \end{matrix}\right| \left|\begin{matrix} y_1 \\ y_2 \\ y_3 \end{matrix}\right| \qquad (-,-,a).$$

Now we may replace y_1 by x_1, for example:

$$\left|\begin{matrix} bax_2 \\ y_2y_3 \end{matrix}\right| \left|\begin{matrix} cax_1 \\ x_2 \end{matrix}\right| \left|\begin{matrix} x_1 \\ y_2 \\ y_3 \end{matrix}\right| \qquad (-,-,a).$$

Combining two steps, we may replace y_2 by bay_2.

$$\left|\begin{matrix} x_2 \\ y_2y_3 \end{matrix}\right| \left|\begin{matrix} cax_1 \\ x_2 \end{matrix}\right| \left|\begin{matrix} x_1 \\ bay_2 \\ y_3 \end{matrix}\right| \qquad (-,-,a).$$

We replace x_2 by y_2x_2

$$\begin{vmatrix} x_2 \\ y_3 \end{vmatrix} \begin{vmatrix} cax_1 \\ y_2x_2 \end{vmatrix} \begin{vmatrix} x_1 \\ bay_2 \\ y_3 \end{vmatrix} \qquad (\text{-},\text{-},a)$$

and y_3 by x_2:

$$\begin{vmatrix} cax_1 \\ y_2x_2 \end{vmatrix} \begin{vmatrix} x_1 \\ bay_2 \\ x_2 \end{vmatrix} \qquad (\text{-},\text{-},a).$$

Now the principal column is completely resolved. We continue with the column associated with x and add the value cax_1 to the substitution list (as x-entry). Combining two steps, we may replace y_2 by ca:

$$\begin{vmatrix} x_1 \\ x_2 \end{vmatrix} \begin{vmatrix} x_1 \\ baca \\ x_2 \end{vmatrix} \qquad (cax_1,\text{-},a).$$

We replace x_2 by x_1:

$$\begin{vmatrix} x_1 \\ baca \\ x_1 \end{vmatrix} \qquad (cax_1,\text{-},a).$$

By (S_1), the matched occurrences of x_1 may be erased, substituting $baca$ for x_1 in the substitution list. The resulting column has only one line $baca$ which is used as substitution entry for y and erased. The final sme-system is empty and the substitution list shows that $(X,Y,Z) = (cabaca, baca, a)$ is a solution of $axbzx == zczyyy$.

Limitation of Simple Transformation

The idea of the sme-systems (which is similar to the idea of generalized equations in Makanin's (or Pécuchet's, Jaffar's) formulation is to simulate the situation of lemma 1.5: if every variable occurs at most twice, then the number of symbols does not grow at a transformation step since at most two new symbols are introduced at the replacement step and two symbols are erased when we take the tail. The concept of flatness shows exactly how far we come with this simple idea. When we reach an open leaf of T_{flat}, then we have only multi-equations left with at least three sides. In this case, the naive transformation strategy leads again to an explosion of the number of symbols: if, for example, a multi-equation has the three-valued head (x,y,z) for paired variables x, y and z and we consider the case where X is assumed to be a proper prefix of Y and of Z, then *four* new occurrences of x are introduced, by the replacement. Then, when we take the tail, only three occurrences of x are erased and the number of symbols grows. Thus it seems that we did not gain anything by artificial duality of variables. Exactly here is the point where we need more theoretical background.

3 Position Equations and Theoretical Background

When we reach an open leaf of the flat part of the search tree, then we cannot longer use the naive transformation strategy. The main idea behind Makanin's algorithm is the following: the basic structures of the algorithm, the *position equations*, represent a geometrical picture of the relative extension of the values of the variables. By artificial duality of variables it is still possible to avoid an enlargement of the number of variable occurrences at a transformation step if a part of the relevant information about corresponding prefixes of variable values is stored in an extra part. In certain situations, this additional part allows to conclude that every solution of the actual structure has a large number of periodical repetitions of the same non-empty subword, in some component. By way of theoretical considerations it is possible to restrict the infinite number of position equations which might occur in a path to a finite number of position equations which possibly have solutions with not too many periodical consecutive repetitions of the same non-empty subword. The content of this section may be found in similar form in [Mak] and in [Péc].

Exponent of Periodicity and Domino Towers

Definition 3.1: The solution (1.2) of a word equation (1.1) has *exponent of periodicity* s if some component X_i has a non-empty subword of the form P^s but no component has a non-empty subword of the form Q^{s+1}.

Definition 3.2: A solution $(X_1, \ldots, X_n)$ of a word equation is called *minimal*, if the length of the word $X_1 X_2 \ldots X_n$ is minimal with respect to the class of all solutions of the equation.

Theorem 3.3 ([Mak], Lemma 1.3): *If d is the notational length of the equation E, then the exponent of periodicity of any minimal solution of E does not exceed the number*

$$(6d)^{2^{(2d^4)}} + 2.$$

The proof of theorem 3.3 may be reconstructed without difficulties from a corresponding proof of a slightly more general result in section 6 (theorem 6.6). V.K.Bulitko [Bul] was the first to give an upper bound for the exponent of periodicity of a minimal solution in terms of the notational length of the equation. Quite recently A.Kościelski and L.Pacholski [KoP] showed that the exponent of periodicity of minimal solutions cannot exceed $2^{2.54d}$.

Definition 3.4: The sequence $(X_1, \ldots, X_n)$ of non-empty words may be arranged to a *domino-tower* $< X_{\lambda_i}, B_i, C_i, S_i >_{1 \le i \le k}$ of height $k > 0$ if the X_i may be ordered to a sequence

$$(X_{\lambda_1}, X_{\lambda_2}, \ldots\ldots\ldots\ldots\ldots\ldots\ldots, X_{\lambda_k})$$

(possibly with multiple occurrences of the X_i) with decompositions $X_{\lambda_i} = B_i C_i$ (for non-empty words B_i, C_i) $(1 \le i \le k)$ such that $B_{i+1} = S_i B_i$ $(1 \le i \le k-1)$ for possibly empty words $S_1, \ldots, S_{k-1}$ and $C_i R_i = C_{i+1} T_i$ for possibly empty words R_i and T_i $(1 \le i \le k-1)$.

The name "domino tower" is motivated by the following figure. Here all parts of consecutive words which have direct contact must be equal. Parts which do not have direct contact are not restricted like that.

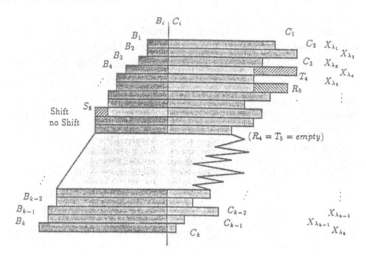

Let $|X|$ denote the length of the word X.

Theorem 3.5 ([Mak], Lemma 1.4): *Suppose the words $X_1, \ldots, X_n$ may be arranged to a domino tower $< X_{\lambda_i}, B_i, C_i, S_i >_{1 \le i \le k}$ of height $k > 0$. If the words $S_1, \ldots, S_{k-1}$ satisfy the "shift condition"*

$$j - i \ge K \Rightarrow |S_i S_{i+1} \ldots S_{j-1} S_j| > 0,$$

then some word X_t has the form $X_t = P^s Q$, where P is non-empty and has $s \ge \frac{k}{Kn^2} - 1$ consecutive repetitions.

Proof: The first aim is to single out a subtower which contains only stones of one type. Unfortunately, it is not possible to simply erase occurrences of other stone types - in general we will not get a domino tower when we omit some stones because parts which get vertical contact would not necessarily be equal. A more tricky procedure will help us. We suppose without loss of generality that the order of the indices of the stone types corresponds to the order of their lengths:

$$|X_1| \leq |X_2| \leq \ldots \leq |X_n|.$$

Between the words X_i of unequal length we distribute all the B_i monotonically with respect to their length. There exists at least one word X_q such that

$$|X_q| \leq |B_{i+1}| \leq |B_{i+2}| \leq \cdots \leq B_{i+p} < |X_{q+1}|$$

for some $p \geq k/n$ and some i. (More exactly: the words B_i could also have smaller length than X_1. This case is treated analogously.) The complete word $X_{\lambda_{i+j}}$ corresponding to B_{i+j} is longer than B_{i+j}, hence, by choice of the distribution, longer than all the words $B_{i+1}, \ldots, B_{i+p}$ $(1 \leq j \leq p)$ and at least as long as X_{q+1}.

If for some $i + 1 \leq r < i + p$ the word C_{r+1} is longer than C_r, for example $C_{r+1} = C_r M$, then the word $X_{\lambda_{r+1}}$ must have the form $X_{\lambda_{r+1}} = X'_{\lambda_{r+1}} M$ where $|X'_{\lambda_{r+1}}| = |B_{r+1} C_r| \geq |B_r C_r| = |X_r|$. The reduced word $X'_{\lambda_{r+1}}$ must still be longer than the words $B_{i+1}, \ldots, B_{i+p}$ and at least as long as X_{q+1}. In the subsequence $(X_{\lambda_{i+1}}, X_{\lambda_{i+2}}, \ldots, X_{\lambda_{i+p}})$ we replace all occurrences of $X_{\lambda_{r+1}}$ by the reduced word $X'_{\lambda_{r+1}}$. The result is a new domino tower with the stone type $X'_{\lambda_{r+1}}$ instead of $X_{\lambda_{r+1}}$. Iterating this we finally get a sequence $X^*_{\lambda_{i+1}}, X^*_{\lambda_{i+2}}, \ldots, X^*_{\lambda_{i+p}}$ with partitions of the following form (note that all the reduced words have length $\geq |X_{q+1}|$, hence iteration is possible):

$$\begin{aligned} X^*_{\lambda_{i+j}} &= B_j^* C_j^* & (1 \leq j \leq p), \\ B_{j+1}^* &= S_j^* B_j^* & (1 \leq i \leq p-1), \\ C_j^* &= C_{j+1}^* T_j^* & (1 \leq j \leq p-1), \end{aligned}$$

for possibly empty words $S_1^*, \ldots, S_{p-1}^*$ and $T_1^*, \ldots T_{p-1}^*$. (Graphically speaking, no stone exceeds his upper predecessor on the right side.) Obviously the new tower satisfies the shift condition. Moreover, we may now omit arbitrary stones of the tower and always get a new domino tower. First we single out the stones $X^*_{\lambda_{i+j}}$ with subindices $i+1, i+1+K, i+1+2K$, etc. The new tower has a shift at every floor and at least $m \geq k/Kn$ stones. Since we only have n types of (reduced) stones one type X_t^* must occur g times, for some $g \geq k/Kn^2$. We single out all stones of this type. The remaining tower

$$(X^*_{\nu_1}, \ldots, X^*_{\nu_g})$$

has length g. We have the equalities $S_i^* X_t^* = X_t^* T_i^*$ $(1 \leq i \leq g-1)$. The solutions of these "conjugacy equations" $sx = xt$ are wellknown (see [Lot], for example). The word X_t^* must have the form $X_t^* = (A_i B_i)^{r_i} A_i$ where $S_i^* = A_i B_i$ and $T_i^* = B_i A_i$ $(1 \leq i \leq g-1)$. Hence

$$(A_i B_i)^{r_i} A_i = (A_{i+1} B_{i+1})^{r_{i+1}} A_{i+1} \quad (1 \leq i \leq g-2)$$

and

$$|A_1 B_1| + \cdots + |A_{g-1} B_{g-1}| < |X_t^*|.$$

It follows from proposition 1.3.5 of [Lot] that the words $A_i B_i$ must be powers of the same simple word P. Since $|S_1 \cdots S_{g-1}| < |X_t^*|$ we have $X_t^* = P^s Q_1$ for $s \geq g - 1 \geq (k/Kn^2) - 1$. The original unreduced word X_t has the form $X_t = P^s Q$. ∎

Position Equations

Let

$$C = \{a_1, \ldots, a_r\} \qquad (r \geq 1)$$
$$X = \{x_1, \ldots, x_n, x_{n+1}, \ldots, x_{2n}\} \qquad (n \geq 0)$$

be two disjoint alphabets of coefficients and variables respectively. The variables x_i and $\bar{x}_{i+n}$ are duals of each other, we write $x_i = \bar{x}_{i+n}, x_{i+n} = \bar{x}_i$ ($1 \leq i \leq n$).

Definition 3.6: A *position equation PE* over (C, X) has the following parts:

1. A set BS of bases. We distinguish the *variable bases* $w_1, \ldots, w_n, w_{1+n}, \ldots, w_{n+n} = x_1, \ldots, x_n, \bar{x}_1, \ldots, \bar{x}_n$ and the *coefficient bases* $w_{2n+1}, \ldots, w_{2n+m}$ ($m > 0$). To every coefficient base w_{2n+i} we assign its type $TYPE(w_{2n+i}) \in C$.
   $TYPE : \{w_{2n+1}, \ldots, w_{2n+m}\} \to C$ is onto.
2. A non-empty initial segment $BD = \{1, \ldots, e\}$ of the natural numbers. BD is the set of *boundaries* of PE. There is a (possibly empty) subset $RF = \{i_1^{(rf)}, \ldots, i_f^{(rf)}\}$ of *right-fossil* boundaries and a similar set $LF = \{i_1^{(lf)}, \ldots, i_g^{(lf)}\}$ of *left-fossil* boundaries.
3. The functions "left-boundary" L und "right-boundary" R with domain $\{w_1, \ldots, w_{2n+m}\}$ and range BD.
5. A (possibly empty) set BC of *boundary connections*. A boundary connection is a sequence of the form

$$(i_0, x_{\lambda(1)}, x_{\lambda(2)}, x_{\lambda(3)}, \ldots, x_{\lambda(k)}, j_0) \qquad (k \geq 1) \qquad (3.1)$$

where $i_0, j_0 \in BD$ and $x_{\lambda(i)} \in BS$, ($1 \leq i \leq k$). The sequence $(x_{\lambda(1)}, \ldots, x_{\lambda(k)})$ is the *path* of length k of (3.1).

The following conditions must be satisfied:
(I) For all bases $w_i \in BS$: $L(w_i) < R(w_i)$.
(II) For every boundary connection (3.1) of PE:
   (IIa) $L(x_{\lambda(1)}) < i_0 < R(x_{\lambda(1)})$,
   (IIb) $L(x_{\lambda(i+1)}) \leq L(\bar{x}_{\lambda(i)})$ ($1 \leq i \leq u - 1$), for some $1 \leq u \leq k$,
   $L(x_{\lambda(u+1)}) > L(\bar{x}_{\lambda(u)})$,
   $L(x_{\lambda(i+1)}) \geq L(\bar{x}_{\lambda(i)})$ ($u < i \leq k - 1$),
   (IIc) $L(\bar{x}_{\lambda(k)}) < j_0 < R(\bar{x}_{\lambda(k)})$.
(3.1) has a non-empty *second part* if $u < k$.
(III) For every boundary connection (3.1) of PE: the final boundary j_0 is right-fossil or there exists a *witness* for (3.1), i.e. a base x_ω such that $j_0 = R(x_\omega)$ which

satisfies the following *convexity condition*: if (3.1) has a non-empty second part, then $L(x_\omega) \geq L(\bar{x}_{\lambda(k)})$.

(IV) Every boundary which is neither the left or right boundary of a base nor fossil is called *inessential*. Every inessential boundary is the first boundary i_0 of a boundary connection (3.1) of PE.

Some additional notions simplify various definitions: a pair (i, j) of boundaries $i < j$ is called a *column* of PE, columns $(i, i+1)$ are called *indecomposable*, the other columns are called *decomposable*. The intersection of two columns is defined in the straightforward way and may be empty or again a column. For every base w_i, $(L(w_i), R(w_i))$ is the column $col(w_i)$ of w_i.

Definition 3.7: The *principal parameters* of a position equation PE are the numbers n, m, furthermore the number $c = 2n + f + n_L + g$ (where f is the number of right-fossil boundaries, g is the number of left-fossil boundaries and n_L is the number of left boundaries) and the maximal length d of a path of a boundary connection of PE.

Definition 3.8: Let PE be a position equation. An assignment S of non-empty words $S((i, i+1))$ over the alphabet $\mathcal{C}$ (shortly written $S(i, i+1)$) to the indecomposable columns $(i, i+1)$ of PE has, by concatenation, a unique extension to an assignment of non-empty words $S((i, j)) = S(i, j)$ to arbitrary columns (i, j) of PE. We identify S with its extension and call S a *presolution* of PE if the following conditions are satisfied:

(PS_1) $S(w_{2n+i}) = TYPE(w_{2n+i})$ $(1 \leq i \leq m)$,
(PS_2) $S(x_i) = S(\bar{x}_i)$, for every variable base x_i of PE.

Here $S(w_i)$ is used as a short notation for $S(col(w_i))$. The *exponent of periodicity* of S is the exponent of periodicity of the words $S(x_1), ..., S(x_n)$ as defined in 3.1. The *index* of S is the number $|S(1, e)|$.

Definition 3.9: Let S be a presolution of the position equation PE. Let v_0 be any integer. For every boundary $i \in BD$ we define its *value* $v_S(i)$ (with respect to S and v_0) as follows:

$v_S(1) = v_0$,
$v_S(i) = v_S(1) + |S(1, i)|$ $(1 < i \leq e)$.

For every $x_i \in BS$ the *dual-difference* $dd_S(x_i)$ of x_i with respect to S is the number $dd_S(x_i) = v_S(L(\bar{x}_i)) - v_S(L(x_i))$. Suppose now that PE has a boundary connection of the form (3.1). For every element $x_{\lambda(i)}$ of (3.1), its *value* $v_S(x_{\lambda(i)})$ (with respect to its position i in (3.1), with respect to S and to v_0) is defined as follows:

$v_S(x_{\lambda(1)}) = v_S(i_0)$,
$v_S(x_{\lambda(i+1)}) = v_S(x_{\lambda(i)}) + dd_S(x_{\lambda(i)})$.

The presolution S *satisfies* (3.1) if $v_S(L(x_{\lambda(i)})) < v_S(x_{\lambda(i)}) < v_S(R(x_{\lambda(i)}))$ $(1 \leq i \leq k)$ and if $v_S(j_0) = v_S(x_{\lambda(k)}) + dd_S(x_{\lambda(k)})$.

Lemma 3.10: (a) The choice of v_0 is irrelevant for satisfaction.
(b) $dd_S(x_i) = v_S(R(\bar{x}_i)) - v_S(R(x_i))$ $(1 \leq i \leq 2n)$.

(c) If S satisfies (3.1), then $v_S(L(\bar{x}_{\lambda(i)})) < v_S(x_{\lambda(i+1)}) < v_S(R(\bar{x}_{\lambda(i)}))$ $(1 \leq i < k)$.

Proof: (a) obvious; (b) follows from (PS_2);
(c) $v_S(L(\bar{x}_{\lambda(i)})) = v_S(L(x_{\lambda(i)})) + dd_S(x_{\lambda(i)}) < v_S(x_{\lambda(i)}) + dd_S(x_{\lambda(i)}) = v_S(x_{\lambda(i+1)}) < v_S(R(x_{\lambda(i)})) + dd_S(x_{\lambda(i)}) = v_S(R(\bar{x}_{\lambda(i)}))$. ∎

Definition 3.11: Let PE be a position equation with presolution S. S is a *solution* of PE if S satisfies all boundary connections of PE.

Lemma 3.12: *If S is a solution of the position equation PE with the boundary connection (3.1), then the words $S(x_{\lambda(i)})$ $(1 \leq i \leq k)$ may be arranged to a domino tower of height at least $k/2$.*

Proof: Step 1: We show that the words $S(x_{\lambda(i)})$ may be decomposed into non-empty subwords $B_i C_i$ such that
$$S(x_{\lambda(i)}) = B_i C_i = S(\bar{x}_{\lambda(i)}),$$
$$S(1, i_0) = S(1, L(x_{\lambda(1)}))B_1,$$
$$S(1, L(\bar{x}_{\lambda(i)}))B_i = S(1, L(x_{\lambda(i+1)}))B_{i+1},$$
$$S(1, L(\bar{x}_{\lambda(k)}))B_k = S(1, j_0).$$
(Here $S(1,1)$ is the empty word.) Since $S(1,e)$ is non-empty we may enumerate the letters of this word with numbers $1, ..., v_S(e)$, where v_S is the evaluation with initial value 0. For every pair p, q of natural numbers $0 \leq p < q \leq v_S(e)$ we denote by $W_S(p,q)$ the letters with numbers $p+1, p+2, ..., q$ of $S(1,e)$. It is trivial that $W_S(v_S(i), v_S(j)) = S(i,j)$ for $1 \leq i < j \leq e$ and that $W_S(v_S(L(x_{\lambda(i)})), v_S(x_{\lambda(i)})) = W_S(v_S(L(\bar{x}_{\lambda(i)})), v_S(x_{\lambda(i+1)}))$ (by (PS_2), definition 3.9 and definition 3.11). Let
$$B_i = W_S(v_S(L(x_{\lambda(i)})), v_S(x_{\lambda(i)})),$$
$$C_i = W_S(v_S(x_{\lambda(i)}), v_S(R(x_{\lambda(i)}))).$$
Since S satisfies the boundary connections B_i and C_i are not empty. We have
$$S(1, i_0) = S(1, L(x_{\lambda(1)}))S(L(x_{\lambda(1)}), i_0)$$
$$= S(1, L(x_{\lambda(1)}))W_S(v_S(L(x_{\lambda(1)})), v_S(i_0))$$
$$= S(1, L(x_{\lambda(1)}))W_S(v_S(L(x_{\lambda(1)})), v_S(x_{\lambda(1)}))$$
$$= S(1, L(x_{\lambda(1)}))B_1,$$
$$S(1, L(\bar{x}_{\lambda(i)}))B_i = W_S(0, v_S(L(\bar{x}_{\lambda(i)})))W_S(v_S(L(x_{\lambda(i)})), v_S(x_{\lambda(i)}))$$
$$= W_S(0, v_S(L(\bar{x}_{\lambda(i)})))W_S(v_S(L(\bar{x}_{\lambda(i)})), v_S(x_{\lambda(i+1)})) = W_S(0, v_S(x_{\lambda(i+1)}))$$
$$= W_S(0, v_S(L(x_{\lambda(i+1)})))W_S(v_S(L(x_{\lambda(i+1)})), v_S(x_{\lambda(i+1)})) = S(1, L(x_{\lambda(i+1)}))B_{i+1},$$
$$S(1, L(\bar{x}_{\lambda(k)}))B_k = W_S(0, v_S(L(\bar{x}_{\lambda(k)})))W_S(v_S(L(\bar{x}_{\lambda(k)})), v_S(x_{\lambda(k)}))$$
$$= W_S(0, v_S(L(\bar{x}_{\lambda(k)})))W_S(v_S(L(\bar{x}_{\lambda(k)})), v_S(j_0)) = W_S(0, v_S(j_0)) = S(1, j_0).$$

Step 2: Now the words
$$S_i = S(L(x_{\lambda(i+1)}), L(\bar{x}_{\lambda(i)}))$$
$$R_i = S(R(\bar{x}_{\lambda(i)}), e)$$
$$T_i = S(R(x_{\lambda(i+1)}), e)$$

for $1 \le i \le u - 1$ (compare definition 3.6, condition (IIb)) show that the decomposition of the words $S(x_{\lambda(i)})$ into $B_i C_i$ ($1 \le i \le u - 1$) defines in fact a domino tower of height $u - 1$. Similarly the decomposition with respect to the second part may be used to get a domino tower of length $k - u + 1$. In any case we get a tower of height at least $k/2$. ∎

The Finite Tree Theorem

The domino towers which we may compose by means of the components $S(x_i)$ of a solution S of a position equation with a boundary connection are yet not directly related to the exponent of periodicity of S: they do not necessarily satisfy the shift condition. Thus we introduce the more rigid concept of normalized position equations where the length of the connections is directly related to the exponent of periodicity of solutions.

Definition 3.13: The subpath $(x_{\lambda(i)}, x_{\lambda(i+1)}, \ldots, x_{\lambda(l)})$ of the boundary connection (3.1) of PE has a *shift* if $L(x_{\lambda(j+1)}) < L(\bar{x}_{\lambda(j)})$ or $L(x_{\lambda(j+1)}) > L(\bar{x}_{\lambda(j)})$ for some $i \le j < l$. In the first (second) case we say that there is a positive (negative) left shift.

Definition 3.14: (a) Let $(\pi_1, x_{\lambda_i}, \pi_2, x_{\lambda_j}, \pi_3)$ be a subpath of the boundary connection (3.1) such that $x_{\lambda_i} = x_{\lambda_j}$ (the π_i are possibly empty subpaths). The subpath (π_2, x_{λ_j}) is called *superfluous* if it does not have any shift.
(b) The variable x_i is *matched* in PE if $col(x_i) = col(\bar{x}_i)$.

Definition 3.15: A position equation PE is *normalized* if the following conditions are satisfied:
(N₁) PE does not have a matched variable.
(N₂) No boundary connection of PE has a superfluous subpath.
(N₃) No boundary connection of PE has a subpath of the form $(x_i, \bar{x}_i)$.

Lemma 3.16: *Suppose the normalized position equation PE with $2n$ variable bases has a boundary connection (3.1) of length k. Then the exponent of periodicity s of any solution of PE is at least*

$$s \ge \frac{k}{4n^3 + 2n^2} - 1.$$

Proof: The path of (3.1) has a shift in every subpath of length $2n + 1$. If S is a solution of PE, then the words $S(x_{\lambda_j})$ may be used to arrange a domino tower of height at least $k/2$ (lemma 3.12). Now theorem 3.5 implies lemma 3.16. ∎

We are almost in the position to prove that we may restrict the search tree to a finite set of (normalized) position equations. First we have to exclude, however, some position equations which are unsolvable in a trivial sense.

Definition 3.17: A position equation over the alphabets $C, \mathcal{X}$ is called *elementary* if C has only one element a_1. If PE is an arbitrary position equation, then we may associate with PE the elementary position equation $PE^{(0)}$ over $\{a_1\}, \mathcal{X}$ where $TYPE$ is replaced by $TYPE^{(0)}$ which assigns a_1 to every coefficient base. It is not difficult to see that the solvability of $PE^{(0)}$ may be expressed as an existential formula of the first order theory for arithmetic without multiplication. As already mentioned, it is well-known that this theory is decidable.

Lemma 3.18: *For every elementary position equation PE there exists an algorithm to decide whether PE has a solution or not.* ∎

Definition 3.19: A position equation PE (with $2n$ variable bases and m coefficient bases) is *trivial* in the following cases:
(1) $\mathcal{X}$ is empty (i.e. $n = 0$),
(2) PE has two coefficient bases with the same column but different type,
(3) $L(w_{2n+i}) < L(x_j)$, $1 \leq i \leq m$ and $1 \leq j \leq 2n$,
(4) PE is elementary.
A nontrivial position equation is called *admissible*.

Lemma 3.20: *If PE is a trivial position equation, then there exists an algorithm to decide whether PE has a solution or not.*

Proof: Cases (1) and (2) are trivial. In case (3) it is simple to decompose PE into two independent parts, the first without variable bases, the second elementary (in the relevant case where the columns of all coefficient bases are indecomposable and the set of variable bases is not empty). ∎

Lemma 3.21: (a) *Suppose the admissible position equation PE has boundary connections*

$$(i_0, \pi, j_0) \tag{1}$$
$$(i_1, \pi, j_0) \tag{2}$$

with path π. Then $i_1 = i_0$ and (1) and (2) are identical.
(b) *If PE is an admissible position equation and $(x_{\lambda(i)}, x_{\lambda(i+1)})$ is a subpath of a boundary connection of PE, then the intersection of the columns of $\bar{x}_{\lambda(i)}$ and of $x_{\lambda(i+1)}$ is a column (thus non-empty).*

Proof: (a) Any solution S of $PE^{(0)}$ satisfies both (1) and (2). If i_0 and i_1 would be different boundaries, then $v_S(i_0) \neq v_S(i_1)$ (for any initial value). As a consequence of definition 3.9 we would have $v_S(j_0) \neq v_S(j_0)$, a contradiction. (b) Otherwise the solution of $PE^{(0)}$ would not allow the decomposition of the proof of lemma 3.12. ∎

Definition 3.22: The position equations PE_1 and PE_2 over $\mathcal{C}, \mathcal{X}$ are isomorphic if PE_1 and PE_2 become identical if we replace (as arguments of the functions L, R and as values of $TYPE$) all variables x_i by $\Phi(x_i)$ and all coefficients a_j by $\Psi(a_j)$, for a permutation Φ of $\mathcal{X}$ preserving duality and a permutation Ψ of $\mathcal{C}$.

In the algorithm, the alphabets which are used for the occurring position equations are subsets of two initial alphabets $\mathcal{C}$ and $\mathcal{X}$. In the following theorem, we assume also that all position equations use alphabets which are subsets of two fixed finite alphabets $\mathcal{C}$ and $\mathcal{X}$. With a generalized notion of isomorphic structures we could dispense with this assumption.

Finite Tree Theorem 3.23: *There exists a recursive function $F(n, m, c, d)$ such that the number of all admissible and normalized position equations PE (up to isomorphism) with parameters n_0, m_0, c_0, d_0 bounded by n, m, c, d does not exceed $F(n, m, c, d)$.*

Proof: Let us ask for the number (up to isomorphism) of all normalized and admissible position equations with principal parameters (n_0, m_0, c_0, d_0), where (n_0, m_0, c_0, d_0) is a fixed quadrucpel satisfying the bound. For these position equations the coefficient alphabet has the form $\{a_1, \ldots, a_{r_1}\}$ for $r_1 \leq m_0$ (since $TYPE$ is onto and we identify isomorphic position equations). The number of possible functions $TYPE$ does not exceed $2^{m_0} + \ldots + m_0^{m_0} \leq m^{m+1}$. We may use at most $2n_0 + f_0 < c_0$ distinct natural numbers as final boundaries of a boundary connection. By lemma 3.21 and condition (IV) of definition 3.6 there are $n_{bc} < (2n_0+1)^{d_0} c_0$ possibilities to introduce boundary connections and inessential boundaries as initial boundaries of such connections. The number of possible lists of boundary connections does not exceed $2^{n_{bc}}$. The total number of boundaries satisfies $n_{bd} \leq n_{bc} + 2(2n_0 + m_0) + f_0 + g_0 < n_{bc} + 2n_0 + 2m_0 + 2c_0$. The number of possible orders of any such number of boundaries does not exceed $(n_{bd})!$. For every such order, the number n_{LF} of possible functions L and R satisfies $n_{LF} \leq (n_{bd})^{4n_0 + 2m_0}$. Combining these results we get a recursive function $F_0(n, m, c, d)$ such that the number of all admissible and normalized position equations PE (up to isomorphism) with parameters n_0, m_0, c_0, d_0 does not exceed $F_0(n_0, m_0, c_0, d_0)$. Theorem 3.23 follows immediately. ∎

4 The Complete Search Tree

We completely describe the algorithm. A new transformation for position equations is given which applies to arbitrary normalized and admissible position equations and has the following properties: a nontrivial left part of the position equation is erased at every transformation step. The number of bases which are transported from the carrier to its dual is as large as possible, in particular we do not necessarily transport only leading bases. In certain situations the complete structure of the carrier is transported to its dual. Boundary connections are only introduced in certain cases.

Index and Exponent of Periodicity in the Flat Tree

Definition 4.1: Suppose the word equation E has a solution S of the form (1.2). The *index* of S is the number $I = |X_1 \ldots X_n|$. If S is a solution of the multi-equation ME of the form (2.1), then the index of S is the number $I = |S(\sigma_{1,1} \ldots \sigma_{1,k_1})|$. If S is a solution of the sme-system $SME = \{ME_1, \ldots, ME_r\}$, then the index of S with respect to SME is the sum of the indices of S with respect to the multi-equations ME_i ($1 \le i \le r$). The exponent of periodicity of S is defined as in definition 3.1. Recall that the index of a solution S of the position equation PE with boundaries $1, \ldots, e$ is $I = |S(1, e)|$.

Remark: Note that there would be another possibility to define the index I of the solution (1.2) of a word equation (1.1), namely $I = |S(\sigma_1 \ldots \sigma_k)|$. Let us use the symbols $I_{(x_1 \ldots x_n)}$ and $I_{(\sigma_1 \ldots \sigma_k)}$ for a moment, in order to distinguish the two possibilities. The point is that $I_{(x_1 \ldots x_n)}$ may be extremely small even if the notational length $nl(E)$ is very large. If x has 437 occurrences in E, then $I_{(x_1 \ldots x_n)} = 2$ if (ab) is a solution of E. But $I_{(\sigma_1 \ldots \sigma_k)} \ge 438$ in the same situation. In our reformulation of the algorithm we will find any solution of E after at most $I_{(x_1 \ldots x_n)}$ transformation steps in the non-flat tree. This is a significant improvement in comparison with the usual formulation of the algorithm where the corresponding number of transformation steps is more or less related with $I_{(\sigma_1 \ldots \sigma_k)}$.

Lemma 4.2: (a) *If the word equation E has a solution S with index I and exponent of periodicity s, then $T_{ini}(E)$ has a successful leaf or an open leaf η labelled with an sme-system SME which has a solution S' with index $I' \le I$ and exponent of periodicity $s' \le s$.*
(b) *Suppose the node η labelled with the flat sme-system SME is an open leaf of $T_{ini}(E)$ or occurs in the non-initial part of $T_{flat}(E)$. If SME has a solution S with index I and exponent of periodicity s, then S leads to an sme system $SME' \in Simpl(Trans(SME))$ which has a solution S' with index $I' < I$ and exponent of periodicity $s' \le s$.*

Proof: (a) Suppose E has the solution $S = (X_1, \ldots, X_n)$. By lemma 2.6 and remark 2.13, S leads to a successful leaf of $T_{ini}(E)$ or to an open leaf of this tree which is labelled with the system SME which has a solution S' corresponding to S. Assume that we are in the second case. Assume first, for simplicity, that $SME(E)$ does not have singles and that no column is erased by the simplification steps which occur along the path from E to SME. The proof of lemma 2.12 shows - in combination with a simple induction - that S has the form $(S'(U^{(1)}), \ldots, S'(U^{(n)}))$ where $U^{(i)}$ is an arbitrary line of the column of SME associated with x_i ($1 \le i \le n$). Thus the index I of S and I' of S' coincide. If, however, simplification erases a column, or if $SME(E)$ has a single, then $I' < I$. If $x_{i,j}$ is a variable occurring in SME, then we may assume, choosing the right lines, that $x_{i,j}$ occurs in some $U^{(l)}$ ($1 \le l \le n$). Thus $S'(x_{i,j})$ is a subword of $S(x_l) = S'(U^{(l)})$ and the exponent of periodicity of S' does not exceed the exponent of periodicity of S.

(b) This follows directly from lemmata 2.12, 2.14 and from remark 2.13. ∎

Corollary 4.3: *If E has a solution with index I and and if no variable occurs more then twice in E, then $T_{flat}(E)$ has a successful leaf in depth bounded by $nl(E) + I$.*

Proof: This follows immediately from theorem 2.17 (c) and lemma 4.2. ∎

Translation of SME-Systems into Position Equations

Definition 4.4: Let SME be a non-flat admissible sme-system with paired variables $x_1, \ldots, x_n$ and set of coefficient symbols C. Suppose SME is given in some order $ME_1, \ldots, ME_r$. The set $Set - PE(SME)$ is the finite set of position equations over C and $\{x_1, \ldots, x_{2n}\}$ which are generated by the following translation algorithm, given SME:

- **Translation** (of the sme-system $SME = \{ME_1, \ldots, ME_r\}$)

Step 1: Suppose that ME_i has the form

$$\sigma_{1,1}^{(i)} \ldots \sigma_{1,k_1^{(i)}}^{(i)} == \sigma_{2,1}^{(i)} \ldots \sigma_{2,k_2^{(i)}}^{(i)} == \ldots == \sigma_{l(i),1}^{(i)} \ldots \sigma_{l(i),k_{l(i)}^{(i)}}^{(i)} \quad (1 \leq i \leq r).$$

For all paired variables x_i of SME, replace the index of one occurrence by $i + n$. If $\sigma_{j,k}^{(i)}$ is the qth occurrence of a coefficient in SME, replace it by a coefficient base w_{2n+q} and define $TYPE(w_{2n+q}) = \sigma_{j,k}^{(i)}$. As a result of step 1, we get the function $TYPE$, the set $BS = \{x_1, \ldots, x_{2n}, w_{2n+1}, \ldots, w_{2n+m}\}$ of bases and the system SME' of multi-equations ME_i' of the form

$$\tau_{1,1}^{(i)} \ldots \tau_{1,k_1^{(i)}}^{(i)} == \tau_{2,1}^{(i)} \ldots \tau_{2,k_2^{(i)}}^{(i)} == \ldots == \tau_{l(i),1}^{(i)} \ldots \tau_{l(i),k_{l(i)}^{(i)}}^{(i)} \quad (1 \leq i \leq r).$$

Step 2: For $1 \leq i \leq r$ and $1 \leq j \leq l(i)$ introduce distinct boundaries

$$b_{left}^{(i)}, b_{j,1}^{(i)}, \ldots, b_{j,k_j^{(i)}-1}^{(i)}, b_{j,k_j^{(i)}}^{(i)} = b_{right}^{(i)},$$

order relations

$$b_{left}^{(i)} < b_{j,1}^{(i)} < b_{j,2}^{(i)} < \ldots < b_{j,k_j^{(i)}-1}^{(i)} < b_{right}^{(i)}$$

and functions (defined for arguments $\tau_{j,1}^{(i)} \in BS$)

$$L(\tau_{j,1}^{(i)}) = b_{left}^{(i)},$$
$$R(\tau_{j,1}^{(i)}) = b_{j,1}^{(i)} = L(\tau_{j,2}^{(i)}), \ldots, R(\tau_{j,k_j^{(i)}-1}^{(i)}) = b_{j,k_j^{(i)}-1}^{(i)} = L(\tau_{j,k_j^{(i)}}^{(i)}),$$

$$R(\tau^{(i)}_{j,k^{(i)}_j}) = b^{(i)}_{right}.$$

Identify $b^{(i)}_{right} = b^{(i+1)}_{left}$ $(1 \le i \le r - 1)$.

Step 3: Compute all linear orders which extend the partial order between boundaries given in step 2. For every such linear order $\preceq$ introduce now a position equation $PE_{\preceq}$ as follows: rename the set of boundaries by consecutive natural numbers $1, 2, \ldots, e$, going from left to right in the linear order. Correct the values of L and R accordingly. $PE_{\preceq}$ does not have fossil boundaries or boundary connections.

Lemma 4.5: *SME has a solution $S = (X_1, \ldots, X_n)$ if and only if some $PE \in Set - PE(SME)$ has a solution S' such that $S'(x_i) = X_i$. Index and exponent of periodicity of the solutions S and S' coincide.*

Proof: Suppose that S is a solution of SME. S is a solution of every multi-equation ME_i. We define

$$b^{(i)}_{j_1,k_1} <_S b^{(i)}_{j_2,k_2}$$

iff $S(\sigma^{(i)}_{j_1,1} \ldots \sigma^{(i)}_{j_1,k_1})$ is a proper prefix of $S(\sigma^{(i)}_{j_2,1} \ldots \sigma^{(i)}_{j_2,k_2})$ and

$$b^{(i)}_{j_1,k_1} =_S b^{(i)}_{j_2,k_2}$$

iff $S(\sigma^{(i)}_{j_1,1} \ldots \sigma^{(i)}_{j_1,k_1}) = S(\sigma^{(i)}_{j_2,1} \ldots \sigma^{(i)}_{j_2,k_2})$. These relations define - in combination with the relations of step 2 - a unique linear order $\preceq$ on the set of boundaries. Now consider $PE_{\preceq} \in Set - PE(SME)$. The indecomposable columns of $PE_{\preceq}$ have - modulo names of boundaries - the form $(b^{(i)}_{left}, b^{(i)}_{j,1})$ or $(b^{(i)}_{j_1,k_1}, b^{(i)}_{j_2,k_2})$. Let

$$S'(b^{(i)}_{left}, b^{(i)}_{j,1}) = S(\sigma^{(i)}_{j,1}),$$
$$S'(b^{(i)}_{j_1,k_1}, b^{(i)}_{j_2,k_2}) = S(\sigma^{(i)}_{j_1,1} \ldots \sigma^{(i)}_{j_1,k_1})^{-1} S(\sigma^{(i)}_{j_2,1} \ldots \sigma^{(i)}_{j_2,k_2}).$$

(Here we use the following notation: if $U = VW$, then $V^{-1}U = W$.) Every symbol $\sigma^{(i)}_{j,k}$ of SME defines a corresponding column

$$(L(\tau^{(i)}_{j,k}), R(\tau^{(i)}_{j,k})) \tag{4.1}$$

of $PE_{\preceq}$. It is not difficult to prove by induction on the number of indecomposable subcolumns that

$$S'(L(\tau^{(i)}_{j,k}), R(\tau^{(i)}_{j,k})) = S(\sigma^{(i)}_{j,k}).$$

Thus it is trivial to verify conditions (PS_1) and (PS_2) of definition 3.8 for S'. Since $PE_{\preceq}$ does not have a boundary connection S' is a solution of $PE_{\preceq}$.
Assume conversely that $PE_{\preceq}$ is a position equation generated by the algorithm and S' is a solution of $PE_{\preceq}$. Since all the $l(i)$ sides of multi-equation ME_i of SME define the same column of $PE_{\preceq}$ it is trivial to show that $S(x_i) = S'(col(x_i))$ defines a solution of SME. ■

Definition 4.6: Let SME and $Set - PE(SME)$ be as in definition 4.4. The set $Set - NPE(SME)$ of *normalized position equations canonically associated with* SME contains all normalized position equations which we get if we erase the matched variables occurring in the elements of $Set - PE(SME)$ and restrict the functions L and R accordingly.

Corollary 4.7: (a) *If SME has a solution $S = (X_1, \ldots, X_n)$ with index I and exponent of periodicity s, then some $PE \in Set - NPE(SME)$ has a solution S' with index $I' = I$ and exponent of periodicity $s' \leq s$ such that $S'(x_i) = X_i$ for all variable bases x_i of PE.*
(b) *If some $PE \in Set - NPE(SME)$ has a solution, then SME has a solution.*

Proof: Trivial. ∎

To describe the complete search tree we need two results in advance:

Transformation Theorem 4.8: *Let PE be a normalized and admissible position equation. The procedure TRANSFORMATION which is described in the following subsection transforms PE into a finite set $TRANS(PE)$ of position equations with the following properties:*
(a) *The principal parameters n, m and c of the position equations in $TRANS(PE)$ do not exceed the corresponding parameters of PE.*
(b) *If PE has a solution S with index I and exponent of periodicity s, then a position equation in $TRANS(PE)$ has a solution S' with index $I' < I$ and exponent of periodicity $s' \leq s$.*
(c) *If a position equation in $TRANS(PE)$ has a solution, then PE has a solution.*

The proof of the transformation theorem may be found in [Sc3].

Normalization Theorem 4.9: *Let PE be a position equation. There exists an algorithm which transforms PE into a finite set $NORM(PE)$ of normalized admissible position equations with the following properties:*
(a) *The principal parameters of the position equations in $NORM(PE)$ do not exceed the corresponding parameters of PE.*
(b) *If PE has a solution S with index I and exponent of periodicity s, then a position equation in $NORM(PE)$ has a solution S' with index $I' = I$ and exponent of periodicity $s' = s$.*
(c) *If a position equation in $NORM(PE)$ has a solution, then PE has a solution.*

A description of the normalization procedure and the proof of the normalization theorem may be found in [Péc]. We are now ready to describe the extension of $T_{flat}(E)$ to $T_{Mak}(E)$.

The Complete Search Tree

Definition 4.10: Let E be a word equation. Let s_{max} denote the upper bound for the exponent of periodicity of a minimal solution of E as given in theorem 3.3. The unordered, finitely branching tree $T_{Mak}(E)$ is the following extension of $T_{flat}(E)$:

Successful Leaves of $T_{Mak}(E)$ are:
- All successful leaves of $T_{flat}(E)$.
- All nodes which are labelled with a trivial and solvable position equation PE.

Blind Leaves of $T_{Mak}(E)$ are:
- All blind leaves of $T_{flat}(E)$.
- All nodes which are labelled with a position equation PE which is isomorphic to the label of a node which has occurred earlier in the same path.
- All nodes which are labelled with a normalized admissible position equation PE with principal parameters n and d showing (compare lemma 3.16) that the exponent of periodicity of any solution exceeds s_{max}.

Successors of other Nodes in $T_{Mak}(E)$:
- Suppose that η is an open leaf of $T_{flat}(E)$ labelled with the non-flat admissible sme-system SME. For every admissible or trivially solvable position equation PE in $Set - NPE(SME)$ the node η has exactly one successor which is labelled with PE. (If there is no such position equation, then η is a blind leaf.)
- Suppose that η is a node of $T_{Mak}(E)$ labelled with the admissible normalized position equation PE. For every admissible or trivially solvable position equation PE' in $NORM(TRANS(PE))$ the node η has exactly one successor which is labelled with PE'. (If there is no such position equation, then η is a blind leaf.)

Main Theorem 4.11: (a) *If E has a solution S with index I, then $T_{Mak}(E)$ has a successful leaf in depth not exceeding $nl(E) + I + 1$.*
(b) *If $T_{Mak}(E)$ has a successful leaf, then E is solvable.*
(c) *$T_{Mak}(E)$ is finite. If E is unsolvable, then every leaf of $T_{Mak}(E)$ is a blind leaf.*

Proof: (a) Without loss of generality S is a minimal solution of E and the exponent of periodicity of S does not exceed s_{max}. By lemma 4.2 (a), either we find a successful leaf of $T_{ini}(E)$ or an open leaf of the initial tree which is labelled with the sme-system SME_1 which has a solution S_1 with index $I_1 \leq I$ and exponent of periodicity $s_1 \leq s_{max}$. Suppose we are in the second case. By minimality of S (which is needed to exclude nodes which are labelled with an sme-system which is isomorphic to an earlier one, compare the proof of lemma 1.5), either we find a successful leaf of $T_{flat}(E)$ or an open leaf of the flat tree which is labelled with the sme-system SME_2 with solution S_2 with index $I_2 \leq I_1 \leq I$ and exponent of periodicity $s_2 \leq s_1 \leq s_{max}$ (lemma 4.2). In the second case, by minimality of S we find a successful node of $T_{Mak}(E)$ labelled with a trivially true position equation PE (theorems 4.8,4.9). On the way from SME_1 to PE, the index of the respective solution decreases at every step (lemma 4.2 (b) and theorems

4.8,4.9). Thus the number of such steps cannot exceed I (since solvable sme-systems or position equations which have a solution of index 1 are trivially solvable). Adding the translation step where we introduce the first position equation and the steps in the initial tree we get at all $w \leq I + 1 + nl(E)$ downward-steps in the search tree.

(b) An immediate consequence of theorems 4.9, 4.8, corollary 4.7 and theorem 2.17.

(c) We stop every path as soon as a boundary connection exceeds a certain length (third type of a blind leaf) or if two isomorphic position equations have occurred (second type). Thus we may give an upper bound for all principal parameters of the position equations which may occur (theorems 4.8 and 4.9). Now (c) follows from theorems 3.23 and 2.17 (a). ∎

A New Transformation for Position Equations

Definition 4.12: A variable base x_i of the position equation PE is called a *leading base* if $L(x_i) \leq L(x_j)$ for all $1 \leq j \leq 2n$. A leading variable base x_i is called a *candidate* if $R(x_i) \geq R(x_j)$ for every leading base x_j. The candidate x_ν with the smallest index ν is called the *carrier* of PE. From now on, the following symbols have a standard meaning: x_ν is always the carrier, $l^* = L(x_\nu), r^* = R(x_\nu), \bar{l}^* = L(\bar{x}_\nu), \bar{r}^* = R(\bar{x}_\nu)$.

Every solution of the position equation PE has to assign the same word to the carrier x_ν and its dual base $\bar{x}_\nu$. Thus, for every base whose column is a subcolumn of (l^*, r^*) there must exist a corresponding part of $(\bar{l}^*, \bar{r}^*)$. We now define the set of all *prints* of x_ν in $\bar{x}_\nu$ which describes all possibilities how the structure of (l^*, r^*) may be reflected in $(\bar{l}^*, \bar{r}^*)$.

Prints

Definition 4.13: Let PE be a normalized admissible position equation. Let

$$l^* = i_1, i_2 = i_1 + 1, ..., i_k = r^*$$

be the complete list of consecutive boundaries between l^* and r^*, let $i_1^{tr}, i_2^{tr}, ..., i_k^{tr}$ be a second copy, using arbitrary new symbols. A *preprint* of x_ν in $\bar{x}_\nu$ is a linear order $\preceq$ on the set $i_1^{tr}, i_2^{tr}, ..., i_k^{tr}, 1, ..., e$ extending the natural order of $1, ..., e$ such that

(i) $\bar{l}^* = l^{*tr} \prec i_r^{tr} \prec i_s^{tr} \prec r^{*tr} = \bar{r}^*$ for $l^* < i_r < i_s < r^*$,

(ii) if there exists a coefficient base w_{2n+i} such that $L(w_{2n+i}) = i, R(w_{2n+i}) = j$, then i and j are consecutive in the order $\preceq$, and if $l^* \leq i < j \leq r^*$, then i^{tr} and j^{tr} are also consecutive.

A *print* for PE is a preprint $\preceq$ which satisfies the following conditions:

(Pr₁) If PE has a boundary connection (i, x_ν, j), then $i^{tr} = j$. (Note that $l^* < i < r^*$, according to condition (IIa) of definition 3.6.)

(Pr₂) If PE has a boundary connection with initial sequence (i, x_ν, x_ψ), then $L(x_\psi) \prec i^{tr} \prec R(x_\psi)$.

(Pr₃) If PE has a boundary connection with subpath $(x_\theta, x_\nu, x_\psi)$, then $L(\bar{x}_\theta)^{tr} \prec R(x_\psi)$ and $L(x_\psi) \prec (min\{r^*, R(\bar{x}_\theta)\})^{tr}$. (Note that $L(\bar{x}_\theta) < r^*$, by lemma 3.21 (b).)

(Pr₄) If PE has a boundary connection with sublist (x_θ, x_ν, j), then $L(\bar{x}_\theta)^{tr} \prec j$. If $R(\bar{x}_\theta) \leq r^*$, then also $j \prec R(\bar{x}_\theta)^{tr}$.

(Pr₅) If PE has a boundary connection $(i, \bar{x}_\nu, j)$, then $j^{tr} = i$.

(Pr₆) If PE has a boundary connection $(i, \bar{x}_\nu, x_\psi)$, then $L(x_\psi)^{tr} \prec i$. If $R(x_\psi) \leq r^*$, then also $i \prec R(x_\psi)^{tr}$.

(Pr₇) If PE has a boundary connection with subpath $(x_\theta, \bar{x}_\nu, x_\psi)$, then $L(x_\psi)^{tr} \prec R(\bar{x}_\theta)$ and $max\{\bar{l}^*, L(\bar{x}_\theta)\} \prec (min\{r^*, R(x_\psi)\})^{tr}$.

(Pr₈) If PE has a boundary connection with sublist $(x_\theta, \bar{x}_\nu, j)$, then $L(\bar{x}_\theta) \prec j^{tr} \prec R(\bar{x}_\theta)$.

Lemma 4.14: *Let S be a solution of the position equation PE with carrier x_ν, boundaries $1, \ldots, e$ and boundaries $l^* = i_1, i_2 = i_1 + 1, \ldots, i_k = r^*$ between l^* and r^*. Define a linear order $\preceq$ on the set $i_1^{tr}, i_2^{tr}, \ldots, i_k^{tr}, 1, \ldots, e$ extending the natural order of $1, \ldots, e$ by $\bar{l}^* = l^{*tr} \prec i_r^{tr} \prec i_s^{tr} \prec r^{*tr} = \bar{r}^*$ if $\bar{l}^* < i_r < i_s < \bar{r}^*$ and by $i_r^{tr} \prec j_s$ (or $i_r^{tr} \succ j_s$, or $i_r^{tr} = js$) if $|S(l^*, i_r)| < |S(\bar{l}^*, j_s)|$ (if $|S(l^*, i_r)| > |S(\bar{l}^*, j_s)|$, if $|S(l^*, i_r)| = |S(\bar{l}^*, j_s)|$). Then $\preceq$ is a print of PE.*

Proof: We show that $\preceq$ satisfies condition (Pr₃), the proof of the other conditions is similar. Suppose that PE has a boundary connection with subpath $(x_\theta, x_\nu, x_\psi)$. By lemma 3.21 (b) it is easy to see that $\bar{l}^* < R(x_\psi)$. We consider any S-evaluation of the connection. Let $v_1 = v_S(l^*)$ and $u = v_S(x_\nu)$. By definition 3.11 and lemma 3.10 we get

$$v_S(L(\bar{x}_\theta)) < u < v_S(R(\bar{x}_\theta)) \tag{1}$$

$$v_S(L(x_\psi)) < u + dd_S(x_\nu) < v_S(R(x_\psi)) \tag{2}$$

Thus $|S(l^*, L(\bar{x}_\theta))| = v_S(L(\bar{x}_\theta)) - v_1 < u - v_1$. Since

$$v_S(R(x_\psi)) = v_S(\bar{l}^*) + |S(\bar{l}^*, R(x_\psi))| = v_1 + dd_S(x_\nu) + |S(\bar{l}^*, R(x_\psi))|$$

we get $|S(l^*, L(\bar{x}_\theta))| < u - v_1 < |S(\bar{l}^*, R(x_\psi))|$, by (2). Thus $L(\bar{x}_\theta)^{tr} \prec R(x_\psi)$, by definition of $\preceq$. Case 1, $\bar{l}^* \leq L(x_\psi)$: by (1), $u < v_S(R(\bar{x}_\theta)) = v_1 + |S(l^*, R(\bar{x}_\theta))|$ and therefore $u - v_1 < |S(l^*, R(\bar{x}_\theta))|$. By (2), $v_S(L(x_\psi)) = v_S(\bar{l}^*) + |S(\bar{l}^*, L(x_\psi))| = v_1 + dd_S(x_\nu) + |S(\bar{l}^*, L(x_\psi))| < u + dd_S(x_\nu)$. Thus $|S(\bar{l}^*, L(x_\psi))| < |S(l^*, R(\bar{x}_\theta))|$. If $R(\bar{x}_\theta) \leq r^*$, then $L(x_\psi) \prec R(\bar{x}_\theta)^{tr}$, by definition of $\preceq$. $L(x_\psi) \leq r^{*tr} = \bar{r}^*$ is trivial, by lemma 3.21 (b). Case 2, $L(x_\psi) < \bar{l}^*$ is trivial. ∎

Definition 4.15: Let x_ν be the carrier of the normalized admissible position equation PE. The *critical boundary* of PE, denoted cr, is the leftmost boundary among the left boundaries of all variable bases x_j such that $L(x_j) < R(x_\nu)$, $R(x_j) > R(x_\nu)$, if there exist such variable bases, and $cr = r^* = R(x_\nu)$ otherwise. A coefficient base w_{2n+i} is called *erasable* if $L(w_{2n+i}) < l^*$. An arbitrary base w_i different from x_ν with

$l^* \leq L(w_i) < cr$ is called a *transport base*. A base w_i with $cr \leq L(w_i)$ which is different from $\bar{x}_\nu$ is called *fixed*. All boundaries i with $l^* < i \leq cr$ and all the boundaries j with $cr < j < r^*$ which occur as right boundary of a transport base are called *transport boundaries*. The remaining boundaries j with $cr < j$ are called *fixed*.

The following figure illustrates the definition for the case $cr \neq r^*$. Bold horizontal (vertical) lines indicate that the base (boundary) is a fixed base (boundary). The new transformation will cut off the left part of the position equation up to cr. The strategy to cut off a maximal left part of the carrier is similarly realized in the string unification algorithm of PROLOG III (A. Colmerauer, personal communication).

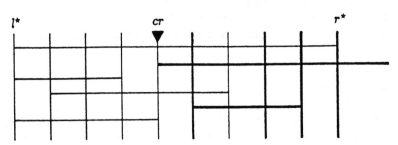

Definition/Remark 4.16: Let PE be a normalized admissible position equation. According to conditions (IIa) and (IIc) of definition 3.6 and condition (N₃) of definition 3.15, every boundary connection (3.1) of PE may be decomposed in the following way:

$$(i_0, \pi_0, x_{\gamma(1)}), (x_{\gamma(1)}, \pi_1, x_{\gamma(2)}), \ldots, (x_{\gamma(k_1)}, \pi_{k_1}, j_0) \qquad (k_1 \geq 0) \qquad (4.2)$$

where the π_i are subsequences of the form $x_\nu, x_\nu, \ldots, x_\nu$ or $\bar{x}_\nu, \bar{x}_\nu, \ldots, \bar{x}_\nu$ of length $l_i \geq 0$ and the boundaries and bases $x_{\gamma(i)}$ are fixed or have transport type (the same holds then for the bases $\bar{x}_{\gamma(i)}$). We call (4.2) the decomposition of (3.1) into *normal subparts*. The sequence $(i_0, \pi_0, x_{\gamma(1)})$ is the *initial normal subpart*, $(x_{\gamma(k_1)}, \pi_{k_1}, j_0)$ is the *final normal subpart* of the boundary connection (4.2).

• **TRANSFORMATION** (of the normalized admissible position equation PE)

The restriction of a print for $PE \preceq$ to the set

$$BD' = \{cr, cr+1, \ldots, e\} \cup \{i^{tr}; l^* < i < r^*, i \text{ transport boundary of } PE\}$$

is called an *r-print* of PE. $TRANS(PE)$ has exactly one element PE', for every r-print $\preceq$ for PE. We sometimes write $PE'_{\preceq}$ in order to stress the dependency from $\preceq$. BD' will be the new set of boundaries. The transformation will be completed after a second, trivial step, where boundaries are renamed in such a way that the set BD'' of renamed boundaries is an initial segment $\{1, \ldots, e'\}$ of the natural numbers. In the following, $\preceq$ is an arbitrary, fixed r-print for PE. We describe how to transform PE into the structure $PE'_{\preceq}$.

Step 0: From the coefficient alphabet C of PE delete a_i if all coefficient bases w_{2n+j} with type a_i are erasable ($1 \leq i \leq r$). From the variable alphabet $\mathcal{X}$ delete $x_\nu, \bar{x}_\nu$ in the case $cr = r^*$. Let C' and $\mathcal{X}'$ denote the new alphabets.

Step 1: If $cr = r^*$, then delete $x_\nu, \bar{x}_\nu$ from the set of variable bases. Delete all erasable coefficient bases. For the remaining coefficient bases, $TYPE'$ and $TYPE$ coincide.

Step 2: BD' is the new set of boundaries. We sometimes write l^{*tr}, r^{*tr} for $\bar{l}^*$ and $\bar{r}^*$.
LF', the new set of left-fossil boundaries contains
- the boundaries $cr < i$ of $BD' \cap LF$,
- all boundaries i^{tr} for $l^* < i \leq cr$, $i \in LF$,
- the boundary $\bar{l}^*$ if $\bar{l}^*$ is neither left nor right in PE'.

RF', the new set of right-fossil boundaries contains
- the boundaries $cr < i$ of $BD' \cap RF$,
- all boundaries i^{tr} for $l^* < i \leq cr$, $i \in RF$,
- the boundaries $r^*, \bar{r}^*$ if they are neither left nor right in PE'.

Step 3: The functions L' and R' are defined as follows:
- $L'(x_\nu) = cr, R'(x_\nu) = r^*$,
  $L'(\bar{x}_\nu) = cr^{tr}, R'(\bar{x}_\nu) = \bar{r}^*$
  if $cr \neq r^*$,
- $L'(w_j) = L(w_j), R'(w_j) = R(w_j)$
  for every fixed base w_j of PE,
- $L'(w_j) = L(w_j)^{tr}, R'(w_j) = R(w_j)^{tr}$
  for every transport base w_j of PE.

Step 4: Consider the normal decomposition (4.2) of a boundary connection (3.1) of PE. PE' will have a corresponding boundary connection with normal decomposition

$$(i'_0, \pi'_0, x_{\gamma(1)}), (x_{\gamma(1)}, \pi'_1, x_{\gamma(2)}), \ldots, (x_{\gamma(k_1)}, \pi'_{k_1}, j'_0) \tag{4.3}$$

where $i'_0 = i_0$ if i_0 is fixed and $i'_0 \in \{i_0, i_0^{tr}\}$ if i_0 is a transport base, similarly for j_0. (Note that besides i_0, j_0 only the sequences π_i are modified.) The transformed boundary connection has the form

$$(i'_0, \pi'_0, x_{\gamma(1)}, \pi'_1, x_{\gamma(2)}, \ldots, x_{\gamma(k_1)}, \pi'_{k_1}, j'_0)$$

and is completely determined by the sequences of (4.3) which we get via transformation of the corresponding normal subparts of (4.2). There are the following *regular transformations* of normal subsequences. Exceptions are described below. Let $\bar{f}_1$ ($\bar{t}_1$) denote a fixed (transport) boundary or a variable base x_j such that $\bar{x}_j$ is a fixed (transport) base. Let f_2 (t_2) denote any fixed (transport) boundary or variable base, $(x_\nu)^k$ (($\bar{x}_\nu)^k$) denote sequences of k elements of the form x_ν ($\bar{x}_\nu$). If t_i is a variable base, then t_i^{tr} denotes an occurrence of t_i in PE'.

Regular Transformation

(f-f) transformation
- A sequence $(\bar{f}_1, (x_\nu)^k, f_2)$ is translated into $(\bar{f}_1, (x_\nu)^k, f_2)$ for $k \geq 0$,
- A sequence $(\bar{f}_1, (\bar{x}_\nu)^k, f_2)$ is translated into $(\bar{f}_1, (\bar{x}_\nu)^k, f_2)$ for $k > 0$.

(t-t) transformation
- A sequence $(\bar{t}_1, (x_\nu)^k, t_2)$ is translated into $(\bar{t}_1^{tr}, (x_\nu)^k, t_2^{tr})$ for $k \geq 0$,
- A sequence $(\bar{t}_1, (\bar{x}_\nu)^k, t_2)$ is translated into $(\bar{t}_1^{tr}, (\bar{x}_\nu)^k, t_2^{tr})$ for $k > 0$.

(f-t) transformation
- A sequence $(\bar{f}_1, (x_\nu)^k, t_2)$ is translated into $(\bar{f}_1, (x_\nu)^{k+1}, t_2^{tr})$ for $k \geq 0$,
- A sequence $(\bar{f}_1, (\bar{x}_\nu)^k, t_2)$ is translated into $(\bar{f}_1, (\bar{x}_\nu)^{k-1}, t_2^{tr})$ for $k > 0$.

(t-f) transformation
- A sequence $(\bar{t}_1, (x_\nu)^k, f_2)$ is translated into $(\bar{t}_1^{tr}, (x_\nu)^{k-1}, f_2)$ for $k > 0$,
- A sequence $(\bar{t}_1, (\bar{x}_\nu)^k, f_2)$ is translated into $(\bar{t}_1^{tr}, (\bar{x}_\nu)^{k+1}, f_2)$ for $k \geq 0$.

New Boundary Connections
- For every transport boundary $cr < i$ of PE we introduce a new boundary connection (i, x_ν, i^{tr}).

Irregular Transformation
- **(Exc 1)** The normal sequence (x_θ, j) is transformed into (x_θ, j) if it is the final part of a boundary connection with an non-empty second part, $\bar{x}_\theta$ is fixed and j is a transport boundary, relativizing the (f-t) transformations.
- **(Exc 2)** The normal sequence (i, x_ψ) is transformed into (i, x_ψ) if x_ψ is fixed and i is a transport boundary, relativizing the (t-f) transformations.
- **(Exc 3)** The normal sequence (x_θ, j) is transformed into $(x_\theta, \bar{x}_\nu, j)$ if $\bar{x}_\theta$ is a transport base, $cr < j$ is a transport boundary and if j is right-fossil or the witness x_ω of the boundary connection of PE is fixed, relativizing the (t-t) transformations.
- **(Exc 4)** The normal sequence (x_θ, x_ν, j) is transformed into (x_θ, j) if it is the final part of a boundary connection with non-empty second part, $\bar{x}_\theta$ is a transport base and j is a transport boundary, relativizing the (t-t) transformations.
- **(Exc 5)** The normal sequence $(i, \bar{x}_\nu, x_\psi)$ is transformed into (i, x_ψ) if i is a transport boundary and x_ψ is a leading base, relativizing the (t-t) transformations.

Finally all trivial boundary connections without variable bases are deleted.

How to Maintain the Substitution List

In section 2 we described how we may augment sme-systems by an additional substitution list which contains a unifier for E when we reach a successful leaf of $T_{flat}(E)$. It is not difficult to have a similar list for the position equations in the non-flat part of the tree. This is not a central point of our work, thus we only indicate the steps which are necessary. Suppose we reach an open leaf of $T_{flat}(E)$ labelled with the sme-system SME and substitution list $(U^{(1)}, \ldots, U^{(n)})$. The variables which occur within this list are either (1) variables $\sigma_{j,k}^{(i)}$ of SME or (2) singles of $SME(E)$ or variables where the corresponding occurrences in SME have been erased by a simplification step. After the translation, we replace variables of the first type by the unique column (4.1) of the position equation which corresponds to $\sigma_{j,k}^{(i)}$. Variables of the second type are not modified. At the normalization steps it is only necessary to use the new names for the columns of the substitution list if boundaries have to be renamed. Before a transformation step, certain columns have to be decomposed: if the substitution list contains a column (i, j) such that $i < l^* < j$, then we replace (i, j) by $(i, l^*)(l^*, j)$. Similarly, if $i < cr < j$, then we replace (i,j) by $(i,cr)(cr,j)$. After these steps there are at most three types of columns in the substitution list: the columns (i,j) where $j \leq l^*$ (type 1), the columns (i,j) where $l^* \leq i < j \leq cr$ (type 2) and the columns (i,j) where $cr \leq i$ (type 3). Columns of type 1 with k indecomposable subcolumns are replaced by a sequence of coefficients and new variable symbols of length k. If the ith indecomposable subcolumn contains a coefficient base of type a_h, then the ith symbol of the sequence is a_h. In the other case, the ith symbol is a new variable symbol. Columns of type 3 are only modified in the trivial sense, renaming boundaries if boundaries are renamed as a result of the transformation. Columns (i,j) of type 2 are replaced by (i^{tr}, j^{tr}). If we reach a trivial position equation with solution S, then we may replace the columns (i,j) of the substitution list by the words $S(i,j)$. In section 6 we will see some examples which demonstrate the main steps.

5 Remarks and Examples

We compare the generation of the initial tree with the generation of a set of position equations corresponding to the given equation in the standard formulation of Makanin's algorithm. The discussion shows that our reformulation does not only concern the flat part of the search tree: the position equations which we get via translation of non-flat sme-systems are (in a sense which will be explained) horizontally shorter than the corresponding position equations in the standard formulation. This substantially reduces the number of transformation steps which are necessary to find a solution of solvable equations. The new transformation is discussed. We explain which technical constraints lead to the details of the definition of a position equation. We show how our decision procedure may be modified in order to generate a minimal and complete set of unifiers for E.

Initial Tree and Position Equations

In the standard formulation of Makanin's algorithm, the word equation E is first translated into a generalized equation. Then a corresponding set of position equations is created which is used as the first level of the search tree. (From this point on, the further development of the search tree follows the definition of the non-flat part of T_{Mak}.) Let us give an example. Here is, roughly, the graphical description of the generalized equation which corresponds to the equation $axbzx = zczyyy$ of example 1.1:

a	x_1	b	z_1	$\bar{x}_1$
$\bar{z}_1$ / $\bar{z}_2$	c	$\bar{z}_2$	y_1	$\bar{y}_1$ / y_2	$\bar{y}_2$

Boundaries separate distinct symbols of both sides of the equation. The first boundaries and the last boundaries of both sides have to be identified. Variables with a single occurrence are erased. Occurrences of variables with two occurrences are replaced by two dual variable bases. If a variable has $k > 2$ occurrences, then the occurrences are replaced by a sequence of $k-1$ pairs of dual variables. According to the different ways in which the partial order of the boundaries may be completed to a total order, a whole set of position equations has to be used for the analysis of a generalized equation. Here is one of several position equations which represent successors of the generalized equation above (with boundaries $1, \ldots, 9$):

a	x_1			b	z_1	$\bar{x}_1$	
$\bar{z}_1$				$\bar{y}_1$			
z_2	c	$\bar{z}_2$	y_1	$\bar{y}_2$			$\bar{y}_2$

Such a position equation without boundary connections stores two types of constraints: the vertical constraints say that we have to assign the same subword to parts which have the same position, and the horizontal constraints say that we have to assign the same subword to the columns of dual variables. In comparison with the standard method to generate the position equations, the new method interchanges the role of these constraints. This may be demonstrated by means of our example. If we use a similar simplification strategy as we did in our formulation of the initial tree, then we may replace the occurrences of z-variables by a:

a	x_1			b	a	$\bar{x}_1$	
				$\bar{y}_1$			
	c	a	y_1	$\bar{y}_2$			$\bar{y}_2$

Now we introduce variables for all columns which do not contain a coefficient base: x_1' denotes the fourth column, y_2' denotes the seventh column and x_2' denotes the last column. By equality of duals, there are the following horizontal constraints: $cax_1' = y_2'x_2'$ (equality of the copies of x_1) and $x_1' = bay_2' = x_2'$ (equality of the y-variables). We get -modulo variable names - exactly the sme-system

$$\begin{vmatrix} cax_1 \\ y_2x_2 \end{vmatrix} \begin{vmatrix} x_1 \\ bay_2 \\ x_2 \end{vmatrix}$$

which we found after the resolution of the principal column in the corresponding example 1.1. It is not hard to see that there is quite general a similar one-to-one correspondence between the position equations which are generated by the standard method and the sme-systems which occur in our approach after the principal column is resolved[*].

---

[*] In last consequence, the algorithm which is used by Abdulrab to compute the "solution schemes" (see [Abd], page 27) corresponds to the resolution of the principal column in our approach.

SME-Systems versus Position Equations

It is worth mentioning that the set of word equations where a solution is found in the flat tree is not at all restricted to equations where variables have only two occurrences. The solvable equations which are mentioned in Abdulrab [Abd], for example, are the following: the equation $xxayby = cayvabd$ with variables x, y, v and solution $(caabd, abd, caabdaabdb)$, the equation $axbyx = ytxt$ with variables x, y, t and solution (ba, a, ba), the equation $xauby = cayvxac$ with variables x, y, u, v and solution (c, cac, cac, b) and the equation $zyxabxyz = ttcyx$ with variables x, y, z, t and solution (c, ab, c, cab). For each of these equations and all solvable examples mentioned in [Rou], the respective solution is found in the flat tree. Let us treat one example.

Example 5.1: Consider the equation

$$zyxabxyz = ttcyx.$$

The corresponding sme-system has the following form:

$$\left|\begin{array}{c} z_1 y_1 x_1 abx_2 y_2 z_2 \\ t_1 t_2 cy_3 x_3 \end{array}\right| \left|\begin{array}{c} z_1 \\ z_2 \end{array}\right| \left|\begin{array}{c} t_1 \\ t_2 \end{array}\right| \begin{array}{|c|} x_1 \\ x_2 \\ x_3 \end{array} \begin{array}{|c|} y_1 \\ y_2 \\ y_3 \end{array} \qquad (-,-,-,-).$$

It is not necessary to have x, y, z and t entries in the substitution list since every line of the corresponding column may be used. Replacing first t_1 by $z_1 t_1$, then t_1 by y_1, t_2 by $x_1 t_2$ and t_2 by ab we get the following sme-system:

$$\left|\begin{array}{c} x_2 y_2 z_2 \\ cy_3 x_3 \end{array}\right| \left|\begin{array}{c} z_1 \\ z_2 \end{array}\right| \left|\begin{array}{c} z_1 y_1 \\ x_1 ab \end{array}\right| \begin{array}{|c|} x_1 \\ x_2 \\ x_3 \end{array} \begin{array}{|c|} y_1 \\ y_2 \\ y_3 \end{array} \qquad (-,-,-,-).$$

Now we replace x_2 by c. Using (S_2) we get

$$\left|\begin{array}{c} y_2 z_2 \\ y_3 c \end{array}\right| \left|\begin{array}{c} z_1 \\ z_2 \end{array}\right| \left|\begin{array}{c} z_1 y_1 \\ cab \end{array}\right| \begin{array}{|c|} y_1 \\ y_2 \\ y_3 \end{array} \qquad (c,-,-,-).$$

Here c is the x-entry of the substitution list. Replacing now y_3 by y_2, we get

$$\left|\begin{array}{c} z_2 \\ c \end{array}\right| \left|\begin{array}{c} z_1 \\ z_2 \end{array}\right| \left|\begin{array}{c} z_1 y_1 \\ cab \end{array}\right| \begin{array}{|c|} y_1 \\ y_2 \\ y_2 \end{array} \qquad (c,-,-,-).$$

By simplification (S_1), the matched occurrences of y_2 are erased. The column associated with y is erased, carrying y_1 to the substitution list.

$$\left|\begin{array}{c} z_2 \\ c \end{array}\right| \left|\begin{array}{c} z_1 \\ z_2 \end{array}\right| \left|\begin{array}{c} z_1 y_1 \\ cab \end{array}\right| \qquad (c, y_1, -, -)$$

We replace z_2 by c. Using (S$_2$) we get

$$\begin{vmatrix} cy_1 \\ cab \end{vmatrix} \qquad (c, y_1, c, \_).$$

Now cy_1 is carried to the substitution list and c is erased:

$$\begin{vmatrix} y_1 \\ ab \end{vmatrix} \qquad (c, y_1, c, cy_1).$$

Eventually y_1 is replaced by ab, the sme-system is empty and the substitution list

$$(c, ab, c, cab)$$

shows that $(X, Y, Z, T) == (c, ab, c, cab)$ is a solution of $zyxabxyz = ttcyx$.

Even if we do not find a solution within the flat tree, the use of sme-systems helps to avoid unnecessary branching of the search tree. As a matter of fact an sme-system corresponds to a set of position equations which may be associated with the system (see definition 4.4). When we postpone the translation as long as possible, then we avoid the use of complicated transformations and we have to treat one simple structure instead of a whole set of complicated ones.

Vertically versus Horizontally Oriented Position Equations

The position equations which are generated by either method have some typical properties:

- In the standard formulation, the position of a base in a position equation reflects exactly its position in the word equation E. If the notational length of E is large, then the corresponding position equations are horizontally large. In our reformulation, position equations are horizontally short. Most constraints are stored in vertical direction.
- In our formulation, position equations are typically decomposed into columns associated with distinct variables of E. These variable columns do not intersect: the column of any base is either completely contained in such a variable column or has a trivial intersection.

If several bases are transported simultaneously by the transformation step (as in the case of our transformation or of the transformation in [Abd]), then the new method will find solutions much faster. For any solution of a word equation which assigns non-empty words $X_1, \ldots, X_n$ of length $l_1, \ldots, l_n$ to the variables $x_1, \ldots, x_n$, the number of transformations (of position equations) where a part of the carrier is erased does not exceed the sum of the l_i, in most cases it is much smaller (theorem 4.11). If the standard method is used, the number of these transformation steps may be arbitrarily large, if variables have many occurrences. It is also clear that the great number of vertical equality constraints helps to exclude unsatisfiable position equations: since there are not so many different positions, distinct coefficient bases are possibly carried to the same position very soon, in the unsolvable case. Let us treat now a rather extreme example where the word equation is very long, as a demonstration:

Example 5.2: Consider the equation

$$xxaybzyzbxz = zyzbxabxzabzy.$$

It is not hard to see that (among others) the following sme-system is generated by resolving the principal column:

$$
\begin{array}{c|c}
\begin{array}{c} ax_1 \\ y_3ab \\ ax_3 \\ ax_4 \\ ax_5 \end{array}
&
\begin{array}{c} x_4a \\ x_5a \\ x_1y_3 \\ x_3a \end{array}
\end{array}
\qquad (ax_1, x_4a, a).
$$

If we translate this system into a set of position equations, then length inconsistencies exclude all position equations but one: since the solutions of x_1 and x_3 have the same length, the solution for y_3 must have length 1 and we get (erasing multiple occurrences of coefficient bases of the same type in the same column):

$$
\begin{array}{c|c|c|c|c}
a & x_1 & & \bar{x}_4 & a \\
\hline
y_3 & a & b & \bar{x}_5 & \\
& x_3 & & \bar{x}_1 & \bar{y}_3 \\
& x_4 & & \bar{x}_3 & \\
& x_5 & & &
\end{array}
\qquad (a(2,4), (4,5)a, a).
$$

Note that variables are replaced by columns in the substitution list. After transformation we get

$$
\begin{array}{c|c|c|c|c}
x_1 & & \bar{x}_4 & \\
\hline
a & b & \bar{x}_5 & a \\
x_3 & & \bar{x}_1 & \\
x_4 & & \bar{x}_3 & \\
x_5 & & &
\end{array}
\qquad (a(1,3), (3,4)a, a).
$$

After the following transformation all remaining variables but the two occurrences of the carrier x_1 are matched and may be erased. Since also x_1 is erased, we get the trivially solvable position equation

$$
\begin{array}{c|c|c}
a & b & a
\end{array}
\qquad (a(1,3), (1,3)a, a) \equiv (aab, aba, a).
$$

Thus the solution $(X, Y, Z) = (aab, aba, a)$ is found after three transformation steps. Clearly the number of transformation steps would be much larger for the horizontal oriented position equations of the standard formulation.

The Uniform Transformation

In order to understand the new transformation strategy it is reasonable to distinguish three levels of growing complication. In the first situation the boundary r^* is not inside the column of another base. Then $cr = r^*$, the complete structure of $col(x_c)$ is transported to $col(\bar{x}_c)$ and $x_c, \bar{x}_c$ are deleted. The transformation does not introduce any new boundary equation. In the following example, x is the carrier of GE.

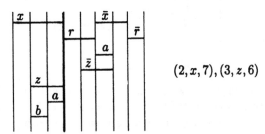

$$(2, x, 7), (3, z, 6)$$

The first boundary connection is used to determine the new position 7 of 2^{tr}. Here is an element of $Transf(GE)$. It is the only successor in this case.

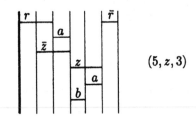

$$(5, z, 3)$$

The second typical situation occurs if there exists a base y, $L(y) < r^*$, which exceeds the carrier, but if there is no transport base whose right boundary falls into the column (cr, r^*). The subpart of $col(x_c)$ up to the critical boundary is transported and cr becomes the new initial boundary. As a consequence of the second condition, no new boundary equations have to be introduced. The following equation GE is an example:

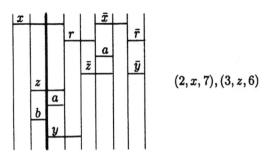

$$(2, x, 7), (3, z, 6)$$

The critical boundary is $cr = 3$. Again $Transf(GE)$ has only one element:

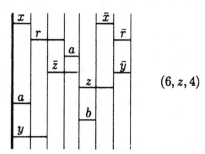

$$(6, z, 4)$$

In the third and most complex situation we have a transport boundary between cr and r^*.

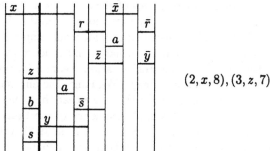

$$(2, x, 8), (3, z, 7)$$

In this case we need a new boundary connection $(2, x, 8)$ after the transformation in order to store all informations on identical subcolumns:

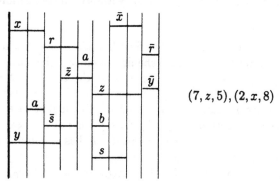

$$(7, z, 5), (2, x, 8)$$

It is simple to see that $l^* < cr$ is always valid. Thus in all three cases a non-trivial left part of the position equation is erased.

Two Technical Remarks on the Choice of Definitions

The rough idea of the concept of a position equation is not difficult to understand: the bases and their relative positions represent the kernel of the position equation. As a matter of fact, the number of non-isomorphic kernels is finite, for a given number of bases. The inessential boundaries and the boundary connections store relevant information on related prefixes of variable bases and are used to indicate unnecessary search in certain situations (if the length of a connection exceeds a critical number). Since the number of non-isomorphic structures has to remain finite it is necessary to link inessential boundaries and boundary connections with the kernel (see conditions (III) and (IV) of definition 3.6). Given this simple idea, the conditions on boundary connections seem perhaps to be more complicated than necessary. Thus let us give an idea of the technical difficulties which led us (better: G.S.Makanin) to the definitions. Let us ask

(1) Why is it necessary to allow that boundary connections have a non-empty second part, in the sense of definition 3.1?

As we have seen, a normal subsequence of the form $(x_\theta, \bar{x}_\nu, j)$ is transformed into $(x_\theta, \bar{x}_\nu, j)$ if $\bar{x}_\theta$ and j are fixed (without any exception). Even if $L(\bar{x}_\nu) \leq L(\bar{x}_\theta)$ it might happen that $L'(\bar{x}_\nu) > L'(\bar{x}_\theta)$. Thus we cannot exclude a non-empty second part of right-directed elements. There is a corresponding situation in Pécuchet's version of Makanin's algorithm and Pécuchet's proof of the transformation theorem for position equations of type IV (in his terminology) contains a gap. Warning: the english translation [Mak] has some misprints which are consistent in the sense that they may lead the reader to the assumption that the shifts in Makanin's boundary connections are also only left-directed. These misprints are eliminated in Makanin's Russian dissertation and in [Ma2] where the notion of a convex boundary connection was introduced.

(2) Why did we use the notion of a witness and of right-fossil boundaries in (III)? Why did we impose the convexity condition for witnesses of connections with non-empty second part?

Suppose that we only require that the final boundary of a connection is "essential" in a certain sense. A transformation step should not enlarge the number of such essential boundaries (finiteness theorem). Now suppose that the transport boundary j, $cr < j$, is such an essential boundary. We might either decide (i) to declare j to be essential in PE' or (ii) to declare j^{tr} to be essential in PE'. In the first case, we have problems to introduce the new connection (j, x_ν, j^{tr}) since j^{tr} is not necessarily essential. If we take possibility (ii), we get a similar problem if PE has a boundary connection with end segment (1) (x_θ, x_ν, j) where (x_θ, x_ν) belongs to the non-empty second part of the connection and $\bar{x}_\theta$ is a transport base. (This might occur for $L(\bar{x}_\theta) = l^*$.) The regular transformation into $(x_\theta, x_\nu, j^{tr})$ leads to a left-directed element (x_θ, x_ν) if $cr < l^*$. Thus the regular transformation cannot be used. We have to transform (1) into (2) (x_θ, j) (compare (Exc 4)). But with the naive idea (ii), j is not necessarily essential now! The notion of a witness is a solution of the problem. But we have to be sure that a witness x_ω for j in (1) is not a transport base, since we want to use x_ω also as witness for j in (2). The convexity condition guarantees that x_ω is in fact fixed, since $cr \leq \bar{l}^* \leq L(x_\omega)$, in our situation. The fossil boundaries are introduced in order to safe connections with final boundary $\bar{r}^*$ and in order to avoid that the essential boundary $\bar{l}^*$ is inessential after the transformation in the case $cr = r^*$. If $\bar{x}_\nu$ is the witness of a connection and if $\bar{x}_\nu$ is completely erased by the transformation step (for $cr = r^\star$), then $r^\star$ and $\bar{r}^\star$ are declared to be right-fossil.

Minimal and Complete Unification

With minor modifications, the search tree $T_{Mak}(E)$ can be used in order to generate a minimal and complete set of unifiers for the equation E. Moreover, as in [Jaf], this can be done in such a way that the algorithm terminates if this set is finite. The proofs are not difficult but long and rather technical, thus we only sketch the main ideas which may be found similarly in [Jaf].

Let us first describe the generation process. The tree $T_{Mak}(E)$ is modified as follows:

(a) We do not stop a path if we reach a node which is labelled with a structure (i.e. a sme-system or a position equation) which is isomorphic to the label of a predecessor.

Now, of course, the search tree might be infinite. But wait for the second step where we ensure termination again.

(b) We do not stop when we reach a trivially solvable position equation unless the set of variable bases is empty (it is straightforward to show that the transformation and normalization algorithms work as long as there is any variable base left.)

As described at the end of section 4, we augment all structures of the search tree by an additional substitution list. When we reach a trivially true position equation without variable bases, then we replace all occurrences of the column (i, j) in the substitution list by a new variable $y_{i,j}$. The set of all substitution list of this form gives, in combination with the set of all substitution lists of successful leaves of the flat tree, a minimal and complete set of unifiers for E. The idea for the proof is, roughly, the following: clearly there exists a minimal and complete set $MCU(E)$ of unifiers for E (see section 1). If U is an element of this set, then U determines a unique path of the modified search tree which leads to a successful leaf. It is possible to show by induction that U is an instance of every substitution list which occurs on this path (modulo translation of columns in sequences of coefficients and variables). The final substitution list V defines a unifier of E. By minimality of $MCU(E)$, V and U are equivalent.

To regain termination (which is, of course, only possible if the set $MCU(E)$ is finite), we have to prove the following theorem:

Theorem 5.3: (a) *There exists a recursive function $g_1(l)$ such that every solvable sme-system SME with notational length l has a solution with exponent of periodicity s not exceeding $g_1(l)$.*
(b) *There exists a recursive function $g_2(n, m, c, d)$ such that every solvable position equation PE with principal parameters n, m, c and d has a solution with exponent of periodicity s not exceeding $g_2(n, m, c, d)$.*

The proof is based on theorem 3.3. Thus sme-systems (position equations) have to be translated into a (set of) word equation(s) in such a way that the notational length of such an equation does not exceed an upper bound which is given in terms of a recursive function in the size (principal parameters) of the sme-system (position equation). See Jaffar's [Jaf] proof of the periodicity lemma for generalized equations where such techniques are used.

Now we consider any node η of the modified search tree and the subtree of all successors of η. We say that a node η' of this subtree is *blind with respect to* η in the following three cases:

(i) if η' is labelled with a position equation PE with principal parameters n and d indicating (according to lemma 3.16) that all solutions of PE have exponent of periodicity which exceeds the bound which we get by applying theorem 5.5 to the label of η, or

(ii) if η' is labelled with a structure which is isomorphic to the label of a predecessor which also belongs to the subtree generated by η, or

(iii) if η' is labelled with a structure which is trivially unsolvable.

Lemma 5.4: η *is labelled with an unsolvable structure if and only if every path of the subtree generated by η leads after a finite number of steps to a node which is blind with respect to η. The maximal number of steps is recursive in the size (principal parameters) of the label of η.*

The proof is straightforward and omitted. Suppose now that E has a finite minimal and complete set of unifiers. Then all but a finite number of nodes of the modified search tree are labelled with unsolvable structures. By a breadth first search the unsolvability of the label of a node may be established after a finite number of steps. We do not extend any subtree with top node η as soon as the unsolvability of the label of η is established. Thus the procedure terminates.

6 Word Equations with Regular Constraints

In this section we prove the following theorem: if E is a word equation with variables $x_1, ..., x_n$ and constants in the alphabet C, and if $L_1, ..., L_n$ are regular languages over C, then it is decidable whether E has a solution $(X_1, ..., X_n)$ such that X_i is in L_i $(i = 1, ..., n)$.

Suppose that we are given a finite deterministic automaton $A = (Q, C, \delta, s, F)$. For any pair (p, q) of states, $\mathcal{L}(A_q^p)$ denotes the language which is accepted by the automaton $A_q^p = (Q, C, \delta, p, \{q\})$. Furthermore we define $\mathcal{L}(A, \emptyset) = C^*$ and $\mathcal{L}(A, \Gamma) = \bigcap_{(p,q) \in \Gamma} \mathcal{L}(A_q^p)$ for $\emptyset \neq \Gamma \subseteq Q \times Q$. An *A-constraint* is a finite set Γ of pairs $(p, q) \in Q \times Q$. An *A-constrained word equation* E is a word equation with coefficients in C and variables $x_1, ..., x_n$ for which the possible values are restricted by means of A-constraints $\Gamma_1, ..., \Gamma_n$. The word $w \in C^*$ is an admissible value for x_i if $w \in \mathcal{L}(A, \Gamma_i)$ $(1 \leq i \leq n)$. In this section, we prove the following

Theorem 6.1: *For every A-constrained word equation E with variables $x_1, \ldots, x_n$ and associated A-constraints $\Gamma_1, \ldots, \Gamma_n$ it is decidable whether E has an A-solution, i.e. a solution $(X_1, \ldots, X_n)$ such that X_i is an admissible value for x_i, $1 \leq i \leq n$.*

The theorem mentioned in the introduction is a corollary. Throughout this section $A = (Q, \mathcal{C}, \delta, s, F)$ denotes a finite deterministic automaton. The number n_A denotes the number of states $|Q|$ of A and $N_A = (n_A!)$. Occasionally we will use the following wellknown fact (†): if $\Gamma \subseteq Q \times Q$, then it is decidable whether the language $\mathcal{L}(A, \Gamma)$ contains at least one non-empty word.

Preliminary Lemmas

General Remark: With the exception of lemma 6.3, the lemmata and proofs of this subsection (up to 6.7) are modifications of similar parts in [Mak].

Definition 6.2: The natural numbers t and t' are called A-equivalent (we write $t \sim_A t'$), if the following two conditions are satisfied:
 (i) $t \equiv t' \ (Mod \ N_A)$,
 (ii) $t > n_A$ if and only if $t' > n_A$.

Lemma 6.3: *Let $v = u_1 w^{t_1} u_2 w^{t_2} .. u_k w^{t_k} u_{k+1}$ and $v' = u_1 w^{t'_1} u_2 w^{t'_2} .. u_k w^{t'_k} u_{k+1}$ $(k \geq 0)$ be two words over the alphabet $\mathcal{C}$, suppose that Γ is any A-constraint. If $t_i \sim_A t'_i$ $(1 \leq i \leq k)$, then $v \in \mathcal{L}(A, \Gamma)$ if and only if $v' \in \mathcal{L}(A, \Gamma)$.*

Proof: We may assume that $\Gamma = \{(p, q)\}$, $k = 1$, $v = u_1 w^t u_2$, $v' = u_1 w^{t'} u_2$ and $t' \leq t$. If $t \leq n_A$, then $t' = t$ and the proof is trivial. If $t > n_A$, then we write

$$v = u_1 w_{(1)} w_{(2)} \ldots w_{(t)} u_2,$$

enumerating the distinguished occurrences of w. To every word $w_{(i)}$ we assign the unique state $q_i \in Q$ with $(p, v) \Rightarrow^*_A (q_i, w_{(i+1)} \ldots w_{(t)} u_2)$. The sequence $(q_1, \ldots, q_t)$ has two identical entries. Suppose b is the minimal index such that $q_a = q_b$, for a number $1 \leq a < b \leq n_A + 1$. We have $a \leq n_A$ and $b - a \leq n_A$. Since A is deterministic we have $q_a = q_b = q_{b+(b-a)} = \ldots = q_{b+l(b-a)}$ for all states $q_{b+l(b-a)}$ in $(q_1, \ldots, q_t)$. In particular we get $q_a = q_{a+N_A} = q_{a+lN_A}$ and

$$q_{a+c} = q_{a+c+lN_A}$$

for all $0 \leq c < N_A$ and q_{a+c+lN_A} in $(q_1, \ldots, q_t)$. Thus $q_t = q_{t'}$ and $(p, u_1 v^t u_2) \Rightarrow^*_A (q, \epsilon)$ if and only if $(p, u_1 v^{t'} u_2) \Rightarrow^*_A (q, \epsilon)$. Therefore $v \in \mathcal{L}(A_q^p)$ if and only if $v' \in \mathcal{L}(A_q^p)$. ∎

Consider a system of linear diophantine equations

$$a_{1,1}z_1 + a_{1,2}z_2 + \cdots + a_{1,n}z_n = b_1$$
$$a_{2,1}z_1 + a_{2,2}z_2 + \cdots + a_{2,n}z_n = b_2$$
$$\cdots$$
$$a_{p,1}z_1 + a_{p,2}z_2 + \cdots + a_{p,n}z_n = b_p$$

$$(S)$$

where the z_i are variables which take natural numbers as values and the coefficients $a_{i,j}$ and b_i are integers $(1 \leq i \leq p, 1 \leq j \leq n)$. We suppose that every row has at least one nonzero coefficient. Let m be the maximum of all the numbers $|a_{i,j}|$ and $|b_i|$, let q be the number of coefficients $a_{1,j}$ of the first row which are different from 0, let I be the index $I = q + (p-1)n$ of the system. The *weight* of a solution is the maximum of the components. The following lemma is a generalization of lemma 1.1 in [Mak].

Lemma 6.4: *If the system (S) has a solution $(z_1^{(0)}, \ldots, z_n^{(0)})$, then we may find a similar solution $(z_1^{(1)}, \ldots, z_n^{(1)})$ with the following properties:*
  (i) $z_r^{(0)} \sim_A z_r^{(1)}$ $(1 \leq r \leq n)$,
  (ii) $z_r^{(1)} \leq 3^{2^I - 2} m^{2^I - 1} 2 N_A$ $(1 \leq r \leq n)$.

Proof: The proof is by induction on the index I of the system. If $I = 1$, then the system has one line

$$a_{1,1}z_1 + a_{1,2}z_2 + \ldots + a_{1,n}z_{1,n} = b_1$$

where only one coefficient $a_{1,i}$ is different from 0. Let $(z_1^{(0)}, \ldots, z_n^{(0)})$ be a solution. Define for $r \neq i$

$$z_r^{(1)} = \begin{cases} z_r^{(0)} - sN_A & \text{if } z_r^{(0)} > 2N_A \\ z_r^{(0)} & \text{if } z_r^{(0)} \leq 2N_A \end{cases}$$

where $N_A < z_r^{(0)} - sN_A \leq 2N_A$ and $z_i^{(1)} = z_i^{(0)}$. Now $(z_1^{(1)}, \ldots, z_n^{(1)})$ satisfies the conditions of the lemma since $z_i^{(0)} = b_1/a_{1,i} \leq m \leq m2N_A$.

Suppose now that $I > 1$. We distinguish two cases:
Case 1: There exist coefficients $a_{1,i} > 0$ and $a_{1,j} < 0$ of the first row.
Suppose $(z_1^{(0)}, \ldots, z_n^{(0)})$ is a solution. Let

$$z_i^{(0)} = (-a_{1,j})z_i^{(0)\prime} + k_j$$
$$z_j^{(0)} = a_{1,i}z_j^{(0)\prime} + k_i$$

where $k_i, k_j, z_i^{(0)\prime}$ and $z_j^{(0)\prime}$ are natural numbers such that $k_j < -a_{1,j}$ and $k_i < a_{1,i}$. Now replace in S the variables as follows:
  $z_r \to z_r'$ for $r \neq i, j$,
  $z_i \to (-a_{1,j})z_i' + k_j$,
  $z_j \to a_{1,i}z_j' + k_i$,

where the z'_r, z'_i and z'_j are new variables. As a result of this transformation we get the system S'. If $z_i^{(0)'} \geq z_j^{(0)'}$, then we replace the variables of S' as follows:

$$z'_r \to z''_r \text{ for } r \neq i,$$
$$z'_i \to z''_i + z''_j.$$

If $z_i^{(0)'} < z_j^{(0)'}$, then we replace the variables of S' as follows:

$$z'_r \to z''_r \text{ for } r \neq j,$$
$$z'_j \to z''_i + z''_j.$$

where the z''_r, z''_i and z''_j are new variables. We get the system S''. Suppose now we are in the first subcase. Then the complete transformation $S \to S''$ may be described as follows:

$$z_r \to z''_r \text{ for } r \neq i, j,$$
$$z_i \to (-a_{1,j})(z''_i + z''_j) + k_j,$$
$$z_j \to a_{1,i}z''_j + k_i.$$

The new coefficients $a''_{s,r}$ and b''_s of S'' are the following ($1 \leq s \leq p$):

$$a''_{s,r} = a_{s,r} \text{ for } r \neq i, j,$$
$$a''_{s,i} = a_{s,i}(-a_{1,j}),$$
$$a''_{s,j} = a_{s,i}(-a_{1,j}) + a_{s,j}a_{1,i},$$
$$b''_s = b_s - a_{s,i}k_j - a_{s,j}k_i.$$

To the solution $(z_1^{(0)}, \ldots, z_n^{(0)})$ of S we have the corresponding solution $(z_1^{(0)''}, \ldots, z_n^{(0)''})$ of S'' which satisfies the following equations (note that by subcase assumption $z_i^{(0)''} \geq 0$):

$$z_r^{(0)} = z_r^{(0)''} \text{ for } r \neq i, j,$$
$$z_i^{(0)} = (-a_{1,j})(z_i^{(0)''} + z_j^{(0)''}) + k_j,$$
$$z_j^{(0)} = a_{1,i}z_j^{(0)''} + k_i.$$

If $(z_1^{(1)''}, \ldots, z_n^{(1)''})$ is any solution of S'', then the same equations (replacing $^{(0)}$ by $^{(1)}$) define a solution of S. The index I_1 of S'' is smaller than I since $a''_{1,j} = 0$. By induction hypothesis S'' has a solution $(z_1^{(1)''}, \ldots, z_n^{(1)''})$ with the following properties:

(j) $z_r^{(0)''} \sim_A z_r^{(1)''}$ ($1 \leq r \leq n$),

(jj) $z_r^{(1)''} \leq 3^{2^{I_1}-2}m_1^{2^{I_1}-1}2N_A$ ($1 \leq r \leq n$).

Here m_1 denotes the maximum of the absolute values of coefficients in S''. Now consider the corresponding solution $(z_1^{(1)}, \ldots, z_n^{(1)})$ of S. Since all transformations are linear it follows from (j) that $z_r^{(1)} \equiv z_r^{(0)} \pmod{N_A}$ ($1 \leq r \leq n$).

Suppose that $z_r^{(0)} > n_A$, for a $1 \leq r \leq n$. If $z_r^{(0)''} > n_A$, then, by induction hypothesis, $z_r^{(1)''} > n_A$ and we have $z_r^{(1)} > n_A$. If $z_r^{(0)''} \leq n_A$, then $z_r^{(1)''} = z_r^{(0)''}$, by (j). If $r \neq i$ we get immediately $z_r^{(1)} = z_r^{(0)}$. If $r = i$, then we have $z_i^{(1)} = (-a_{1,j})(z_i^{(1)''} + z_j^{(1)''}) + k_j = (-a_{1,j})(z_i^{(0)''} + z_j^{(1)''}) + k_j$. If $z_j^{(0)''} \leq n_A$ we get $z_j^{(1)''} = z_j^{(0)''}$. In the other case we have $z_j^{(1)''} > n_A$, by induction hypothesis. Thus, in any case $z_i^{(1)} > n_A$.

Suppose that $z_r^{(0)} \leq n_A$, for a $1 \leq r \leq n$. If $r \neq i$, then we get $z_r^{(0)''} \leq n_A$, thus, by induction hypothesis $z_r^{(1)''} = z_r^{(0)''}$ and $z_r^{(1)} = z_r^{(0)} \leq n_A$. If $r = i$, then it follows that $z_i^{(0)''} \leq n_A$ and $z_j^{(0)''} \leq n_A$. By induction hypothesis, $z_i^{(1)''} = z_i^{(0)''}$ and $z_j^{(1)''} = z_j^{(0)''}$. Thus $z_i^{(1)} = z_i^{(0)} \leq n_A$. This proves condition (i) of the lemma.

Let h_1 denote the weight of $(z_1^{(1)''}, \ldots, z_n^{(1)''})$ and h the weight of $(z_1^{(1)}, \ldots, z_n^{(1)})$. By induction hypothesis,

$$h_1 \leq 3^{2^{l-1}-2} m_1^{2^{l-1}-1} 2N_A.$$

We have $m_1 \leq 3m^2$ and $h \leq m2h_1 + m = m(2h_1 + 1)$, thus $h_1 = 0$ or $h_1 \leq 3mh_1$. Therefore

$$\begin{aligned}
h &\leq 3m3^{2^{l-1}-2} m_1^{2^{l-1}-1} 2N_A \\
&\leq 3m3^{2^{l-1}-2} (3m^2)^{2^{l-1}-1} 2N_A \\
&= 3^{(1+2^{l-1}-2+2^{l-1}-1)} m^{(1+2^l-2)} 2N_A \\
&= 3^{2^l-2} m^{2^l-1} 2N_A.
\end{aligned}$$

This proves condition (ii). In the second subcase the proof is similar.

Case 2: All coefficients $a_{1,1}, \ldots, a_{1,n}$ are non-negative or all are non-positive. We may assume that the coefficients are non-negative. Let $a_{1,i} \neq 0$. Now $z_i^{(0)} \leq |b_1| \leq m$. We apply the transformation $z_i \to z_i^{(0)}$ and get the system S'. For every solution $(z_1^{(1)'}, \ldots, z_{i-1}^{(1)'}, z_{i+1}^{(1)'}, \ldots, z_n^{(1)'})$ of S' we get a corresponding solution $(z_1^{(1)}, \ldots, z_n^{(1)})$ of S where $z_r^{(1)} = z_r^{(1)'}$ for $r \neq i$ and $z_i^{(1)} = z_i^{(0)}$. The rest is as in case 1. ∎

A word P over the coefficient alphabet C is *simple* if it does not have the form Q^m with $m > 1$. The distinguished occurrence of P in a word $W = W_1 P W_2$ is called *stable*, if W_1 ends with P and W_2 starts with P. It is not difficult to show that every word W has (with respect to P) a unique *stable representation* of the form

$$W = D_1 P D_2 P D_3 \ldots D_h P D_{h+1},$$

where the D_i are (possibly empty) words and all the stable occurrences of the simple word P are distinguished. The representation

$$C_1 P^{l_1} C_2 P^{l_2} C_3 \ldots C_m P^{l_m} C_{m+1} \tag{6.1}$$

of a parametric word (where P is simple, the C_i are arbitrary words and the l_i variables which may take natural numbers as values) is called stable if the words C_1 and $P C_2, \ldots, P C_m$ end with P and the words $C_2 P, \ldots, C_m P$ and C_{m+1} start with P. It is not difficult to show that every parametric word of the form (6.1) has the following unique *complete representation*

$$C_1' P^{L_1} C_2' P^{L_2} C_3' \ldots C_q' P^{L_q} C_{q+1}', \tag{6.2}$$

where all C_i' are non-empty words and the L_i are nontrivial linear forms in the variables $l_1, \ldots, l_m$ such that for every assignment of positive values to the variables the resulting word has the stable representation

$$C_1' P^{L_1^{(0)}} C_2' P^{L_2^{(0)}} C_3' \ldots C_q' P^{L_q^{(0)}} C_{q+1}'.$$

Lemma 6.5: *If the words W and V have p and q stable occurrences of the simple word P respectively, then the word WV has r stable occurrences of P with $p + q \leq r \leq p + q + 3$.*

Proof: trivial. ∎

Theorem 6.6: *Let E be an A-constrained word equation with coefficient alphabet C and notational length $nl(E) = d$. If E has any A-solution, then it has also an A-solution S' with exponent of periodicity s' such that*

$$s' \leq (9d)^{2^{2^{d^3}}} 2N_A + 2.$$

Proof: Let E be an equation with A-solution S. Suppose that S has exponent of periodicity s, let P be a simple word which has s consecutive repetitions in a component X_i of S. In every component of S we distinguish all stable occurrences of P:

$$X_1 = C_{1,1} P^{u_{1,1}} C_{1,2} P^{u_{1,2}} \ldots C_{1,q_1} P^{u_{1,q_1}} C_{1,q_1+1}$$
$$\ldots \tag{6.3}$$
$$X_n = C_{n,1} P^{u_{n,1}} C_{n,2} P^{u_{n,2}} \ldots C_{n,q_n} P^{u_{n,q_n}} C_{n,q_n+1}.$$

Here the words $C_{i,j}$ are non-empty and the numbers $u_{i,j}$ are positive naturals. To get a graphical idea of the following, suppose we want to find further solutions just by varying the numbers $u_{i,j}$ in the components without changing other parts. We replace the $u_{i,j}$ by variables $z_{i,j}$ and get stable parametric words

$$X_1^{(par)} = C_{1,1} P^{z_{1,1}} C_{1,2} P^{z_{1,2}} \ldots C_{1,q_1} P^{z_{1,q_1}} C_{1,q_1+1}$$
$$\ldots$$
$$X_n^{(par)} = C_{n,1} P^{z_{n,1}} C_{n,2} P^{z_{n,2}} \ldots C_{n,q_n} P^{z_{n,q_n}} C_{n,q_n+1}.$$

Substituting these words into the original equation $\Phi = \Psi$, we get parametric words $\Phi^{(par)}$ and $\Psi^{(par)}$ which we write down in the complete representation

$$\Phi^{(par)} = D_1 P^{L_{1,1}} D_2 P^{L_{1,2}} D_3 \ldots D_{r_1} P^{L_{1,r_1}} D_{r_1+1}$$
$$\Psi^{(par)} = D'_1 P^{L_{2,1}} D'_2 P^{L_{2,2}} D'_3 \ldots D'_{r_2} P^{L_{2,r_2}} D'_{r_2+1}. \tag{6.4}$$

Here the D_i und D'_i are non-empty and the exponents $L_{i,j}$ are nontrivial linear forms in the variables $z_{i,j}$ with non-negative coefficients. (For example, a coefficient of $z_{i,j}$ is greater than 1 if $X_i^{(par)} = P^{z_{i,1}}$ and the variable x_i occurs consecutively several times in E. The linear forms have constant terms if we get new stable occurrences of P by concatenating some words $X_i^{(par)}$. By lemma 6.5 we have at most 3 new stable occurrences for every concatenation. Hence the sum of all constant terms does not exceed $3d$.) Now we replace in (6.4) the variables $z_{i,j}$ by $u_{i,j}$ and get words

$$\Phi^{(0)} = D_1 P^{L_{1,1}^{(0)}} D_2 P^{L_{1,2}^{(0)}} D_3 \ldots D_{r_1} P^{L_{1,r_1}^{(0)}} D_{r_1+1}$$
$$\Psi^{(0)} = D'_1 P^{L_{2,1}^{(0)}} D'_2 P^{L_{2,2}^{(0)}} D'_3 \ldots D'_{r_2} P^{L_{2,r_2}^{(0)}} D'_{r_2+1}$$

which are identical and distinguish all stable occurrences of P (compare the definition of the complete representation). Since this representation is unique, $r_1 = r_2$ and $D_i^{(0)} = D_i'^{(0)}$ $(1 \leq i \leq r_1 + 1)$. We see that the numbers $u_{i,j}$ are a solution of the system of linear diophantine equations

$$L_{1,i} = L_{2,i} \qquad (1 \leq i \leq r_1). \tag{6.5}$$

We now want to estimate the relevant parameters m, q, n of (6.5) in terms of the notational length d of the word equation. But still the number of equations may cause problems and we have to start with some trivial simplifications of the system. We erase every equation of the form $z_{i,j} = z_{k,l}$ and replace in the remaining system $z_{k,l}$ by $z_{i,j}$. If the resulting system has two equations $z_{i,j} = L_{p,r}$ and $z_{k,l} = L_{p,r}$, we also replace $z_{k,l}$ by $z_{i,j}$. As a result the two equations coincide and one copy is erased. In the remaining system S we have equations of type "z-variable = proper linear form" (type 1) and equations of type "proper linear form A = proper linear form B" (type 2). The number of proper linear forms does not exceed d, hence we have $p \leq d$ equations. (Here Makanin gives a weaker estimate $p \leq (d(d-1)/2) + d \leq d^2$.) Every proper linear form contains at most d variables. In the equations occur the proper linear forms, furthermore for every equation of type 1 an additional z-variable. The number of these isolated z-variables does not exceed d, hence we have $n \leq d + d^2$ variables in the system. The number q of nonzero coefficients of the first row of the system does not exceed d. The coefficients of the z-variables do not exceed the maximal number of occurrences of a variable in the equation E which is at most d. The constant terms of the linear forms do not exceed $3d$, hence $m \leq 3d$. The index $I = q + (p-1)n$ satisfies $I \leq d^3$. Now remember that the numbers $z_{i,j}^{(0)} = u_{i,j}$ define a solution of S. By lemma 6.4, we may find another solution $z_{i,j}^{(1)}$ $(1 \leq i \leq n, 1 \leq j \leq q_i)$ with weight h bounded by

$$3^{2^I - 2} m^{2^I - 1} 2N_A \leq (3m)^{2^I} 2N_A \leq (9d)^{2^I} 2N_A \leq (9d)^{2^{d^3}} 2N_A.$$ Moreover, by lemma 6.3 and condition (i) of lemma 6.4 this solution may be chosen in such a way that the words

$$X_1^{(1)} = C_{1,1} P^{z_{1,1}^{(1)}} C_{1,2} P^{z_{1,2}^{(1)}} \dots C_{1,q_1} P^{z_{1,q_1}^{(1)}} C_{1,q_1+1}$$

$$\dots$$

$$X_n^{(1)} = C_{n,1} P^{z_{n,1}^{(1)}} C_{n,2} P^{z_{n,2}^{(1)}} \dots C_{n,q_n} P^{z_{n,q_n}^{(1)}} C_{n,q_n+1}$$

(which define a solution S' of E) satisfy again all A-constraints imposed on the variables $x_1, \dots, x_n$ in E. The exponent of periodicity of S' satisfies $s' = h + 2 \leq (9d)^{2^{d^3}} 2N_A + 2$.

The Flat Tree with Regular Constraints

Definition 6.7: A *word expression* over the alphabets C, X is a non-empty word

$$\sigma_1 \dots \sigma_k$$

in $(C \cup X)^+$. An *A-constrained word expression* is an expression Σ of the form,

$$(\sigma_1, \Gamma_1) \dots (\sigma_k, \Gamma_k)$$

where $\sigma_1 \ldots \sigma_k$ is a word expression (called the *kernel* of Σ) and each Γ_i is an A-constraint which is empty if σ_i is a coefficient $(1 \leq i \leq k)$. We say that (σ_i, Γ_i) *occurs* in Σ. Similarly, if $(p, q) \in \Gamma_i$, then we say that (p, σ_i, q) *occurs* in Σ $(1 \leq i \leq k)$. An *A-constrained l-sided multi-equation* is an expression CME of the form

$$\Sigma_1 == \ldots == \Sigma_l \qquad (l \geq 1)$$

where the Σ_i are A-constrained word expressions. Note that we allow $l = 1$ now. The kernel of CME is the l−sided multi-equation whose sides are the kernels of the Σ_i, $1 \leq i \leq l$. An *A-constrained sme-system*, or a *csme-system* is a finite set $CSME$ of A-constrained multi-equations such that every variable which occurs in $CSME$ has exactly two occurrences.

Definition 6.8: The csme-system $CSME$ over the alphabets $\mathcal{C}, \mathcal{X}$ is *A-admissible* if the kernel of $CSME$ is admissible in the sense of definition 2.9 and if the following condition is satisfied:
for all $x_i \in \mathcal{X}$: if (x_i, Γ_1) and (x_i, Γ_2) occur in $CSME$, then the language $\mathcal{L}(A, \Gamma_1 \cup \Gamma_2)$ contains at least one non-empty word.

Definition 6.9: The sequence $S = (X_1, \ldots, X_n)$ is an *A-solution* of the csme-system $CSME$ over the alphabets $\mathcal{C}$ and $\mathcal{X} = \{x_1, \ldots, x_n\}$ if it is a solution of the kernel of $CSME$ and if $X_i \in \mathcal{L}(A_q^p)$, for every variable $x_i \in \mathcal{X}$ and all states $p, q \in Q$ such that (p, x_i, q) occurs in $CSME$. The index I and the exponent of periodicity s of S are defined as in definition 4.1.

Lemma 6.10: (a) *There exists an algorithm which decides whether $CSME$ is A-admissible or not, for any csme-system $CSME$.*
(b) *If all A-constrained multi-equations of the csme-system $CSME$ are one-sided, then it is decidable whether $CSME$ is A-solvable or not.*

Proof: (a) It is decidable whether the kernel is admissible in the sense of definition 2.9. By (†) it is decidable whether $CSME$ satisfies the condition of definition 6.8. To prove (b), note that in this case $CSME$ is A-solvable if and only if $CSME$ is A-admissible. ∎

In this section, the sme-system $SME(E)$ canonically associated with E is defined similarly as in section 2, but has in addition a one-sided multi-equation $x_{i,1}$ associated with the variable x_i, for every variable x_i with only one occurrence in E. Thus every variable of $SME(E)$ has exactly two occurrences.

Definition 6.11: Let E be an A-constrained word equation with variables $x_1, .., x_n$ and associated A-constraints $\Gamma_1, \ldots, \Gamma_n$. The csme-system $CSME(E)$ *canonically associated* with E is defined as follows:
- $CSME(E)$ has kernel $SME(E)$,

- for every variable x_i with $n_i > 0$ occurrences in E: the A-constrained multi-equation associated with x_i has the form

$$(x_{i,1}, \Gamma_i) == (x_{i,2}, \emptyset) == \ldots == (x_{i,n_i}, \emptyset),$$

- the principal equation of $CSME(E)$ has the form

$$(\sigma_1, \emptyset) \ldots (\sigma_k, \emptyset) == (\sigma_{k+1}, \emptyset) \ldots (\sigma_{k+l}, \emptyset). \tag{6.6}$$

Lemma 6.12: *The A-constrained word equation E is A-solvable iff $CSME(E)$ is A-solvable.*

Proof: obvious. ■

From now on we shall sometimes omit the prefix "A-constrained" if misunderstandings are excluded. In any case, the letter C indicates that the structure is A-constrained. We shall assume that $CSME(E)$ is always A-admissible.

Definition 6.13: The csme-systems $CSME_1$ and $CSME_2$ over the alphabets $C, \mathcal{X}$ are *isomorphic* if $CSME_2$ and $CSME_1$ become graphically identical when we replace all occurrences of variables x in SME_2 by $\Psi(x)$, for a permutation Ψ of $\mathcal{X}$.

Definition 6.14: The *head* of the two-sided multi-equation

$$(\sigma_1, \Gamma_1) \ldots (\sigma_k, \Gamma_k) == (\sigma_{k+1}, \Gamma_{k+1}) \ldots (\sigma_{k+l}, \Gamma_{k+l}) \tag{6.7}$$

is the expression $((\sigma_1, \Gamma_1), (\sigma_{k+1}, \Gamma_{k+1}))$, the *tail* of (6.7) is the expression

$$(\sigma_2, \Gamma_2) \ldots (\sigma_k, \Gamma_k) == (\sigma_{k+2}, \Gamma_{k+2}) \ldots (\sigma_{k+l}, \Gamma_{k+l}).$$

Definition 6.15: Let $\Gamma_1, \Gamma_2, \Gamma = \{(p_1, q_1), \ldots, (p_k, q_k)\}$ be A-constraints. The pair (Γ_1, Γ_2) is an *unfolded instance* of Γ if there exist $t_1, \ldots, t_k \in Q$ such that

$$\Gamma_1 = \{(p_1, t_1), \ldots, (p_k, t_k)\}$$
$$\Gamma_2 = \{(t_1, q_1), \ldots, (t_k, q_k)\}.$$

If $a_j \in C$ and $\delta(p_i, a_j) = t_i$ $(1 \leq i \leq k)$, then (Γ_1, Γ_2) is the (unique) a_j-*unfolded* instance of Γ.

Definition 6.16: For every word equation E, the *initial search tree* $T_{ini}^A(E)$ is defined as follows:
- The top node of $T_{ini}^A(E)$ is labelled with $CSME(E)$. We introduce in addition marks which indicate the type (principal equation, multi-equation associated with x_1,...) of any multi-equation in $CSME(E)$.

Suppose η is any node of $T_{ini}^A(E)$, labelled with the A-admissible system $CSME$. In the following cases, η is a leaf of $T_{ini}^A(E)$:

(a) If $CSME$ is empty, then η is a successful leaf.

(b) If all multi-equations of $CSME$ are one-sided, then we decide whether $CSME$ has an A-solution or not. In the first case, η is a successful leaf, in the second case η is blind.

(c) If the principal column has been completely resolved and $CSME$ contains a multi-equation which is at least two-sided, then η is an *open leaf* of $T_{ini}^A(E)$ (as in section 2, the term "open" indicates that η is not a leaf of $T_{Mak}^A(E)$).

If $CSME$ has non-empty principal equation, the successors of η are defined by means of A-transformation and A-simplification which are defined below. We get the finite set $Simpl^A(Trans^A(CSME))$ of A-constrained sme-systems. For every A-admissible csme-system $CSME_i \in Simpl^A(Trans^A(SME))$ the node η has one successor η_i. If, however, $Simpl^A(Trans^A(CSME))$ does not have any A-admissible element, then η is blind. As in section 2 it might happen that only one side of the principal equation is empty, as a result of the transformation. Such a structure is immediately deleted.

The following rules are formulated more generally as it would be necessary for the initial tree. For the resolution of the principal column, the distinguished equation CME is always the principal equation and the following transformation rules may be simplified: the sets Γ_1 and Γ_2 which occur in the description are always empty. The rule (T_4^A) does not apply. Later we need the rules in full generality.

• **A-Transformation** (of the A-admissible csme-system $CSME$ with distinguished equation CME)

(T_1^A) Suppose that CME has head $((a_i, \emptyset), (a_i, \emptyset))$ with two identical coefficient symbols. Take the tail of CME, leave the other multi-equations unmodified. The resulting system is admissible and the label of the unique successor of η.

(T_2^A) Suppose that CME has head $((x_{i_1,j_1}, \Gamma_1), (x_{i_2,j_2}, \Gamma_2))$ with two distinct variables. Suppose that the second occurrences of the variables have the form (x_{i_1,j_1}, Δ_1) and (x_{i_2,j_2}, Δ_2).

  (1) To get the first element of $Trans^A(CSME)$,
    replace (x_{i_2,j_2}, Δ_2) by $(x_{i_1,j_1}, \Delta_2 \cup \Gamma_1 \cup \Gamma_2)$.
    Then take the tail of the distinguished equation and leave the other multi-equations unmodified.

  (2) For all unfolded instances $(\Gamma_{1,1}, \Gamma_{1,2})$ of Γ_1 and $(\Delta_{1,1}, \Delta_{1,2})$ of Δ_1, the set $Trans^A(CSME)$ has an element constructed as follows:
    replace (x_{i_1,j_1}, Γ_1) by $(x_{i_2,j_2}, \Gamma_{1,1})\ (x_{i_1,j_1}, \Gamma_{1,2})$ and
    replace (x_{i_1,j_1}, Δ_1) by $(x_{i_2,j_2}, \Delta_{1,1} \cup \Gamma_{1,1} \cup \Gamma_2)\ (x_{i_1,j_1}, \Delta_{1,2})$.
    Then take the tail of the distinguished equation and leave the other multi-equations unmodified.

  (3) For all unfolded instances $(\Gamma_{2,1}, \Gamma_{2,2})$ of Γ_2 and $(\Delta_{2,1}, \Delta_{2,2})$ of Δ_2, the set $Trans^A(CSME)$ has an element constructed as follows:
    replace (x_{i_2,j_2}, Γ_2) by $(x_{i_1,j_1}, \Gamma_{2,1})\ (x_{i_2,j_2}, \Gamma_{2,2})$ and
    replace (x_{i_2,j_2}, Δ_2) by $(x_{i_1,j_1}, \Delta_{2,1} \cup \Gamma_{2,1} \cup \Gamma_1)\ (x_{i_2,j_2}, \Delta_{2,2})$.

Then take the tail of the distinguished equation and leave the other multi-equations unmodified.

(T_3^A) Suppose that CME has head $((a_h, \emptyset), (x_{i,j}, \Gamma))$ or $((x_{i,j}, \Gamma), (a_h, \emptyset))$ where $a_h \in C$. Suppose the second occurrences of the variable $x_{i,j}$ has the form $(x_{i,j}, \Delta)$. $Trans^A(CSME)$ has - at most - two elements $CSME_1$ and $CSME_2$:

  (1) If $\delta(p, a_h) = q$ for all pairs $(p, q) \in \Gamma \cup \Delta$, then there exists $CSME_1$: to get $CSME_1$, replace $(x_{i,j}, \Delta)$ by $(a_h, \emptyset)$. Then take the tail of the distinguished equation and leave the other multi-equations unmodified.

  (1) To get $CSME_2$, replace $(x_{i,j}, \Gamma)$ by $(a_h, \emptyset)$ $(x_{i,j}, \Gamma_2)$ where (Γ_1, Γ_2) is the a_h-unfolded instance of Γ and replace $(x_{i,j}, \Delta)$ by $(a_h, \emptyset)$ $(x_{i,j}, \Delta_2)$ where (Δ_1, Δ_2) is the a_h-unfolded instance of Δ. Then take the tail of the distinguished equation and leave the other multi-equations unmodified.

(T_4^A) Suppose that CME has head $((x_{i,j}, \Gamma_1), (x_{i,j}, \Gamma_2))$. Then take the tail of CME, leave the other multi-equations unmodified. The resulting system is admissible and the single element of $Trans^A(CSME)$.

Lemma 6.17: *Let $CSME$ be an A-admissible csme-system with distinguished non-empty equation CME. $CSME$ is solvable iff a csme-system in $Trans^A(CSME)$ is solvable.*

Proof: Suppose that $CSME$ has the A-solution S which assigns the words $X_{i,j}$ to the variables $x_{i,j}$ occurring in $CSME$. Suppose, for example, that CME has the form which is assumed for (T_2^A). Suppose that X_{i_2,j_2} is a proper prefix of $X_{i_1,j_1} = X_{i_2,j_2} X'_{i_1,j_1}$. Then $\hat{\delta}(p, X_{i_1,j_1}) = q$ for all $(p, q) \in \Gamma_1 \cup \Delta_1$. We say shortly that X_{i_1,j_1} satisfies $\Gamma_1 \cup \Delta_1$. Similarly X_{i_2,j_2} satisfies $\Gamma_2 \cup \Delta_2$. To find $CSME'$, unfold $\Gamma_1 = \{(p_1, q_1), \ldots, (p_k, q_k)\}$ with the states $t_i = \hat{\delta}(p_i, X_{i_2,j_2})$ with result $(\Gamma_{1,1}, \Gamma_{1,2})$ and unfold in the same way Δ_1 with result $(\Delta_{1,1}, \Delta_{1,2})$. Now follow the steps of (2) in (T_2^A) to get $CSME'$. Define $X'_{i,j} = X_{i,j}$ for $(i, j) \neq (i_1, j_1)$. Similarly as in the corresponding proof of section 2, the words $X'_{i,j}$ (where $x_{i,j}$ occurs in $CSME$) define a solution S' of $CSME'$. By choice of the unfolded instance, X'_{i_1,j_1} satisfies $\Gamma_{1,2} \cup \Delta_{1,2}$ and $X'_{i_2,j_2} = X_{i_2,j_2}$ satisfies $\Gamma_{1,1} \cup \Delta_{1,1} \cup \Gamma_2$. It follows easily that S' is an A-solution of $CSME'$.

Suppose conversely that $CSME'$ is defined as in (2) of (T_2^A) and has an A-solution S' which assigns the words $X'_{i,j}$ to the variables occurring in $CSME$. Define $X_{i,j} = X'_{i,j}$ for $(i, j) \neq (i_1, j_1)$ and $X_{i_1,j_1} = X'_{i_2,j_2} X'_{i_1,j_1}$. Since X'_{i_2,j_2} satisfies $\Gamma_{1,1}$ and X'_{i_1,j_1} satisfies $\Gamma_{1,2}$ now X_{i_1,j_1} satisfies Γ_1. The rest of the subcase is trivial. The other cases are proved similarly. ∎

Remark: The proof shows that an A-solution S of an csme-system occurring in $T_{ini}^A(E)$ (not labelling a leaf) leads to a unique successor node. We shall see immediately that the A-simplification of an A-solvable csme-system $CSME$ leads to a unique A-solvable csme-system $CSME'$. Thus the argument may be iterated.

- **A-Simplification** (of the csme-system $CSME$)

The following A-simplification rules are applied until the system is erased or a system $CSME'$ is reached which cannot be further simplified by the rules.

(S_1^A) If a two-sided multi-equation CME of $CSME$ has the form $(x_{i,j}, \Gamma_1) == (x_{i,j}, \Gamma_2)$, then check whether $\mathcal{L}(A, \Gamma_1 \cup \Gamma_2)$ contains at least one non-empty word (see (†)). In the positive case, erase CME. In the negative case, delete $CSME$.

(S_2^A) If a multi-equation CME of $CSME$ has at least three sides, if two sides of CME have the form $(x_{i,j}, \Gamma_1)$ respectively $(x_{i,j}, \Gamma_2)$ and another side of CME has the form $(x_{i',j'}, \Gamma_3)$, then erase $(x_{i,j}, \Gamma_1)$ and $(x_{i,j}, \Gamma_2)$ and replace $(x_{i',j'}, \Gamma_3)$ by $(x_{i',j'}, \Gamma_3 \cup \Gamma_1 \cup \Gamma_2)$.

(S_3^A) If a multi-equation CME of $CSME$ has a side $(a_h, \emptyset)$ and all other sides have the form (x_{i_r,j_r}, Γ_r) $(1 \le r \le l-1)$, then check whether $\delta(p, a_h) = q$ for all (p, q) in $\bigcup_{0 < r \le l-1} \Gamma_r$ and in the constraints assigned to the other occurrences of these variables. In the negative case, delete $CSME$. In the positive case, erase CME and replace all occurrences of the variables x_{i_r,j_r} by a_h $(1 \le r \le l-1)$.

Lemma 6.18: *Suppose that we get the csme-system $CSME'$ from $CSME$ by applying one of the rules (S_1^A) - (S_3^A). Then $CSME$ is A-solvable if and only if $CSME'$ is A-solvable.*

Proof: This is trivial for (S_1^A). Suppose that (S_2^A) was applied where the other side has the form $(x_{i',j'}, \Gamma_3)$. If $CSME$ has an A-solution S, then $S(x_{i,j}) = S(x_{i',j'})$ is in $\mathcal{L}(A, \Gamma_3 \cup \Gamma_1 \cup \Gamma_2)$ and $CSME'$ has an A-solution which coincides for the remaining variables with S. The converse direction is similar. The case (S_3^A) is simple. ∎

Theorem 6.19: (a) *If E has an A-solution, then $T_{ini}^A(E)$ has a succesful leaf or an open leaf which is labelled with an A-solvable csme-system. If $T_{ini}^A(E)$ has a successful leaf or an open leaf which is labelled with an A-solvable csme-system, then E is A-solvable.*
(b) *If all variables of E occur only once, then $T_{ini}^A(E)$ does not have an open leaf. E has a solution if and only if $T_{ini}^A(E)$ has a successful leaf.*
(c) *The length of the paths in $T_{ini}^A(E)$ does not exceed $nl(E)$.*

Proof: (a) By the remark, lemma 6.12, lemma 6.17 and lemma 6.18.
(b) is trivial since A-transformation does not change the number of lines of the multi-equations and A-simplification may only decrease the number of lines.
(c) is trivial since the notational length of the principal equation decreases at every A-transformation step. ∎

Definition 6.20: The unordered, finitely branching tree $T_{flat}^A(E)$ is an extension of $T_{ini}^A(E)$. Suppose η is any node of $T_{flat}^A(E)$ labelled with the A-admissible csme-system $CSME$. In the following cases, η is a leaf of $T_{flat}^A(E)$:

- If all multi-equations of $CSME$ are one-sided. Then we decide whether $CSME$ is solvable or not (lemma 6.10 (b)). In the first case, η is a successful leaf of $T_{flat}^A(E)$, in the second case η is blind.
- If η has a predecessor η' which is labelled with the csme-system $CSME'$ which is isomorphic to $CSME$, then η is a blind leaf of $T_{flat}^A(E)$.
- If $CSME$ does not have a two-sided multi-equation, but has a multi-equation which is at least three-sided, then η is an open leaf of $T_{flat}^A(E)$.

In the other case, $CSME$ has at least one two-sided multi-equation CME. Then η has one successor η_i for every A-admissible system SME_i in $Simpl^A(Trans^A(CSME))$. If $Simpl^A(Trans^A(CSME))$ does not have an A-admissible element, then η is a blind leaf.

Theorem 6.21: (a) *The maximal length of a path in $T_{flat}^A(E)$ does not exceed the number $(d!)^3 d 2^{n_a^2}$, where $d = 2nl(E)$.*
(b) *If E has an A-solution with index I and exponent of periodicity s, then $T_{flat}^A(E)$ has a successful leaf in depth not exceeding $nl(E) + I$ or $T_{flat}^A(E)$ has an open leaf in depth D which is labelled with an A-constrained csme-system $CSME$ which has an A-solution S' with index $I' \leq I - t$ where $t \geq D - nl(E)$ denotes the number of transformation steps in the non-initial part of the path. The exponent of periodicity s' of S' satisfies $s' \leq s$. Conversely, if $T_{flat}^A(E)$ has a successful leaf or an open leaf which is labelled with an A-solvable csme-system $CSME$, then E is A-solvable.*
(c) *If no variable occurs more than twice in E, then $T_{flat}^A(E)$ does not have an open leaf. E is solvable if and only if $T_{flat}^A(E)$ has a successful leaf.*

Proof: (a) The number of possible constraints for a given symbol of an sme-system is $2^{n_a^2}$. The number of symbols of the systems which occur in the flat tree does not exceed $2nl(E)$. Thus for any any sme-system there are at most $2nl(E)2^{n_a^2}$ possibilities to introduce constraints. The rest is as in the proof of theorem 2.17.
(b) Suppose E is A-solvable. The A-solution S of E leads either to a successful leaf of $T_{ini}^A(E)$ or to an open leaf of this tree which is labelled with an A-solvable csme-system $CSME_1$. In the first case the depth of the leaf is bounded by $nl(E)$, by theorem 6.19. In the second case it follows easily that the A-solution S_1 of the label $CSME_1$ of the open leaf has index $I_1 \leq I$ and exponent of periodicity $s' \leq s$. Now the A-solution of $CSME_1$ leads to a successor with a corresponding A-solution etc. At every step the index of the respective A-solution decreases. This process stops only in three cases: in the first case, we find a successful leaf of $T_{flat}^A(E)$. Then the bound given in the theorem is obviously correct. In the second case, we find a open leaf of $T_{flat}^A(E)$. Again, the bound is obviously correct. In the third case, we find a node η_3 which is labelled with the A-solvable csme-system $CSME_3$ which is isomorphic to the label $CSME_2$ of a predecessor node. Now suppose that we continue the path at η_3, completely ignoring isomorphic labels. Then we will find a successful leaf η_5 of $T_{ini}^A(E)$ or an open leaf η_5 of $T_{flat}^A(E)$ which is labelled with the A-solvable csme-system $CSME_5$ and the bounds of the theorem are correct. Let η_4 be the last node on the path from η_3 to η_5 such that the label $CSME_4$ of η_4 is isomorphic to $CSME_3$. Now we may imitate the steps

which lead from η_4 to η_5, but starting at η_2 and using the isomorphism. It is obvious that we find a successful or open leaf η_5' of $T_{flat}^A(E)$ which is labelled with an A-solvable csme-system $CSME_5'$ which satisfies the bounds of the theorem. Thus we may stop at η_3.

(c) If no variable has more than two occurrences in E, then all multi-equations in $SME(E)$ are two-sided or one-sided. A-transformation does not change the number of sides. A-simplification may only decrease the number of sides. Now (c) follows from (b).

Remark: In view of theorem 6.7 one might have the following idea: we formulate the search tree for the regular language exactly as in sections 2 and 4, omitting all A-constraints. For all transformation steps, the correspondence of solutions is really only a correspondence of general solutions now, not of A-solutions. If a solution is found, we test, calculating backwards, whether the corresponding solution of E is an A-solution or not (this is no problem, as may be shown, even if we get a nonground unifier). If we find (in $T_{Mak}(E)$) a position equation such that all solutions have to large exponent of periodicity, then we stop. If we find isomorphic structures in the same path, then we stop. But exactly this last point causes problems: suppose that we have a situation as in part (b) of the preceeding proof. Suppose that η_5' is labelled with a solvable sme-system but that the corresponding solution of E is not an A-solution. Even in this case it might very well happen that the solution of E which corresponds to the solution of the label of η_5 is an A-solution. Thus it would not be correct just to stop at η_3. But if we ignore isomorphic structures we get an infinite search tree and do not have a decision procedure.

A-Constrained Position Equations

Definition 6.22: An *A-constrained position equation CPE* is a position equation PE together with a function $Con : Col(PE) \to \mathcal{P}(Q \times Q)$ (where $Col(PE)$ is the set of columns of PE) which assigns to every column (i, j) of PE an A-constraint $Con(i, j)$. PE is called the *kernel* of CPE. In contrast to definition 3.6 we allow now that PE does not have any coefficient base. The solution S of PE is an A-solution of CPE if $S(i, j) \in \mathcal{L}(A, Con(i, j))$ for all columns (i, j) of PE. Index and exponent of periodicity of S are defined as usual. Let $Con(w_i)$ denote the set $Con(col(w_i))$, for every base w_i of PE.

We shall sometimes omit the prefix "A-constrained".

Definition 6.23: The position equation CPE is *unfolded* if $Con(i, j) = \emptyset$ for all decomposable columns of CPE.

Lemma 6.24: *There exists an algorithm which assigns to every position equation CPE a finite set $Unfold(CPE)$ of unfolded position equations such that CPE is A-solvable if and only if an element of $Unfold(CPE)$ is A-solvable.*

Proof: The algorithm is as follows: if CPE is unfolded there is nothing to do. If CPE has a column (i,k) such that $Con(i,k) = \{(p_1,q_1,\ldots,(p_l,q_l)\}$ $(l \geq 1)$ and $i < j < k$ for a boundary j of CPE, then erase CPE and introduce an A-constrained position equation CPE_{Γ_1,Γ_2} for every unfolded instance of $Con(i,k)$ as follows: Add Γ_1 to $Con(i,j)$ and Γ_2 to $Con(j,k)$. Then delete all entries of $Con(i,k)$. If one of the structures CPE_{Γ_1,Γ_2} is not unfolded, then the procedure is iterated. Since the set of decomposable columns is finite eventually a set $Unfold(CPE)$ is reached where all elements are unfolded. The rest of the proof is simple and similar as the proof of lemma 6.17. ■

Definition 6.25: The A-constrained position equation CPE is *trivial* in the following cases:
(i) if CPE does not have any variable base.
(ii) If CPE has two coefficient bases with the same column but different type,
(ii) if CPE has kernel PE and the elementary position equation $PE^{(0)}$ (which coincides with PE if PE does not have any coefficient base) is unsolvable.
CPE is *admissible* if it is not trivial.

Lemma 6.26: *If the A-constrained position equation CPE is trivial, then it is decidable whether CPE is A-solvable or not.*

Proof: In cases (ii) and (iii), CPE is unsolvable. If CPE satisfies condition (i) (and neither (ii) nor (iii)) we may assume, by lemma 6.24, that CPE is unfolded. Then CPE has an A-solution if the following two conditions are satisfied:
(j) for all coefficient bases w_i of CPE: if w_i has type $a_j \in C$, then $\delta(p, a_j) = q$ for all pairs $(p,q) \in Con(w_i)$,
(jj) for all indecomposable columns $(i, i+1)$ of CPE, the language $\mathcal{L}(A, Con(i, i+1))$ contains at least one non-empty word.
In fact, we just assign a_j to every column of a coefficient base of type a_j and any non-empty word $w_{i,j}$ of $\mathcal{L}(A, Con(i, i+1))$ to the other columns (i,j) in order to find an A-solution of CPE. By (†), both (j) and (jj) are decidable. ■

Definition 6.27: The A-constrained position equation CPE is *A-admissible* if CPE is admissible and if the following conditions are satisfied:
- For every column (i,j) of CPE, the language $\mathcal{L}(A, Con(i,j))$ has at least one non-empty word,
- For all variable bases x_i of CPE: the language $\mathcal{L}(A, Con(x_i) \cup Con(\bar{x}_i))$ contains at least one non-empty word.
- For all coefficient bases w_i of PE: if w_i has type $a_j \in C$, then $\delta(p, a_j) = q$ for all pairs $(p,q) \in Con(w_i)$.

Lemma 6.28: *Let CPE be an A-constrained position equation. It is decidable whether CPE is A-admissible or not. If CPE is admissible but not A-admissible, then CPE has no A-solution.*

Proof: As mentioned already earlier, admissibility may be decided. By (†), A-admissibility may be decided. The rest is trivial. ∎

Lemma 6.29: *There exists an algorithm which translates every A-admissible csme-system $CSME$ into a finite set $Set - CNPE(CSME)$ of A-admissible, A-constrained and normalized position equations such that $CSME$ is A-solvable if and only if CPE is A-solvable for a CPE in $Set - CNPE(CSME)$.*

Proof: The algorithm is just a trivial generalization of the corresponding algorithm in definition 4.4 and definition 4.6: suppose SME is the kernel of $CSME$. For every $PE \in Set - NPE(SME)$ we introduce an A-constrained position equation CPE with kernel PE as follows: every symbol $\sigma_{j,k}^{(i)}$ of SME defines a unique column c of the form (4.1) of PE. If $(\sigma_{j,k}^{(i)}, \Gamma_{j,k}^{(i)})$ is the corresponding entry of $CSME$, then $\Gamma_{j,k}^{(i)}$ is used as a subset of $Con(c)$. All constraints of columns are introduced in this way. As a matter of fact, we finally test A-admissibility. ∎

Definition 6.30: The A-constrained position equations CPE_1 and CPE_2 with variables in $\mathcal{X}$ are *isomorphic* if CPE_1 and CPE_2 become identical when we replace in CPE_1 all variables bases x_i by $\Phi(x_i)$, for a permutation Φ of $\mathcal{X}$ which respects duality.

Definition 6.31: Suppose the A-constrained position equation CPE has kernel PE. The principal parameters of CPE are the principal parameters of PE.

Finite Tree Theorem 6.32: *There exists a recursive function $F^A(n, m, c, d)$ such that the number (up to isomorphism) of all A-constrained, A-admissible and normalized position equations CPE with coefficient bases of type C and variables in $\mathcal{X}$ with parameters n_0, m_0, c_0, d_0 bounded by n, m, c, d does not exceed $F^A(n, m, c, d)$.*

Proof: Follows immediately from 3.23 since there is a bound for the number of columns. The modified notion of isomorphic structures is obviously inessential. ∎

Transformation Theorem 6.33: *Let CPE be a normalized and A-admissible position equation. The procedure A-$TRANSFORMATION$ which is described below transforms CPE into a finite set $TRANS^A(CPE)$ of A-constrained position equations with the following properties:*
(a) The principal parameters n, m and c of the elements of $TRANS^A(CPE)$ do not exceed the corresponding parameters of CPE.
(b) If CPE has an A-solution S of index I and exponent of periodicity s, then an A-constrained position equation in $TRANS^A(CPE)$ has an A-solution S' of index $I' < I$ and exponent of periodicity $s' \leq s$.

(c) *If an A-constrained position equation in* $TRANS^A(PE)$ *has an A-solution, then* CPE *has an A-solution.*

Normalization Theorem 6.34: *Let* CPE *be an A-constrained position equation. There exists an algorithm which transforms* CPE *into a finite set* $NORM^A(CPE)$ *of normalized and A-admissible or trivial and A-solvable position equations with the following properties:*
(a) *The principal parameters of the elements of* $NORM^A(CPE)$ *do not exceed the corresponding parameters of* CPE.
(b) *If* CPE *has an A-solution* S *of index* I *and exponent of periodicity* s, *then an element of* $NORM^A(CPE)$ *has an A-solution* S' *of index* $I' = I$ *and exponent of periodicity* $s' = s$.
(c) *If an A-constrained position equation in* $NORM^A(CPE)$ *has an A-solution, then* CPE *has an A-solution.*

The normalization procedure is just a trivial modification of the normalization procedure given in [Péc]: by normalization columns are never erased. In some cases, the procedure introduces new boundaries. By choice of BD, some of the old boundaries may change their names and in this case Con has to be updated with the new column names. As a matter of fact, we finally have to check A-solvability for the resulting trivial structures and A-admissibility for the non-trivial structures. Nothing else has to be modified.

Remark: It is easy to see that the transformation of position equations defined in section 4 may be generalized to position equations PE which do not have coefficient bases, as allowed in this section. Instead of a proof consider the following (unnecessary) modification of PE: introduce a new boundary $e + 1$, a new coefficient base w_{2n+1} of type a_1 (where $2n$ is the number of variable bases of PE) and define $L(w_{2n+1}) = e$, $R(w_{2n+1}) = e + 1$. Then we get a position equation in the old sense, but the transformation is completely independent from the new part. Thus TRANSFORMATION fails to be defined only if PE does not have any variable base. This is excluded by A-admissibility as defined above.

Definition 6.35: The procedure A-TRANSFORMATION is defined as follows:
Step 1: First, if there is any column (i,j) in CPE such that $Con(i,j) = \Gamma \neq \emptyset$ and $i < l^* < j$ (or $i < cr < j$), then we unfold CPE as follows: for every unfolded instance (Γ_1, Γ_2) of Γ we introduce the A-constrained position equation CPE_{Γ_1,Γ_2} where the constraints for (i,j) are deleted and Γ_1 respectively Γ_2 are added to $Con(i,l^*)$ and $Con(l^*,j)$ (or to $Con(i,cr)$ and $Con(cr,j)$). By iteration, eventually the finite set $Unfold_{(l^*,cr)}(CPE)$ is reached where the set of constraints is empty for all columns (i,j) as above. (It is straightforward to show that CPE is A-solvable if and only if an element of $Unfold_{(l^*,cr)}(CPE)$ is A-solvable.)
Step 2: Suppose now that CPE' is in $Unfold_{(l^*,cr)}(CPE)$. The coefficient bases w_i and boundaries j such that $R(w_i) \leq l^*$ and $j \leq l^*$, the columns (i,j) such that $j \leq l^*$

and their constraints define a subpart of CPE' which may be regarded as a trivial A-constrained position equation. It is exactly this part which is erased by the transformation of position equations as defined in section 4. Thus, before we erase this part, we check the A-solvability. If this subpart is not A-solvable, then CPE' is A-unsolvable and we deleted it.

Step 3: Suppose that CPE' is not deleted by step 2. Then the kernel of CPE' is transformed as described in section 4. As a result, every column (i,j) with $l^* \le i < j \le cr$ has an image (i^{tr}, j^{tr}) in the transformed structures. Now the constraints $Con(i,j)$ are transformed. If at least one boundary of (i^{tr}, j^{tr}) is new, then we define $Con(i^{tr}, j^{tr}) = Con(i,j)$. If both transported boundaries are identified with boundaries of CPE', then we add $Con(i,j)$ to $Con(i^{tr}, j^{tr})$. (The proof of the transformation theorem in [Sc3] shows that the correspondence of solutions may be lifted to a correspondence of A-solutions.)

The Complete Search Tree with Regular Constraints

Definition 6.36: Let E be an A-constrained word equation. If E has any A-solution, then E has an A-solution with exponent of periodicity not exceeding s_{max} where s_{max} is the bound of theorem 6.6. The unordered, finitely branching tree $T^A_{Mak}(E)$ is the following extension of $T^A_{flat}(E)$:

Successful Leaves of $T^A_{Mak}(E)$ are:
- All successful leaves of $T^A_{flat}(E)$.
- All nodes which are labelled with a trivial and A-solvable position equation CPE.

Blind Leaves of $T^A_{Mak}(E)$ are:
- All blind leaves of $T^A_{flat}(E)$.
- All nodes which are labelled with an A-constrained position equation CPE which is isomorphic to the label of a node which has occurred earlier in the same path.
- All nodes which are labelled with a normalized A-admissible position equation CPE with principal parameters n and d showing (lemma 3.16) that the exponent of periodicity of any solution exceeds s_{max}.

Successors of other Nodes in $T^A_{Mak}(E)$:
- Suppose that η is an open leaf of $T^A_{flat}(E)$ labelled with the A-admissible csme-system $CSME$. For every A-admissible or trivial and A-solvable position equation CPE in $Set - CNPE(CSME)$ the node η has exactly one successor which is labelled with CPE. Note that CPE is normalized. (If there is no such position equation, then η is a blind leaf.)
- Suppose that η is a node of $T^A_{Mak}(E)$ labelled with the A-admissible normalized position equation CPE. For every A-admissible or trivial and A-solvable position

equation CPE' in $NORM^A(TRANS^A(CPE))$ the node η has exactly one successor which is labelled with CPE'. (If there is no such position equation, then η is a blind leaf.)

Theorem 6.37: (a) *If E has an A-solution S with index I, then $T^A_{Mak}(E)$ has a successful leaf in depth not exceeding $nl(E) + I$.*
(b) *If $T^A_{Mak}(E)$ has a successful leaf, then E is A-solvable.*
(c) *$T^A_{Mak}(E)$ is finite and may effectively be computed.*

Proof: (a) Theorem 6.21, lemma 6.29, theorem 6.33 and theorem 6.34. The exclusion of nodes which are labelled with a position equation which is isomorphic to the label of a predecessor node is justifiable exactly as in the proof of 6.213 or lemma 1.5.
(b) Theorem 6.21, lemma 6.29, theorem 6.33 and theorem 6.34.
(c) Theorem 6.32 and previous results. ∎

As a matter of fact, theorem 6.1 is an immediate consequence. Moreover, we get the following corollaries:

Theorem 6.38: *For every word equation E with variables $x_1, \ldots, x_n$ and coefficients in C and for arbitrary regular languages $\mathcal{L}_1, \ldots, \mathcal{L}_n$ over C it is decidable whether E has a solution $(X_1, \ldots, X_n)$ such that $X_i \in \mathcal{L}_i$ $(1 \leq i \leq n)$.*

Proof: Suppose that for every language $\mathcal{L}_i$ a finite automaton A_i is specified which accepts $\mathcal{L}_i$. The automata are assumed to be disjoint. Introducing a new initial state and vacuous transitions to the initial states of the A_i we get a finite automaton where every language $\mathcal{L}_i$ is characterized by means of a corresponding set of final states. Employing the usual construction, we compute a finite deterministic automaton A which is equivalent. Since this construction is independent from the set of final states which is specified it is clear that again every language $\mathcal{L}_i$ may be recognized with A, specifying an appropriate set of final states depending from $\mathcal{L}_i$. It is now trivial to introduce a finite number of A-constrained word equations such that E has a solution $(X_1, \ldots, X_n)$ such that $X_i \in \mathcal{L}_i$ $(1 \leq i \leq n)$ if and only if some constrained equation has an A-solution. ∎

Theorem 6.39: *For n boolean combinations $B_1, \ldots, B_n$ of finitely generated submonoids of C^* it is decidable whether E has a solution $(X_1, \ldots, X_n)$ such that $X_i \in B_i$ $(1 \leq i \leq n)$.*

Proof: Every boolean combination of finitely generated submonoids of C^* is a regular language over C^*. ∎

As a matter of fact all results may be lifted to systems of equations, compare the remark on page 15.

Example 6.40: Let us conclude with a rather complex example which demonstrates the main steps of the algorithm for regular languages. We ask whether the equation

$$cxyzzbxcy == zbaxzcybayzba$$

has a solution (X, Y, Z) such that all components are in the regular language $\mathcal{L}((aba)^* \cup (aba)^*(ca))$. Here is a finite deterministic automaton which accepts this language - s is the initial state, s and f are the final states:

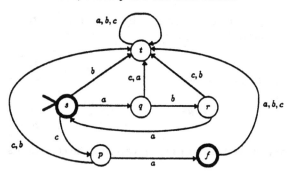

Let us treat the subcase where we assume that X and Y lead to s and Z leads to f. Accordingly we assign the A-constraint (s, s) to x, y and (s, f) to z. The following csme-system is canonically associated with the resulting A-constraint word equation:

$$\left| \begin{array}{c} cx_1y_1z_1z_2bx_2cy_2 \\ z_3bax_3z_4cy_3bay_4z_5ba \end{array} \right| \left| \begin{array}{c} x_1(s, s) \\ x_2 \\ x_3 \end{array} \right| \left| \begin{array}{c} y_1(s, s) \\ y_2 \\ y_3 \\ y_4 \end{array} \right| \left| \begin{array}{c} z_1(s, f) \\ z_2 \\ z_3 \\ z_4 \\ z_5 \end{array} \right| \qquad (-, -, -).$$

We replace z_3 by cz_3 and get

$$\left| \begin{array}{c} x_1y_1z_1z_2bx_2cy_2 \\ z_3bax_3z_4cy_3bay_4z_5ba \end{array} \right| \left| \begin{array}{c} x_1(s, s) \\ x_2 \\ x_3 \end{array} \right| \left| \begin{array}{c} y_1(s, s) \\ y_2 \\ y_3 \\ y_4 \end{array} \right| \left| \begin{array}{c} z_1(s, f) \\ z_2 \\ cz_3 \\ z_4 \\ z_5 \end{array} \right| \qquad (-, -, -).$$

Now we want to replace x_1 by z_3x_1. Thus we unfold (s, s) choosing the unfolded instance $((s, q), (q, s))$. We get

$$\left| \begin{array}{c} x_1y_1z_1z_2bx_2cy_2 \\ bax_3z_4cy_3bay_4z_5ba \end{array} \right| \left| \begin{array}{c} z_3(s, q)x_1(q, s) \\ x_2 \\ x_3 \end{array} \right| \left| \begin{array}{c} y_1(s, s) \\ y_2 \\ y_3 \\ y_4 \end{array} \right| \left| \begin{array}{c} z_1(s, f) \\ z_2 \\ cz_3 \\ z_4 \\ z_5 \end{array} \right| \qquad (-, -, -).$$

The next step is to replace x_1 by bx_1. The unique b-unfolded instance of (q,s) is the pair $((q,r),(r,s))$. Thus we get

$$
\begin{vmatrix} x_1 y_1 z_1 z_2 bx_2 cy_2 \\ ax_3 z_4 cy_3 bay_4 z_5 ba \end{vmatrix}
\begin{vmatrix} z_3(s,q)b(q,r)x_1(r,s) \\ x_2 \\ x_3 \end{vmatrix}
\begin{vmatrix} y_1(s,s) \\ y_2 \\ y_3 \\ y_4 \end{vmatrix}
\begin{vmatrix} z_1(s,f) \\ z_2 \\ cz_3 \\ z_4 \\ z_5 \end{vmatrix}
\qquad (\text{-},\text{-},\text{-}).
$$

After similar steps (replacing x_1 by ax_1, x_3 by $x_1 x_3$, then x_3 by y_1 and z_4 by z_1 we reach the following csme-system

$$
\begin{vmatrix} z_2 bx_2 cy_2 \\ cy_3 bay_4 z_5 ba \end{vmatrix}
\begin{vmatrix} z_3(s,q)b(q,r)a(r,s)x_1(s,s) \\ x_2 \\ x_1 y_1 \end{vmatrix}
\begin{vmatrix} y_1(s,s) \\ y_2 \\ y_3 \\ y_4 \end{vmatrix}
\begin{vmatrix} z_1(s,f) \\ z_2 \\ cz_3 \\ z_1 \\ z_5 \end{vmatrix}
\qquad (\text{-},\text{-},\text{-}).
$$

Now we erase the matched occurrences of z_1, carrying the constraints (s,f) to z_2 (compare (S_2^A)):

$$
\begin{vmatrix} z_2 bx_2 cy_2 \\ cy_3 bay_4 z_5 ba \end{vmatrix}
\begin{vmatrix} z_3(s,q)b(q,r)a(r,s)x_1(s,s) \\ x_2 \\ x_1 y_1 \end{vmatrix}
\begin{vmatrix} y_1(s,s) \\ y_2 \\ y_3 \\ y_4 \end{vmatrix}
\begin{vmatrix} z_2(s,f) \\ cz_3 \\ z_5 \end{vmatrix}
\qquad (\text{-},\text{-},\text{-}).
$$

Then we replace z_2 by cz_2, y_3 first by $z_2 y_3$, then by by_3, x_2 first by $y_3 x_2$, then by bx_2 and ax_2, y_4 by x_2, z_5 by cz_5 and y_2 first by $z_5 y_2$, then by by_2 and by a. Now we have reached the following open leaf of the initial (and of the flat) tree:

$$
\begin{vmatrix} z_3(s,q)b(q,r)a(r,s)x_1(s,s) \\ y_3 bax_2 \\ x_1 y_1 \end{vmatrix}
\begin{vmatrix} y_1(s,s) \\ z_5 ba \\ z_2 by_3 \\ x_2 \end{vmatrix}
\begin{vmatrix} c(s,p)z_2(p,f) \\ cz_3 \\ cz_5 \end{vmatrix}
\qquad (z_3 bax_1, y_1, cz_2).
$$

By translation we get (for example) the following A-constrained position equation (similar as in example 6.4 we choose the y-column as the leftmost part):

$$
\begin{array}{|c|c c|c|c|c|c|}
\hline
\bar{y}_1 & & & z_3 & b & a & x_1 \\
\hline
z_5 & b & a & y_3 & & & x_2 \\
\hline
z_2 & & \bar{y}_3 & \bar{x}_1 & & & y_1 \\
\hline
\bar{x}_2 & & & & & & \\
\hline
\end{array}
\; c \;
\begin{array}{|c|}
\hline
\bar{z}_2 \\
\hline
\bar{z}_3 \\
\hline
\bar{z}_5 \\
\hline
\end{array}
\qquad
\begin{bmatrix}
Con(1,4) = \{(s,s)\} \\
Con(4,5) = \{(s,q)\} \\
Con(5,6) = \{(q,r)\} \\
Con(6,7) = \{(r,s)\} \\
Con(7,8) = \{(s,s)\} \\
Con(9,10) = \{(p,f)\}
\end{bmatrix}
$$

The substitution list is $((4,5)ba(7,8),(1,4),c(9,10))$. After the first transformation (carrier x_2), the matched occurrences of y_1 and x_2 are erased.

$$
\begin{array}{|c|c|c|c|c|c|c|c|}
\hline
z_3 & b & a & x_1 & & & c & \bar{z}_2 \\
\hline
\end{array}
\quad
\left[
\begin{array}{l}
Con(1,2) = \{(s,q)\} \\
Con(2,3) = \{(q,r)\} \\
Con(3,4) = \{(r,s)\} \\
Con(4,7) = \{(s,s)\} \\
Con(8,9) = \{(p,f)\}
\end{array}
\right]
\quad ((1,2)ba(4,7),(4,7),c(8,9)).
$$

After the second transformation the matched occurrences of x_1 are erased.

$$
\begin{array}{|c|c|c|c|c|}
\hline
z_5 & b & a & c & \bar{z}_2 \\
\hline
\end{array}
\quad
\left[
\begin{array}{l}
Con(1,2) = \{(s,q)\} \\
Con(2,3) = \{(q,r)\} \\
Con(3,4) = \{(r,s)\} \\
Con(1,4) = \{(s,s)\} \\
Con(5,6) = \{(p,f)\}
\end{array}
\right]
\quad ((1,2)ba(1,4),(1,4),c(5,6)).
$$

Now the substitution list is replaced by $((1,2)ba(1,2)(2,4),(1,2)(2,4),c(5,6))$ and the constraint (s,s) for $(1,4)$ has to be unfolded over the critical boundary $cr = 2$. We get $Con(1,2) = \{(s,q)\}$, $Con(2,4) = \{(q,s)\}$. After the transformation with carrier y_3 the matched occurrences of y_3 are erased.

$$
\begin{array}{|c|c|c|c|}
\hline
b & a & c & \bar{z}_2 \\
\hline
\end{array}
\quad
\left[
\begin{array}{l}
Con(1,2) = \{(q,r)\} \\
Con(2,3) = \{(s,q),(r,s)\} \\
Con(1,3) = \{(q,s)\} \\
Con(4,5) = \{(p,f)\}
\end{array}
\right]
\quad ((4,5)ba(4,5)(1,3),(4,5)(1,3),c(4,5))
$$

Now the substitution list is replaced by $((4,5)ba(4,5)b(2,3),(4,5)b(2,3),c(4,5))$. The constraint for the part which will be erased by the next transformation is checked for satisfiability; in fact b satisfies (q,r) and may be erased. The constraint (q,s) for $(1,3)$ has to be unfolded over the left boundary 2 of the carrier z_2. We get $Con(1,2) = \{(q,r)\}$, $Con(2,3) = \{(r,s)\}$. The b satisfies also (q,r) and this constraint may be forgotten. After the next transformation, the position equation has the trivial form

$$
\begin{array}{|c|c|}
\hline
c & a \\
\hline
\end{array}
\quad
\left[
\begin{array}{l}
Con(2,3) = \{(s,q),(r,s)\} \\
Con(2,3) = \{(p,f)\}
\end{array}
\right]
\quad ((2,3)ba(2,3)b(2,3),(2,3)b(2,3),c(2,3)).
$$

The A-constrained position equation is trivial since no variable base is left. Since all constraints are satisfied it is A-solvable. Carrying the values of the columns to the substitution list we find the A-solution $(X,Y,Z) = (abaaba, aba, ca)$ of the equation $cxyzzbxcy == zbaxzcybayzba$.

References

[Abd] H.Abdulrab, *Résolution d'équations sur les mots: Etude et implémentation LISP de l'algorithme de MAKANIN*, Thèse de doctorat - Laboratoire d'informatique, Rouen 1987.

[Bul] V.K.Bulitko, *Equations and Inequalities in a Free Group and a Free Semigroup*, Tul. Gos. Ped. Inst. Učen. Zap. Mat. Kafedr Vyp.2, Geometr. i Algebra (1970), pp. 242-252 (Russian).

[Coo] D.C.Cooper, *Theorem-proving in Arithmetic without Multiplication*, Machine Intelligence **7** (1972), 82-95.

[Far] Z.Farkas, *Listlog - A Prolog Extension for List Processing*, Proc. TAPSOFT 1987, March 1987, LNCS 250, Vol. 2, 82-95.

[Fri] L.Fribourg, *List Concatenation via Extended Unification*, Programmation en Logique, Actes du Séminaire 1987, Trégastel 1987, 45-59.

[Hme] J.I.Hmelevskiĭ, *Equations in Free Semigroups*, Trudy Mat. Inst. Steklov. **107** (1971); English transl., Proc. Steklov Inst. Math. **107** (1971).

[Jaf] J.Jaffar, *Minimal and Complete Word Unification*, JACM **37** (1), 47-85.

[KoP] A.Kościelski,L.Pacholski, *Complexity of Unification in free groups and free semigroups*, Proc. 3st Annual IEEE Symposium on Foundations of Computer Science, Los Alamitos 1990, 824-829.

[Len] A.Lentin, *Equations in Free Monoids*, in Automata Languages and Programming, (M.Nivat, ed.) North Holland Publishers, Amsterdam 1972, 67-85.

[LeS] A.Lentin,M.P.Schützenberger, *A Combinatorial Problem in the Theory of Free Monoids*, Proc. of the University of North-Carolina (1967), 128-144.

[LiS] M.Livesey,J.Siekmann, *Termination and Decidability Results for String Unification*, Essex University, Memo, 1975.

[Lot] M.Lothaire, *Combinatorics on words*, Addison Wesley, 1983.

[Mak] G.S.Makanin, *The problem of solvability of equations in a free semigroup*, Math. USSR Sbornik **32**, 2 (1977), 129-198.

[Ma2] G.S.Makanin, *Recognition of the Rank of Equations in a Free Semigroup*, Math. USSR Izvestija **14**, 3 (1980), 499-545.

[Péc] J.P.Pécuchet, *Equations avec constantes et algorithme de Makanin*, Thèse de doctorat, Laboratoire d' informatique, Rouen 1981.

[Plo] G.Plotkin, *Building-in Equational Theories*, Machine Intelligence **7** (1972), 73-90.

[Rob] J.A.Robinson, *A Machine-Oriented Logic based on the Resolution Principle*, Journal of the ACM **12** (1965), 32-41.

[Rou] A.Roussel, *Programmation de l'algorithme de Makanin en PROLOG II*, Memoire de D.E.A., Université Aix-Marseille II, 1987.

[Sc1] K.U.Schulz, *A Guide to Makanin's Algorithm*, SNS-Bericht 88-39, Seminar für natürlich-sprachliche Systeme, University of Tübingen 1989.

[Sc2] K.U.Schulz, *A Note on Makanin's Algorithm*, SNS-Bericht 89-53, Seminar für natürlich-sprachliche Systeme, University of Tübingen 1989.

[Sc3] K.U.Schulz, *Makanin's Algorithm - Two Improvements and a Generalization*, CIS-Report 91-39, Centrum für Informations- und Sprachverarbeitung, University of Munique,1991.

[Si1] J.Siekmannn, *A Modification of Robinson's Unification Procedure*, M.Sc.Thesis, 1972.

[Si2] J.Siekmannn, *Unification and Matching Problems*, Ph.D.Thesis, Essex University, Memo CSA-4-78, 1978.

Unification Theory

Franz Baader

German Research Center for AI (DFKI)

Postfach 2080

W-6750 Kaiserslautern, Germany

e-mail: baader@dfki.uni-kl.de

Abstract

The purpose of this paper is not to give an overview of the state of art in unification theory. It is intended to be a short introduction into the area of equational unification which should give the reader a feeling for what unification theory might be about. The basic notions such as complete and minimal complete sets of unifiers, and unification types of equational theories are introduced and illustrated by examples. Then we shall describe the original motivations for considering unification (in the empty theory) in resolution theorem proving and term rewriting. Starting with Robinson's first unification algorithm it will be sketched how more efficient unification algorithms can be derived.

We shall then explain the reasons which lead to the introduction of unification in non-empty theories into the above mentioned areas theorem proving and term rewriting. For theory unification it makes a difference whether single equations or systems of equations are considered. In addition, one has to be careful with regard to the signature over which the terms of the unification problems can be built. This leads to the distinction between elementary unification, unification with constants, and general unification (where arbitrary free function symbols may occur). Going from elementary unification to general unification is an instance of the so-called combination problem for equational theories which can be formulated as follows: Let E, F be equational theories over disjoint signatures. How can unification algorithms for E, F be combined to a unification algorithm for the theory $E \cup F$.

1 What is E-unification?

E-unification is concerned with solving term equations modulo an equational theory E. The theory is called "unitary" ("finitary") if the solutions of an equation can always be represented by one (finitely many) "most general" solutions. Otherwise the theory is of type "infinitary" or "zero." Equational theories which are of unification type unitary or finitary play an important rôle in automated theorem provers with "built in" theories (see e.g., [Pl72,Ne74,Sl74,St85]), in generalizations of the Knuth-Bendix algorithm (see e.g., [Hu80,PS81,JK86,Bc87]), and in logic programming with equality (see e.g., [JL84]).

The first two applications will be considered in subsequent sections. In the present section we shall introduce the basic notions of unification theory such as complete and minimal complete sets of unifiers and unification types of equational theories, and illustrate them by examples.

Let Ω be a signature, i.e., a set of function symbols with fixed arity, and let V be a countable set of variables. The set of Ω-terms with variables in V is denoted by $T(\Omega, V)$. A set of identities $E \subseteq T(\Omega, V) \times T(\Omega, V)$ defines an *equational theory* $=_E$, i.e., the equality of terms induced by E. The quotient algebra $T(\Omega, V)/=_E$ is the *E-free algebra* with generators V, i.e., the free algebra with countably many generators over the class of all models of E.

Example 1.1 Let Ω be the signature consisting of one binary function symbol f. The set of identities $A := \{f(x, f(y, z)) = f(f(x, y), z)\}$ defines the theory of semigroups. Obviously, the $=_A$-classes may be considered as words over the alphabet V, and the A-free algebra $T(\Omega, V)/=_A$ is isomorphic to the free semigroup V^+.

Informally, we can now say that *E-unification* is just solving equations in the E-free algebra $T(\Omega, V)/=_E$. To be more precise, we have to define what is meant by equation and by solution of the equation.

For this reason we consider *substitutions* which are mappings $\theta: V \to T(\Omega, V)$ such that $\{x \in V \mid x\theta \neq x\}$ is finite. Since $T(\Omega, V)$ is the free Ω-algebra with generators V, this mapping can uniquely be extended to a homomorphism $\theta: T(\Omega, V) \to T(\Omega, V)$. A *unification problem* (the equation) is a pair of terms s, t, and an *E-unifier* of the problem (the solution of the equation $s = t$ in $T(\Omega, V)/=_E$) is a substitution θ such that $s\theta =_E t\theta$. The set of all E-unifiers of s, t will be denoted by $U_E(s, t)$.

Example 1.2 Let Ω be the signature consisting of a binary function symbol f and a constant symbol a. We consider the terms $s = f(x, a)$ and $t = f(a, y)$.

$E = \emptyset$: In this case, the substitution $\theta = \{x \mapsto a, y \mapsto a\}$ is the only $\emptyset$-unifier of the terms s, t.

$E = C := \{f(x, y) = f(y, x)\}$: Obviously, θ is also a C-unifier of s, t. But since f is now commutative, there exists another C-unifier, namely $\sigma = \{x \mapsto y\}$. These two solutions of our equation $s = t$ are however not independent of each other. In fact, θ is an instance of σ because $\theta = \sigma \circ \{y \mapsto a\}$.

For most applications, one does not need the set of all E-unifiers. A complete set of E-unifiers, i.e., a set of E-unifiers from which all E-unifiers can be generated by E-*instantiation*, is usually sufficient. More precisely, we extend the relation $=_E$ to $U_E(s, t)$, and define the quasi-ordering $\leq_E$ on $U_E(s, t)$ by

$$\sigma =_E \theta \quad \text{iff} \quad x\sigma =_E x\theta \text{ for all variables } x \text{ occurring in } s \text{ or } t.$$

$$\sigma \leq_E \theta \quad \text{iff} \quad \text{there exists a substitution } \lambda \text{ such that } \theta =_E \sigma \circ \lambda.$$

If $\sigma \leq_E \theta$ then θ is called an E-*instance* of σ, and σ is said to be *more general* than θ.

A *complete set* $cU_E(s,t)$ *of* E-*unifiers of* s,t has to satisfy the conditions

- $cU_E(s,t) \subseteq U_E(s,t)$, and

- for all $\theta \in U_E(s,t)$ there exists $\sigma \in cU_E(s,t)$ such that $\sigma \leq_E \theta$.

For reasons of efficiency, such a set should be as small as possible. Thus one is interested in *minimal complete sets* $\mu U_E(s,t)$ *of* E-*unifiers of* s,t, that is, complete sets satisfying the additional condition

- For all $\sigma, \theta \in \mu U_E(s,t)$, $\sigma \leq_E \theta$ implies $\sigma = \theta$.

Example 1.3 As in Example 1.2 we consider the terms $s = f(x,a)$ and $t = f(a,y)$.

$E = A := \{f(x,f(y,z)) = f(f(x,y),z)\}$: The substitutions $\theta = \{x \mapsto a, y \mapsto a\}$ and $\tau = \{x \mapsto f(a,z), y \mapsto f(z,a)\}$ are A-unifiers of s,t, and it is easy to see that the set $\{\theta, \tau\}$ is complete. In addition, θ and τ are not comparable with respect to $\leq_A$, which shows that $\{\theta, \tau\}$ is a minimal complete set of E-unifiers of s,t.

A minimal complete set of E-unifiers may not always exist, but if it exists it is unique up to the equivalence defined by $\sigma \equiv_E \theta$ iff $\sigma \leq_E \theta$ and $\theta \leq_E \sigma$. For this reason, the *unification type* of an equational theory E can be defined with reference to the cardinality and existence of minimal complete sets.

Type 1 (unitary):   A set $\mu U_E(s,t)$ exists for all s,t and has cardinality ≤ 1.

Type ω (finitary):   A set $\mu U_E(s,t)$ exists for all s,t and is of finite cardinality.

Type ∞ (infinitary): A set $\mu U_E(s,t)$ exists for all s,t, but may be infinite.

Type 0 (zero):   There are terms s,t such that a set $\mu U_E(s,t)$ does not exist.

For example, the empty theory $\emptyset$ is unitary (see [Ro65]), commutativity $C = \{f(x,y) = f(y,x)\}$ is finitary (see e.g., [Si76]), associativity $A = \{f(x,f(y,z)) = f(f(x,y),z)\}$ is infinitary (see [Pl72]), and the theory $B = A \cup \{f(x,x) = x\}$ of idempotent semigroups (bands) is of type zero (see [Ba86,Sc86]).

If a theory E is unitary, then a minimal complete set $\mu U_E(s,t)$ is either empty, if s,t are not unifiable, or it consists of a single E-unifier of s,t. This unifier is called *most general E-unifier* of s,t. It is unique up to $\equiv_E$-equivalence. For the empty theory, this means that most general unifiers are unique up to variable renaming, but in general the relation $\equiv_E$ may be more complex.

As already mentioned above, most applications of E-unification presuppose that the theory E is unitary or finitary. Of course, for these applications it is not enough to just know that a given theory is of type finitary. One also needs an E-*unification algorithm*.

Such an algorithm should be able to decide whether a given pair s, t of terms is unifiable; and if the answer is "yes" it should compute a finite complete set of E-unifiers of s, t. This notion of a "unification algorithm" should be distinguished from the notion "unification procedure" which is only required to enumerate a (possibly infinite) complete set of E-unifiers, without necessarily yielding a decision procedure for E-unifiability (see e.g., [Pl72] for an example of such a procedure for A-unification).

In order to get efficient applications, the complete set computed by the unification algorithm should be as small as possible; but for some theories, computing a minimal complete set as opposed to just computing a finite complete set may cause too much overhead compared to what is gained by having a smaller set. As an example of a theory for which this phenomenon occurs one can take commutativity $C = \{f(x, y) = f(y, x)\}$. It is very easy to devise an algorithm computing finite complete sets of C-unifiers, but it is much harder to get minimal complete sets (see e.g., [Si76,He87]).

2 Unification in the empty theory

The earliest references for unification of terms (which in the framework of the previous section is called $\emptyset$-unification) date back to E. Post in the 1920s and J. Herbrand in 1930 (see [Si89] for an account of the early history of unification theory). But its real importance became clear only when $\emptyset$-unification was independently rediscoverd in J.A. Robinson's paper on the resolution principle [Ro65] and in D. Knuth's paper on completion of term rewriting systems [KB70]. Both papers were seminal for their respective fields, namely automated theorem proving and term rewriting.

Robinson and Knuth show that two unifiable terms always have a most general $\emptyset$-unifier, i.e., that the empty theory is unitary, and they describe an algorithm which computes this most general $\emptyset$-unifier.

2.1 An informal description of Robinson's algorithm

We shall now explain Robinson's algorithm with the help of two examples. A formal description of a very similar algorithm can be found in the next section.

Example 2.1 Assume that we want to unify $s = f(x, g(a, z))$ and $t = f(g(a, y), x)$, where f, g are binary function symbols, a is a constant symbol, and x, y, z are variables.

In the first step, one reads the terms simultaneously from left to right until the first disagreement occurs. In our example, this disagreement occurs at the variable x in s and the function symbol g in t. These places of disagreement define the so-called disagreement terms, which are in our example x and $g(a, y)$. To unify s and t one has to unify these disagreement terms, and this can obviously be done with the help of the substitution $\sigma_1 := \{x \mapsto g(a, y)\}$.

Now one applies this substitution to s and t, and carries on with reading the obtained terms—which are $s\sigma_1 = f(g(a, y), g(a, z))$ and $t\sigma_1 = f(g(a, y), g(a, y))$ in our example—

from left to right until the first disagreement occurs. This process has to be iterated until the terms are unified. In the example, we get the terms z and y as the next pair of disagreement terms. After applying the substitution $\sigma_2 := \{y \mapsto z\}$ to $s\sigma_1$ and $t\sigma_1$, we have obtained the unified term $s\sigma_1\sigma_2 = f(g(a,z), g(a,z)) = t\sigma_1\sigma_2$. The composition $\sigma := \sigma_1 \circ \sigma_2$ is a most general $\emptyset$-unifier of s, t.

Obviously, we could also have used the substitution $\{z \mapsto y\}$ instead of $\sigma_2 = \{y \mapsto z\}$. This explains why most general $\emptyset$-unifiers are unique only up to variable renaming.

Until now we have only treated the case where the two terms are unifiable. The next example considers all the possible reasons for non-unifiability of terms.

Example 2.2 First, assume that we want to unify the terms $s = f(g(a,y), z)$ and $t = f(f(x,y), z)$. In this case, the disagreement occurs at the function symbol g in s and at the second symbol f in t. This means that the disagreement terms—namely $g(a,y)$ and $f(x,y)$—have different function symbols as top level symbol. Obviously, this means that the disagreement terms, and thus also the terms s, t, are not unifiable. This kind of reason for non-unifiability is called *clash failure*.

Second, assume that we want to unify the terms $s = f(g(a,x), z)$ and $t = f(x,z)$. Here we obtain disagreement terms $g(a,x)$ and x. These two terms cannot be unified because the variable x occurs in the term $g(a,x)$. In fact, for any substitution σ the size of the term $x\sigma$ is strictly smaller than the size of $g(a,x)\sigma = g(a, x\sigma)$. This kind of reason for non-unifiability is called *occur-check failure*.

2.2 Motivations for using $\emptyset$-unification

In the remainder of this section we shall shortly sketch the reason why unification is important for resolution-based theorem proving and completion of term rewriting systems.

The aim of resolution-based theorem proving is to refute a given set of clauses. In the propositional case, the *resolution principle* can be described roughly as follows. Suppose that one already has derived clauses $A \lor p$ and $B \lor \neg p$ where A, B are clauses and p is a propositional variable. Then one can also deduce $A \lor B$.

In the first order case, the rôle of propositional variables is played by atomic formulae. For example, assume that we have clauses of the form $A \lor P(x,a)$ and $B \lor \neg P(a,y)$. Before the resolution rule can be applied one has to instantiate the variables x, y in a suitable way. The appropriate instantiations can be found via unification (where the predicate symbols are treated like function symbols). In the example, we can apply the $\emptyset$-unifier $\theta := \{x \mapsto a, y \mapsto a\}$, which yields the clauses $A\theta \lor P(a,a)$ and $B\theta \lor \neg P(a,a)$. After applying the resolution rule we thus get $A\theta \lor B\theta$.

In the present example, there was only one $\emptyset$-unifier of the given pair of atomic formulae, but in general there may exist infinitely many unifiers. However, it can be shown that one can restrict oneself to most general unifiers without losing refutation completeness.

The aim of a *completion procedure for term rewriting systems* is to transform a given system into an equivalent complete (i.e., confluent and terminating) system, which then can be used to decide the word problem for the corresponding equational theory.

If a rewrite system is terminating, then confluence is equivalent to local confluence, and this property can be decided by considering finitely many critical pairs (see e.g., [KB70,Hu80]). For local confluence, one has to consider triples s, t_1, t_2 of terms where t_1 is obtained from s by applying some rule $g \to d$ of the system, and t_2 is obtained from s by applying some rule $l \to r$. The system is *locally confluent* iff for all such triples there exists a common descendant t of t_1 and t_2 (see the picture below). The picture also shows

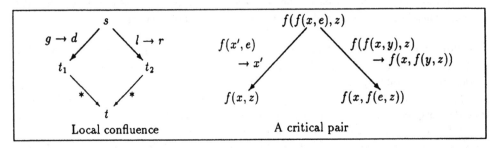

Local confluence   A critical pair

an example for such a triple where the rule for a right neutral element e applied to the subterm $f(x, e)$ of $s = f(f(x, e), z)$ yields $t_1 = f(x, z)$, whereas the associativity rule for f applied to s yields $t_2 = f(x, f(e, z))$. The term s of this example was generated from the two rules $f(x', e) \to x'$ and $f(f(x, y), z) \to f(x, f(y, z))$ as follows:[1] We have applied the unifier $\theta := \{x' \mapsto x, y \mapsto e\}$ of the left hand side $f(x', e)$ of the first rule and the subterm $f(x, y)$ of the other left hand side to this other left hand side. The *critical pair* t_1, t_2 was then obtained from s by applying the two rules at the appropriate positions. As for resolution it can be shown that it suffices to use most general $\emptyset$-unifiers in the computation of critical pairs.

3   Efficient algorithms for $\emptyset$-unification

The naive unification algorithm described in the previous section is of exponential time and space complexity. This is demonstrated by the following example.

Example 3.1 We consider the terms

$$
\begin{aligned}
s_n &= f(f(x_0, x_0), f(f(x_1, x_1), f(f(x_2, x_2), f(\ldots, f(x_{n-1}, x_{n-1}))\ldots))) \text{ and} \\
t_n &= f(x_1, f(x_2, f(x_3, f(\ldots, x_n)\ldots)))
\end{aligned}
$$

where f is a binary function symbol and $x_0, \ldots, x_n$ are variables. The most general $\emptyset$-unifier of s_n, t_n computed by the naive algorithm is of the form

$$
\sigma_n = \{x_1 \mapsto f(x_0, x_0),
$$

---

[1]Please note that the variables in the two rules have been made disjoint.

$$x_2 \mapsto f(f(x_0, x_0), f(x_0, x_0)),$$
$$x_3 \mapsto f(f(f(x_0, x_0), f(x_0, x_0)), f(f(x_0, x_0), f(x_0, x_0))),$$
$$\vdots$$
$$\}.$$

This means that $x_i \sigma_n$ contains the variable x_0 2^i times, and hence x_0 is contained in the unified term $\sum_{i=1}^{n} 2^i = 2^{n+1} - 1$ times. Since the size of s_n, t_n is linear in n, this shows that we need space—and thus also time—which is exponential in the size of the input terms.

Until now we have represented terms as strings of symbols. The example shows that more efficient unification algorithms depend on a better representation of terms. Robinson himself [Ro71] proposed a more succinct representation of terms by tables which improves the space complexity, but his algorithm is still exponential with respect to time complexity. Algorithms having almost linear time complexity were e.g. discovered by Huet [Hu76] and by Baxter [Bx76]; and finally Paterson and Wegman [PW78], and Martelli and Montanari [MM77] developed algorithms which are of linear time complexity. Later on, an algorithm which is of quadratic time complexity, but shows a better behaviour than the linear ones for most applications, was proposed by Bidoit and Corbin [BC83] (for a more complete survey of the history of efficient algorithms for $\emptyset$-unification see e.g. [Kn89,Si89]).

3.1 A recursive version of Robinson's algorithm working on dags

The algorithms of Paterson and Wegman and of Bidoit and Corbin use *directed acyclic graphs (dags)* for the representation of terms. This representation differs from the usual tree representation in that variables have to be shared and other subterms may be shared. The following picture shows the terms s_3, t_3 and the unified term $s_3 \sigma_3 = t_3 \sigma_3$ of Example 3.1 in dag-representation. This example shows that the unified term—which in string or

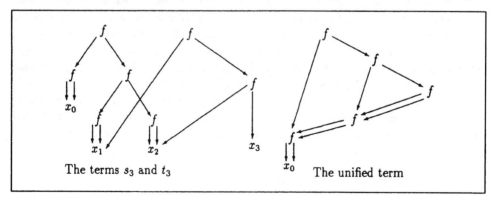

The terms s_3 and t_3 The unified term

tree representation would have been exponential in the size of the input terms—can be represented by a dag which is not larger than the input terms.

Now we shall give a recursive version of the naive algorithm which works on dags. This algorithm will be linear with respect to space complexity, but still exponential with respect to time complexity. Then it will be shown how this algorithm can be modified, first to a quadratic algorithm, and then to an almost linear algorithm.

We assume that dags consist of nodes. Any node in a given dag defines a unique (sub)dag (consisting of the nodes which can be reached from this node), and thus a unique (sub)term. There are two different types of nodes, namely variable nodes and function nodes. Function nodes carry the following information: the name of the function symbol, the arity n of this symbol, and a list (of length n) of the nodes corresponding to the arguments of the function, the so-called successor list. Both function and variable nodes may be equipped with one additional pointer to another node.

The *input* of the unification procedure (see Figure 1) is a pair of nodes in a dag. The *output* is "true" or "false," depending on whether the corresponding terms are unifiable or not. As a *side effect* the procedure creates an additional pointer structure which allows us to read off the unified term and the most general $\emptyset$-unifier.

These additional pointers are manipulated or used in the following three auxiliary procedures:

find: This procedure gets a node of the dag as input, and follows the additional pointers until it reaches a node without such a pointer. This node is the output of *find.*

union: This procedure gets a pair u, v of nodes (which do not have additional pointers) as input, and it creates an additional pointer from u to v.

occur: This procedure gets a variable node u and another node v (both of which do not have additional pointers) as input, and it performs the *occur check*, i.e., it tests whether the variable is contained in the term corresponding to v. This test is performed on the virtual term expressed by the additional pointer structure, i.e., one first applies *find* to all nodes reached during this test.

The unification algorithm described in Figure 1 requires only linear space since it does not create new nodes, and it creates at most one additional pointer for each variable node. However, the time complexity is still exponential. To see this one can consider the behaviour of the procedure *unify1* for the input terms $f(s_n, f(t'_n, x_n))$ and $f(t_n, f(s'_n, y_n))$ where s_n, t_n are defined as in Example 3.1 and s'_n, t'_n are obtained from s_n, t_n by replacing the x_i's by y_i's. The procedure needs exponentially many calls of *unify1* to finally unify the node corresponding to x_n with the node corresponding to y_n. To be more precise, these nodes are already unified after n calls of *unify1* (when x_1 and y_1 are unified), but the procedure needs exponentially many additional calls of *unify1* to recognize this fact.

3.2 A quadratic algorithm

As a solution to this problem, Bidoit and Corbin propose to not only replace variable nodes during the unification process, but also function nodes, provided that one unifies

```
        procedure unify1(k_1, k_2)

        if      k_1 = k_2 then return true
        else    %k_1 and k_2 are physically different nodes

                if      function-node(k_2)      %if one of the nodes is a
                then    u := k_1; v := k_2        %variable node then u
                else    u := k_2; v := k_1        %is now a variable node
                fi

                if      variable-node(u)
                then    if      occur(u,v)
                        then    return false    %occur-check failure
                        else    union(u,v);     %replace variable u by the
                                return true     %term corresponding to v
                        fi

                else    % u and v are function nodes
                        if      function-symbol(u) != function-symbol(v)
                        then    return false    %clash failure

                        else    n := arity(function-symbol(u));
                                (u_1,...,u_n) := successor-list(u);
                                (v_1,...,v_n) := successor-list(v);
                                i := 0; bool := true

                                while i < n and bool do
                                i := i + 1; bool := unify1(find(u_i),find(v_i))
                                od

                                return bool
                        fi
                fi
        fi
        end procedure unify1
```

Figure 1: A recursive version of Robinson's algorithm working on dags

the corresponding arguments. This can be achieved by a very simple modification of our procedure *unify1*. One simply has to insert the statement "union(u,v)" immediately in front of the while-loop. Thereby, one obtains a *procedure unify2* which is of quadratic time complexity. Since each call of *unify2* either returns "true" immediately (if the nodes were physically identical) or makes one more node virtually unreachable (i.e., it can no longer be the result of a find operation), there can only be linearly many recursive calls of *unify2*. This also shows that there are only linearly many calls of *find*, *union*, and *occur*.

The quadratic time complexity comes from the fact that the complexity of both *find* and *occur* is not constant, but may be linear. This should be obvious for *occur*. As an example for the linearity of *find*, consider the unification problem for the terms $s_1 := f(x_2, f(x_3, \ldots, f(x_n, y) \ldots))$ and $s_2 := f(x_1, f(x_1, \ldots, f(x_1, x_1) \ldots))$. Let k_1, k_2 be the nodes corresponding to s_1, s_2 in a dag-representation of this problem. During the execution of unify2(k_1, k_2), *find* is called n times with the node corresponding to x_1, and for $i = 1, \ldots n$, the ith call has to follow a pointer chain of length $i - 1$.

3.3 An almost linear algorithm

Thus we have detected two sources of non-linearity of *unify2*, namely *occur* and *find*. The first source can easily be circumvented by just omitting the occur check during the execution of the unification procedure. Since occur-check failures are not detected immediately, the procedure may return "true" even if the terms are not unifiable. But in this case a cyclic structure has been generated, and this can be recognized by a linear search. The complexity of *find* can be reduced by employing a more efficient union-find algorithm as e.g. described in [Tr75]. In this way one gets an almost linear unification algorithm (see Figure 2) which is very similar to Huet's algorithm. To be more precise, the algorithm is of time complexity $O(n \cdot \alpha(n))$ where the function α is an extremely slow-growing function, which for practical purposes (i.e., for all terms representable at all on a computer) never exceeds the value 5.

The algorithm uses three additional auxiliary procedures, namely:

collapsing-find: Like *find*, this procedure gets a node k of the dag as input, and follows the additional pointers until the node find(k) is reached. In addition, *collapsing-find* relocates the pointer of all the nodes reached during this process to find(k).

union-with-weight: This procedure gets a pair u, v of nodes (which do not have additional pointers) as input. If the set $\{k \mid k$ is a node with find(k) $= u\}$ is larger than the set $\{k \mid k$ is a node with find(k) $= v\}$, then it creates an additional pointer from v to u, otherwise it creates an additional pointer from u to v.

not-cyclic: This procedure gets a node k as input, and it tests the graph which can be reached from k for cycles. The test is performed on the virtual graph expressed by the additional pointer structure, i.e., one first applies *collapsing-find* to all nodes reached during this test.

Please note that we cannot apply the weighted union procedure in the case where we have a variable node and a function node. In this case the pointer has to go from the variable to the function node. Otherwise we should lose important information such as the name of the function symbol and the argument list. However, it is easy to see that the use of this non-optimal *union* can increase the time complexity at most by a summand $O(m)$ where m is the number of different variable nodes occurring in the dag.

procedure unify3(k_1,k_2)

if	cyclic-unify(k_1,k_2) and
	not-cyclic(k_1)

then **return** true
else **return** false
fi end procedure unify3

procedure cyclic-unify(k_1,k_2)

if　　$k_1 = k_2$ **then return** true
else　　%k_1 *and* k_2 *are physically different nodes*

　　　　if　　function-node(k_2)　　　%*if one of the nodes is a*
　　　　then　$u := k_1; v := k_2$　　　%*variable node then u*
　　　　else　$u := k_2; v := k_1$　　　%*is now a variable node*
　　　　fi

　　　　if　　variable-node(u)
　　　　then　**if**　　variable-node(v)
　　　　　　　　then　union-with-weight(u,v)
　　　　　　　　else　union(u,v)　　　　　%*no weighted union*
　　　　　　　　fi
　　　　　　　return true　　　　　　　%*no occur-check*

　　　　else　% *u and v are function nodes*
　　　　　　　if　　function-symbol(u) $\neq$ function-symbol(v)
　　　　　　　then　**return** false　%*clash failure*

　　　　　　　else　$n :=$ arity(function-symbol(u));
　　　　　　　　　　$(u_1,...,u_n) :=$ successor-list(u);
　　　　　　　　　　$(v_1,...,v_n) :=$ successor-list(v);
　　　　　　　　　　$i := 0$; bool $:=$ true;
　　　　　　　　　　union-with-weight(u,v)

　　　　　　　　　　while $i < n$ **and** bool **do**
　　　　　　　　　　$i := i + 1$;
　　　　　　　　　　bool $:=$ cyclic-unify(collapsing-find(u_i),collapsing-find(v_i))
　　　　　　　　　　od

　　　　　　　　　　return bool
　　　　　　　fi
　　　　fi
fi end procedure unify3

Figure 2: An almost linear unification algorithm

4 Unification in non-empty theories

In this section we shall first sketch by two examples why unification in equational theories was introduced into the fields automated theorem proving and term rewriting. Then we shall give some examples for new problems—i.e., problems not occurring for the empty theory—which arise in theory unification. These examples will show that one has to be very careful when trying to generalize definitions and results from $\emptyset$-unification to unification in non-empty theories.

4.1 Motivations for using E-unification

Plotkin [Pl72] observed that resolution theorem provers may waste a lot of time by applying axioms like associativity and commutativity. As a solution to this problem he proposed to build such equational axioms into the theorem proving mechanism. As a consequence one has to use unification modulo theses axioms in place of unification in the empty theory. Plotkin's paper was seminal for unification theory since, for example, the important notion of minimal complete set of unifiers (which Plotkin called a maximally general set of unifiers) was formally introduced for the first time.

Example 4.1 Assume that we have the axioms $f(f(x,y),z) = f(x,f(y,z))$ for associativity and $f(x,x) = x$ for idempotence, and that we should like to apply idempotency to the term

$$f(x_0, f(x_1, \ldots, f(x_{n-1}, f(x_n, f(x_0, \ldots, f(x_{n-1}, x_n) \ldots))) \ldots))$$

There are exponentially many ways of rearranging the parentheses with the help of associativity, and it takes a lot of time if the theorem prover has to search for the right one.

To solve this problem one can consider what a human mathematician would do in this case. (S)he would of course use words instead of terms, i.e., (s)he would work modulo associativity. In this framework one could at once apply idempotency $xx = x$ to the word

$$\underbrace{x_0 \cdots x_n}_{x} \underbrace{x_0 \cdots x_n}_{x}.$$

If we want to adopt this proceeding in a resolution theorem prover, then we have to replace $\emptyset$-unification in the resolution step by A-unification.

In term rewriting one comes very soon to the point where one would like to work modulo an equational theory. This is a consequence of the fact that certain identities cannot be oriented into terminating rewrite rules. As a solution to this problem one can leave some identities unoriented, and then pursue rewriting modulo these equational axioms. But then one must also use unification modulo these axioms when computing critical pairs (see e.g., [PS81,JK86] for details).

Example 4.2 In [KB70], the equational theory of groups was used as the motivating example. If one tries to apply the completion method of [KB70] to the theory of abelian groups, then one has to face the following problem: obviously, the commutativity axiom $f(x,y) = f(y,x)$ cannot be oriented into a terminating rule. A solution to this problem is to use rewriting modulo

$$AC := \{f(f(x,y),z) = f(x,f(y,z)), f(x,y) = f(y,x)\}.$$

Rewriting modulo $C := \{f(x,y) = f(y,x)\}$ is not possible because modulo C associativity cannot be oriented into a C-terminating rule. Modulo AC one can for example make the following rewrite step:

$$f(x, f(y, i(x))) \xrightarrow[\;f(x,i(x)) \to e\;]{\text{modulo } AC} f(e, y)$$

When computing critical pairs for the resulting system one has to use an AC-unification algorithm.

In the following subsections we shall consider some of the particular problems which arise when one goes from unification in the empty theory to unification in non-empty theories.

4.2 Single equations versus systems of equations

Until now we have considered E-unification problems for pairs s, t of terms, i.e., we have considered a single equation $s = t$ which has to be solved in the E-free algebra $T(\Omega, V)/_{=_E}$. For many applications, one has to solve systems $\Gamma = \{s_1 = t_1, \ldots, s_n = t_n\}$ of equations, i.e., one wants to find a substitution σ satisfying $s_1\sigma =_E t_1\sigma, \ldots, s_n\sigma =_E t_n\sigma$. The substitution σ is then called E-unifier of the system Γ. Now the notions complete and minimal complete set of unifiers and unification type of a theory can also be defined with respect to solving systems of equations.

For the empty theory, solving systems of equations can trivially be reduced to solving single equations. In fact, it is easy to see that σ is an $\emptyset$-unifier of the system $\Gamma = \{s_1 = t_1, \ldots, s_n = t_n\}$ iff it is an $\emptyset$-unifier of the pair $f(s_1, f(s_2, \cdots f(s_{n-1}, s_n) \cdots))$, $f(t_1, f(t_2, \cdots f(t_{n-1}, t_n) \cdots))$ of terms.

More general, for unitary and finitary theories E, a finite E-unification algorithm for single equations can always be used to get a finite E-unification algorithm for finite systems of equations. This is an immediate consequence of the following fact: Let Γ be a finite unification problem, and let s, t be terms. If $cU_E(\Gamma)$ is a finite complete set of E-unifiers of Γ, and if for all substitutions $\sigma \in cU_E(\Gamma)$, the sets $cU_E(s\sigma, t\sigma)$ are finite complete sets of E-unifiers of $s\sigma, t\sigma$ then

$$\bigcup_{\sigma \in cU_E(\Gamma)} \sigma \circ cU_E(s\sigma, t\sigma) := \{\sigma \circ \tau \mid \sigma \in cU_E(\Gamma) \text{ and } \tau \in cU_E(s\sigma, t\sigma)\}$$

is a finite complete set of E-unifiers of $\Gamma \cup \{s = t\}$ (see e.g., [He87]).

There are also non-finitary equational theories where solving finite systems of equations can be reduced to solving single equations; but the reduction may often be more complex. As an example for this case one can take associativity (see e.g., [Pé81]). However, a reduction cannot be achieved for all equational theories. This is demonstrated by the following two results:

- Schmidt-Schauß has shown that there exists an equational theory E (see [BH89]) such that

    - E-unification for single equations is infinitary, but
    - E-unification for systems of equations is of type zero.

- Narendran and Otto have shown that there exists an equational theory F (see [NO90]) such that

    - testing for unifiability is decidable for single equations, but
    - it is undecidable for systems of equations.

4.3 A closer look at the signature

It is important to note that the signature over which the terms of the unification problems may be built has considerable influence on the unification type and on the existence of unification algorithms.

To make this remark more precise, we define the *signature of an equational theory E* (for short $sig(E)$) as the set of function symbols occurring in the identities of E. When talking about unification in the theory E, often one only thinks of E-unification problems where the terms to be unified are built over $sig(E)$, i.e., are elements of $T(sig(E), V)$. However, the applications of E-unification in theorem proving and term rewriting usually require that these terms may contain additional constant symbols, or even function symbols of arbitrary arity. Because the interpretation of these symbols is not constrained by the equation theory, they are called *free symbols*.

Example 4.3 The theory $A = \{f(f(x,y), z) = f(x, f(y, z))\}$ has signature $sig(A) = \{f\}$. When talking about A-unification, one first thinks of unifying modulo A terms built by using just the symbol f and variables, or equivalently, of unifying words over the alphabet V.

However, suppose that our resolution theorem prover—which has built in the theory A—gets the formula

$$\exists x : (\forall y : f(x,y) = y \ \wedge \ \forall y \exists z : f(z,y) = x)$$

as axiom. In a first step, this formula has to be Skolemized, i.e., the existential quantifiers have to be replaced by new function symbols. In our example, we need a nullary symbol

e and a unary symbol i in the Skolemized form

$$\forall y : f(e,y) = y \ \wedge \ \forall y : f(i(y),y) = e$$

of the axiom. This shows that, even if we start with formulae containing only terms built over f, our theorem prover has to handle terms containing additional free symbols.

The same situation occurs in term rewriting modulo equation theories. In Example 4.2 we have proposed to use rewriting modulo AC for the theory of abelian groups. Obviously, the remaining rules (for the neutral element and the inverse) also contain symbols not contained in $sig(AC) = \{f\}$.

To sum up, one should distinguish between three types of E-unification, namely

Elementary E-unification: the terms of the problem may contain only symbols of $sig(E)$;

E-unification with constants: the terms of the problem may contain additional free constant symbols;

General E-unification: the terms of the problem may contain additional free function symbols of arbitrary arity.

For the empty theory, we have of course considered general $\emptyset$-unification because elementary unification and unification with constants are trivial in this case. The following facts show that there really is a difference between the three types of E-unification.

- There exist theories which are unitary with respect to elementary unification, but finitary with respect to unification with constants. An example for such a theory is the theory of abelian monoids, i.e., $AC \cup \{f(x,1) = x\}$ (see e.g., [He87,Ba89]), or the theory of idempotent abelian monoids, i.e., $AC \cup \{f(x,1) = x, f(x,x) = x\}$ (see e.g., [BB86]).

- There exists an equational theory for which elementary unification is decidable, but unification with constants is undecidable (see [Bü86]).

- From the developement of the first algorithm for AC-unification with constants [St75,LS75] it took almost a decade until the termination of an algorithm for general AC-unification was shown by Fages [Fa84].

4.4 The combination problem for unification algorithms

Motivated by the previous section, one can now ask: How can algorithms for elementary unification (or for unification with constants) be used to get algorithms for general unification? This leads to the more general question of how to combine unification algorithms for equational theories with disjoint signatures.

More formally, this *combination problem* can be described as follows. Assume that two finitary equational theories E, F with $sig(E) \cap sig(F) = \emptyset$ are given. How can unification algorithms for E and F be combined to a unification algorithm for $E \cup F$? Recall that by a unification algorithm we mean an algorithm which computes a finite complete set of unifiers.

This combination problem was first considered in [St75,St81,Fa84,HS87] for the case where several AC-symbols and free symbols may occur in the terms to unified. More general combination problems were for example treated in [Ki85,Ti86,He86,Ye87], but the theories considered in these papers always had to satisfy certain restrictions (such as collapse-freeness or regularity[2]) on the syntactic form of their defining identities.

The problem was finally solved in its most general form by Schmidt-Schauß [Sc89]. His combination algorithm imposes no restriction on the syntactic form of the defining identities. The only requirements are:

- There exist unification algorithms for unification with constants for E and F.

- All constant elimination problems must be finitary solvable in E and F.

A *constant elimination problem* in a theory E is a finite set $\{(c_1, t_1), \ldots, (c_n, t_n)\}$ where the c_i's are free constants (i.e., constant symbols not occurring in $sig(E)$) and the t_i's are terms (built over $sig(E)$, variables, and some free constants). A solution to such a problem is called a *constant eliminator*. It is a substitution σ such that for all $i, 1 \leq i \leq n$, there exists a term t_i' not containing the free constant c_i with $t_i' =_E t_i\sigma$. The notion *complete set of constant eliminators* is defined analogously to the notion complete set of unifiers. The requirement that all constant elimination problems must be finitary solvable in E means that one can always compute a finite complete set of constant eliminators for E.

A more efficient version of this combination method has been described in [Bo90]. It should be noted that the method of Schmidt-Schauß can also handle theories which are not finitary. In this case, procedures which enumerate complete sets of unifiers for E and F can be combined to a procedure enumerating a complete set of unifiers for $E \cup F$.

5 Some topics in unification theory

The purpose of this paper was not to give an overview of the state of art in unification theory, and for this reason there are many important topics we have not touched. Now we shall give a (certainly not complete) list of some of the research problems of current interest in unification theory.

- Determine unification types of equational theories (see e.g., [Sc86,Ba87,Fr89]).

---

[2]A theory E is called collapse-free if it does not contain an identity of the form $x = t$ where x is a variable and t is a non-variable term, and it is called regular if the left and right hand sides of the identities contain the same variables.

- Investigate the decidability of the unification problem, and the complexity of this decision problem (see e.g., [SS86,KN89]; in [KN89] there is a table summarizing many of the known results in this direction).

- Devise unification algorithms for specific unitary and finitary theories. For example, a lot of work was—and still is—devoted to finding efficient algorithms for AC-unification (see e.g., [St81,Fa84,Ki85,Bt86,He87,HS87,Fo85,CF89,BD90]).

- Devise universal unification algorithms, i.e., algorithms where the equational theory also belongs to the input of the algorithm. Examples are methods based on narrowing (see e.g., [Fy79,Hl80,Fi84,NR89]), or on Martelli and Montanari's decomposition technique (see e.g., [GS87,KK90]).

For more information on these and other topics in unification theory one can consult Siekmann's overview of the state of art in unification theory [Si89], or Jouannaud and Kirchner's survey of unification [JK90]. In particular, these papers contain an almost complete list of references on unification theory.

References

[Ba86] F. Baader, "The Theory of Idempotent Semigroups is of Unification Type Zero," *J. Automated Reasoning* **2**, 1986.

[Ba87] F. Baader, "Unification in Varieties of Idempotent Semigroups," *Semigroup Forum* **36**, 1987.

[Ba89] F. Baader, "Unification in Commutative Theories," in C. Kirchner (ed.), *Special Issue on Unification, J. Symbolic Computation* **8**, 1989.

[BB86] F. Baader, W. Büttner, "Unification in Commutative Idempotent Monoids," *Theoretical Computer Science* **56**, 1986.

[Bc87] L. Bachmair, *Proof Methods for Equational Theories*, Ph.D. Thesis, Dep. of Comp. Sci., University of Illinois at Urbana-Champaign, 1987.

[Bx76] L. Baxter, *The Complexity of Unification*, Ph.D. Thesis, University of Waterloo, Waterloo, Ontario, Canada, 1976.

[BC83] M. Bidoit, J. Corbin, "A Rehabilitation of Robinson's Unification Algorithm," In R.E.A. Pavon, editor, *Information Processing 83*, North Holland, 1983.

[Bo90] A. Boudet, "Unification in a Combination of Equational Theories : An Efficient Algorithm," *Proceedings of the 10th Conference on Automated Deduction, LNCS* **449**, 1990.

[BD90] A. Boudet, E. Contejean, H. Devie, "A New AC-unification Algorithm with a New Algorithm for Solving Diophantine Equations," *Proceedings of the 5th IEEE Symposium on Logic in Computer Science, Philadelphia*, 1990.

[Bü86]   H.-J. Bürckert, "Some Relationships Between Unification, Restricted Unification, and Matching," *Proceedings of the 8th Conference on Automated Deduction, LNCS* **230**, 1986.

[BH89]   H.-J. Bürckert, A. Herold, M. Schmidt-Schauß, "On Equational Theories, Unification, and Decidability," in C. Kirchner (ed.), *Special Issue on Unification, J. Symbolic Computation* **8**, 1989.

[Bt86]   W. Büttner, "Unification in the Data Structure Multiset," *J. Automated Reasoning* **2**, 1986.

[CF89]   M. Clausen, A. Fortenbacher, "Efficient Solution of Linear Diophantine Equations," in C. Kirchner (ed.), *Special Issue on Unification, J. Symbolic Computation* **8**, 1989.

[Fa84]   F. Fages, "Associative-Commutative Unification," *Proceedings of the 7th Conference on Automated Deduction, LNCS* **170**, 1984.

[Fy79]   M. Fay, "First Order Unification in an Equational Theory," *Proceedings of the 4th Workshop on Automated Deduction*, Austin, Texas, 1979.

[Fo85]   A. Fortenbacher, "An Algebraic Approach to Unification under Associativity and Commutativity," *Proceedings of the 1st Conference on Rewriting Techniques and Applications*, Dijon, France, *LNCS* **202**, 1985.

[Fr89]   M. Franzen, "Hilbert's Tenth Problem Has Unification Type Zero," Preprint, 1989. To appear in *J. Automated Reasoning*.

[Fi84]   L. Fribourg, "A Narrowing Procedure with Constructors," *Proceedings of the 7th Conference on Automated Deduction, LNCS* **170**, 1984.

[GS87]   J.H. Gallier, W. Snyder, "A General Complete E-Unification Procedure," *Proceedings of the Second Conference on Rewriting Techniques and Applications*, Bordeaux, France, *LNCS* **256**, 1987.

[He86]   A. Herold, "Combination of Unification Algorithms," *Proceedings of the 8th Conference on Automated Deduction, LNCS* **230**, 1986.

[He87]   A. Herold, *Combination of Unification Algorithms in Equational Theories*, Dissertation, Fachbereich Informatik, Universität Kaiserslautern , 1987.

[HS87]   A. Herold, J.H. Siekmann, "Unification in Abelian Semigroups," *J. Automated Reasoning* **3**, 1987.

[Hu76]   G.P. Huet, *Résolution d'équations dans des langages d'ordre* $1, 2, ..., \omega$, Thèse d'État, Université de Paris VII, 1976.

[Hu80]   G.P. Huet, "Confluent Reductions: Abstract Properties and Applications to Term Rewriting Systems," *J. ACM* **27**, 1980.

[Hl80]   F.M. Hullot, "Canonical Forms and Unification," *Proceedings of the 5th Confer-*
         ence on Automated Deduction, LNCS **87**, 1980.

[JL84]   J. Jaffar, J.L. Lassez, M. Maher, "A Theory of Complete Logic Programs with
         Equality," *J. Logic Programming* **1**, 1984.

[JK86]   J.P. Jouannaud, H. Kirchner, "Completion of a Set of Rules Modulo a Set of
         Equations," *SIAM J. Computing* **15**, 1986.

[JK90]   J.P. Jouannaud, C. Kirchner, "Solving Equations in Abstract Algebras: A Rule-
         Based Survey of Unification," Preprint, 1990. To appear in the Festschrift to
         Alan Robinson's birthday.

[KN89]   D. Kapur, P. Narendran, "Complexity of Unification Problems with Associative-
         Commutative Operators," Preprint, 1989. To appear in *J. Automated Reasoning*.

[Ki85]   C. Kirchner, *Méthodes et Outils de Conception Systématique d'Algorithmes*
         d'Unification dans les Théories equationnelles, Thèse d'Etat, Univ. Nancy,
         France, 1985.

[KK90]   C. Kirchner, F. Klay, "Syntactic Theories and Unification," *Proceedings of the*
         5th IEEE Symposium on Logic in Computer Science, Philadelphia, 1990.

[Kn89]   K. Knight, "Unification: A Multidisciplinary Survey," *ACM Computing Surveys*
         21, 1989.

[KB70]   D.E. Knuth, P.B. Bendix, "Simple Word Problems in Universal Algebras," In
         J. Leech, editor, *Computational Problems in Abstract Algebra*, Pergamon Press,
         Oxford, 1970.

[LS75]   M. Livesey, J.H. Siekmann, "Unification of AC-Terms (bags) and ACI-Terms
         (sets)," Internal Report, University of Essex, 1975, and Technical Report 3-76,
         Universität Karlsruhe, 1976.

[MM77]   A. Martelli, U. Montanari, "Theorem Proving with Structure Sharing and Ef-
         ficient Unification," *Proceedings of International Joint Conference on Artificial*
         Intelligence, 1977.

[NO90]   P. Narendran, F. Otto, "Some Results on Equational Unification," *Proceedings*
         of the 10th Conference on Automated Deduction, LNCS **449**, 1990.

[Ne74]   A.J. Nevins, "A Human Oriented Logic for Automated Theorem Proving,"
         J. ACM **21**, 1974.

[NR89]   W. Nutt, P. Réty, G. Smolka, "Basic Narrowing Revisited," *J. Symbolic Com-*
         putation **7**, 1989.

[PW78]   M.S. Paterson, M.N. Wegman, "Linear Unification," *J. Comput. Syst. Sci.* **16**,
         1978.

[Pé81]   J.P. Pécuchet, *Équation avec constantes et algorithme de Makanin*, Thèse de Doctorat, Laboratoire d'informatique, Rouen, 1981.

[PS81]   G. Peterson, M. Stickel, "Complete Sets of Reductions for Some Equational Theories," *J. ACM* **28**, 1981.

[Pl72]   G. Plotkin, "Building in Equational Theories," *Machine Intelligence* **7**, 1972.

[Ro65]   J.A. Robinson, "A Machine-Oriented Logic Based on the Resolution Principle," *J. ACM* **12**, 1965.

[Ro71]   J.A. Robinson, "The Unification Computation," *Machine Intelligence* **6**, 1971.

[Sc86]   M. Schmidt-Schauß, "Unification under Associativity and Idempotence is of Type Nullary," *J. Automated Reasoning* **2**, 1986.

[Sc89]   M. Schmidt-Schauß, "Combination of Unification Algorithms," *J. Symbolic Computation* **8**, 1989.

[Si76]   J.H. Siekmann, "Unification of Commutative Terms," SEKI-Report, Universität Karlsruhe 1976.

[Si89]   J.H. Siekmann, "Unification Theory: A Survey," in C. Kirchner (ed.), *Special Issue on Unification, Journal of Symbolic Computation* **7**, 1989.

[SS86]   J.H. Siekmann, P. Szabo, "The Undecidability of the D_A-Unification Problem," SEKI-Report SR-86-19, Universität Kaiserslautern, 1986, and *J. Symbolic Logic* **54**, 1989.

[Sl74]   J.R. Slagle, "Automated Theorem Proving for Theories with Simplifiers, Commutativity and Associativity," *J. ACM* **21**, 1974.

[St75]   M. Stickel, "A Complete Unification Algorithm for Associative-Commutative Functions," *Proceedings of the International Joint Conference on Artificial Intelligence*, 1975.

[St81]   M.E. Stickel, "A Unification Algorithm for Associative-Commutative Functions," *J. ACM* **28**, 1981.

[St85]   M.E. Stickel, "Automated Deduction by Theory Resolution," *J. Automated Reasoning* **1**, 1985.

[Ti86]   E. Tiden, "Unification in Combinations of Collapse Free Theories with Disjoint Sets of Function Symbols," *Proceedings of the 8th Conference on Automated Deduction, LNCS* **230**, 1986.

[Tr75]   E.T. Trajan, "Efficiency of a Good But Not Linear Set Union Algorithm," *J. ACM* **22**, 1975.

[Ye87]   K. Yelick, "Unification in Combinations of Collapse Free Regular Theories," *J. Symbolic Computation* **3**, 1987.

Algebraic and Logical Aspects of Unification

Alexander Bockmayr
Max-Planck-Institut für Informatik
Im Stadtwald
D-6600 Saarbrücken
e-mail: bockmayr@cs.uni-sb.de

During the last years unification theory has become an important subfield of automated reasoning and logic programming. The aim of the present paper is to relate unification theory to classical work on equation solving in algebra and mathematical logic. We show that many problems in unification theory have their counterpart in classical mathematics and illustrate by various examples how classical results can be used to answer unification-theoretic questions.

1 Introduction

During the last years unification theory has become an important subfield of automated reasoning and logic programming [Sie89, JK91]. As a matter of fact, various problems of unification theory have been studied earlier in algebra and mathematical logic. However, many of the interesting results that have been obtained in these classical disciplines are widely unknown in unification theory.

One reason may be that much of the relevant work has been done by Russian mathematicians. Although most of their papers have been translated into English they are usually not read by computer scientists.

The aim of the present paper is to relate unification theory to classical work on equation solving in algebra and mathematical logic with the main emphasis on the decidability of equational problems. We show that many problems in unification theory have their counterpart in classical mathematics and illustrate by various examples how algebraic and logical results can be used to answer unification-theoretic questions.

A preliminary version of this paper appeared in [Boc89].

2 Preliminaries

Let us recall briefly the basic definitions. For more details see the survey [HO80].

$\Sigma = (S, F)$ denotes a *recursive signature* with a recursive set S of sort symbols and a recursive set F of function symbols together with an arity function. A sort $s \in S$ is *strict* iff there exists $f \in F$ such that either $f :\rightarrow s$ or $f : s_1 \times \ldots \times s_n \rightarrow s$ and s_i is strict for $i = 1, \ldots, n$. A signature Σ is *strict* iff each $s \in S$ is strict.

A Σ-*algebra* A consists of a family of non-empty sets $(A_s)_{s \in S}$ and a family of functions $(f^A)_{f \in F}$ such that if $f : s_1 \times \ldots \times s_n \rightarrow s$ then $f^A : A_{s_1} \times \ldots \times A_{s_n} \rightarrow A_s$.

X represents a family $(X_s)_{s\in S}$ of countably infinite sets X_s of *variables* of sort s. $T(F, X)$ is the Σ-algebra of *terms* with variables over Σ. If Σ is strict, $T(F)$ is the Σ-algebra of *ground terms* over Σ.

Let A and B be Σ-algebras. A Σ-homomorphism $h : A \to B$ is a family $h = (h_s)_{s\in S}$ of mappings $h_s : A_s \to B_s$ such that $h_s(f^A(t_1, \ldots, t_n)) = f^B(h_{s_1}(t_1), \ldots, h_{s_n}(t_n))$, for all $f : s_1 \times \ldots \times s_n \to s$ in F and all $t_i \in A_{s_i}$.

For a set E of conditional Σ-equations $\equiv_E$ denotes the Σ-congruence on $T(F, X)$ generated by E. By $Mod_\Sigma(E)$ we denote the class of all Σ-algebras A that are models of E. By Birkhoff's theorem we know that $s \equiv_E t$ iff $s \doteq t$ is valid in $Mod_\Sigma(E)$, for any $s, t \in T(F, X)$. Given a recursively enumerable set E of conditional Σ-equations we say that the *word problem for E is decidable* iff we can decide for any $s, t \in T(F, X)$ whether $s \equiv_E t$.

The Σ-algebra $T_E(X) \stackrel{\text{def}}{=} T(F, X)/\equiv_E$ is *free* in $Mod_\Sigma(E)$ over X. This means that for any model A of E and any variable assignment $\alpha : X \to A$ there is a unique extension of α to a Σ-homomorphism $\overline{\alpha} : T_E(X) \to A$. Mostly, we will not distinguish α and $\overline{\alpha}$.

If the signature Σ is strict, then the Σ-algebra $T_E \stackrel{\text{def}}{=} T(F)/\equiv_E$ is *initial* in $Mod_\Sigma(E)$. This means that for any model A of E there is a unique homomorphism $h : T_E \to A$.

A system of Σ-equations $S : t_1 \doteq u_1 \wedge \ldots \wedge t_n \doteq u_n$, with $t_i, u_i \in T(F, X), i = 1, \ldots, n, n \geq 1$, is called *E-unifiable* iff there is a substitution $\sigma : X \to T(F, X)$ with $\sigma(t_i) \equiv_E \sigma(u_i)$, for all $i = 1, \ldots, n$. By $\exists S$ we denote the *existential closure* of the first-order formula S.

Given a recursively enumerable set E of conditional Σ-equations we say that the *E-unification problem is decidable* iff we can decide for any system of Σ-equations S whether it is E-unifiable.

3   Unification and the Diophantine Theory

In this paper we investigate the decidability of the E-unification problem from the viewpoint of mathematical logic. The key to this approach is the following proposition:

Proposition 1 *Let E be a set of conditional Σ-equations. For any system of Σ-equations $S : t_1 \doteq u_1 \wedge \ldots \wedge t_n \doteq u_n, n \geq 1$, the following are equivalent:*

1. *S is E-unifiable.*

2. *The existential closure $\exists S$ is valid in the free algebra $T_E(X)$.*

3. *The existential closure $\exists S$ is valid in the model class $Mod_\Sigma(E)$.*

If Σ is strict, then (1) to (3) are equivalent to

4. *The existential closure $\exists S$ is valid in the initial algebra T_E.*

Proof: $(1) \Rightarrow (2)$: Assume that the system of Σ-equations $S : t_1 \doteq u_1 \wedge \ldots \wedge t_n \doteq u_n$ is E-unifiable. Then there exists a substitution $\sigma : X \to T(F, X)$ with $\sigma(t_i) \equiv_E \sigma(u_i)$, for all $i = 1, \ldots, n$. Composing σ with the quotient mapping $\rho : T(F, X) \to T_E(X)$ yields $(\rho \circ \sigma)(t_i) = (\rho \circ \sigma)(u_i)$, for all $i = 1, \ldots, n$. This implies $T_E(X) \models \exists S$.

$(2) \Rightarrow (3)$: If $\exists S$ is valid in $T_E(X)$, then there exists a variable assignment $\tau : X \to T_E(X)$ such that $\tau(t_i) = \tau(u_i)$, for all $i = 1, \ldots, n$. Let A a be model of E and $\alpha : X \to A$

a variable assignment. Then $(\overline{\alpha} \circ \tau)(t_i) = (\overline{\alpha} \circ \tau)(u_i)$, for all $i = 1, \ldots, n$. This implies $A \models \exists S$ and since A was arbitrary we get $Mod_\Sigma(E) \models \exists S$.

$(3) \Rightarrow (2)$: Obvious, since $T_E(X)$ is a model of E.

$(2) \Rightarrow (1)$: Assume $\tau(t_i) = \tau(u_i), i = 1, \ldots, n$, for a variable assignment $\tau : X \to T_E(X)$. By choosing a substitution $\sigma : X \to T(F, X)$ with $\rho \circ \sigma = \tau$, where again $\rho : T(F, X) \to T_E(X)$ is the quotient mapping, we get $\sigma(t_i) \equiv_E \sigma(u_i)$, for all $i = 1, \ldots, n$, which means that S is E-unifiable.

If Σ is strict, then the initial algebra T_E is a model of E. This yields the implication $(3) \Rightarrow (4)$. Since T_E is also a subalgebra of $T_E(X)$ we get finally $(4) \Rightarrow (2)$. □

This proposition shows that unification problems correspond to certain first-order formulas. In mathematical logic the following classification of formulas is common:

Definition 2 Let $\mathcal{K}$ be a class of Σ-algebras.

The *elementary theory* of $\mathcal{K}$ is the set of all closed formulas in first-order predicate logic with equality that are valid in $\mathcal{K}$.

Any such formula can be represented in prenex disjunctive normal form

$$Q_1 x_1 \ldots Q_m x_m((t_{11} \doteq\kern-1.2ex\neq u_{11} \wedge \ldots \wedge t_{1j_1} \doteq\kern-1.2ex\neq u_{1j_1}) \vee \ldots \vee (t_{q1} \doteq\kern-1.2ex\neq u_{q1} \wedge \ldots \wedge t_{qj_q} \doteq\kern-1.2ex\neq u_{qj_q})),$$

where each Q_i is a universal or existential quantifier, $x_1, \ldots, x_m$ are variables, $m \geq 0$, and each atomic formula has the form $t_{ij} \doteq u_{ij}$ or $t_{ij} \neq u_{ij}$ with terms $t_{ij}, u_{ij} \in T(F, X)$.

The *universal theory* of $\mathcal{K}$ is the collection of those closed formulas valid in $\mathcal{K}$ that are of the form

$$\forall x_1 \ldots \forall x_m((t_{11} \doteq\kern-1.2ex\neq u_{11} \wedge \ldots \wedge t_{1j_1} \doteq\kern-1.2ex\neq u_{1j_1}) \vee \ldots \vee (t_{q1} \doteq\kern-1.2ex\neq u_{q1} \wedge \ldots \wedge t_{qj_q} \doteq\kern-1.2ex\neq u_{qj_q}))$$

The *existential theory* of $\mathcal{K}$ is the collection of those closed formulas valid in $\mathcal{K}$ that are of the form

$$\exists x_1 \ldots \exists x_m((t_{11} \doteq\kern-1.2ex\neq u_{11} \wedge \ldots \wedge t_{1j_1} \doteq\kern-1.2ex\neq u_{1j_1}) \vee \ldots \vee (t_{q1} \doteq\kern-1.2ex\neq u_{q1} \wedge \ldots \wedge t_{qj_q} \doteq\kern-1.2ex\neq u_{qj_q}))$$

The *positive theory* of $\mathcal{K}$ is the collection of those closed formulas valid in $\mathcal{K}$ that are of the form

$$Q_1 x_1 \ldots Q_m x_m((t_{11} \doteq u_{11} \wedge \ldots \wedge t_{1j_1} \doteq u_{1j_1}) \vee \ldots \vee (t_{q1} \doteq u_{q1} \wedge \ldots \wedge t_{qj_q} \doteq u_{qj_q})),$$

The *diophantine theory* of $\mathcal{K}$ is the collection of those closed formulas valid in $\mathcal{K}$ that are of the form

$$\exists x_1 \ldots \exists x_m(t_1 \doteq u_1 \wedge \ldots \wedge t_n \doteq u_n)$$

The *equational theory* of $\mathcal{K}$ is the collection of those closed formulas valid in $\mathcal{K}$ that are of the form

$$\forall x_1 \ldots \forall x_m(t \doteq u)$$

Let E be a set of conditional equations over a signature Σ. In the sequel, we will consider both the theory of the full class $Mod_\Sigma(E)$ and the theory of the class consisting of the single model $T_E(X)$. For positive formulas, this does not make any difference.

Proposition 3 *Let E be a set of conditional Σ-equations. Then the positive theory of $Mod_\Sigma(E)$ and the positive theory of $T_E(X)$ coincide.*

Proof: It is enough to prove that the positive theory of $T_E(X)$ is included in the positive theory of $Mod_\Sigma(E)$. We will show by induction on positive formulae that $T_E(X) \models \phi [\rho|X]$, where $\rho|X$ is the variable assignment induced by the quotient mapping $\rho : T(F,X) \to T_E(X)$, implies $A \models \phi [\alpha]$, for all models A of E and all variable assignments $\alpha : X \to A$. For the rest of the proof, we will not distinguish ρ and $\rho|X$.

Since $T_E(X)$ is free in $Mod_\Sigma(E)$ over X, any assignment $\alpha : X \to A$ can be extended to a Σ-homomorphism $\overline{\alpha} : T_E(X) \to A$ with $\overline{\alpha} \circ \rho = \alpha$.

If ϕ is an equation $s \doteq t$, with terms $s, t \in T(F,X)$, we have $T_E(X) \models (s \doteq t) [\rho] \iff \rho(s) = \rho(t) \implies \overline{\alpha}(\rho(s)) = \overline{\alpha}(\rho(t)) \iff A \models (s \doteq t) [\overline{\alpha} \circ \rho] \iff A \models (s \doteq t) [\alpha]$, for all models A of E and all variable assignments $\alpha : X \to A$.

If ϕ_1, ϕ_2 are two positive formulae such that $T_E(X) \models (\phi_1 \wedge \phi_2) [\rho]$ (resp. $T_E(X) \models (\phi_1 \vee \phi_2) [\rho]$), then $T_E(X) \models \phi_1 [\rho]$ and (resp. or) $T_E(X) \models \phi_2 [\rho]$. By induction hypothesis, we get for all models A of E and all variable assignments $\alpha : X \to A$ that $A \models \phi_1 [\alpha]$ and (resp. or) $A \models \phi_2 [\alpha]$, which implies $A \models (\phi_1 \wedge \phi_2) [\alpha]$ (resp. $A \models (\phi_1 \vee \phi_2) [\alpha]$).

Now let ϕ be of the form $\exists x \psi$, with a positive formula ψ.

If $T_E(X) \models (\exists x \psi) [\rho]$, then $T_E(X) \models \psi [\rho_{\overline{t}}^x]$, for some $\overline{t} \in T_E(X)$ and the variable assignment $\rho_{\overline{t}}^x : X \to A$ defined by $\rho_{\overline{t}}^x(y) = \begin{cases} \rho(y) & \text{if } y \neq x \\ \overline{t} & \text{if } y = x. \end{cases}$

We select a term $t \in T(F,X)$ such that $\rho(t) = \overline{t}$. By renaming bounded variables in ψ we may assume that t is free for x in ψ. It follows that $T_E(X) \models \psi(x/t) [\rho]$.

By the induction hypothesis, $A \models \psi(x/t) [\alpha]$, for all models A of E and all variable assignments $\alpha : X \to A$. Again, since t is free for x in ψ we get $A \models \psi [\alpha_{\overline{\alpha}(\overline{t})}^x]$ for the variable assignment $\alpha_{\overline{\alpha}(\overline{t})}^x$ defined by $\alpha_{\overline{\alpha}(\overline{t})}^x(y) = \begin{cases} \alpha(y) & \text{if } y \neq x \\ \overline{\alpha}(\overline{t}) & \text{if } y = x \end{cases}$, and this implies $A \models \exists x \psi [\alpha]$.

Finally, let ϕ be of the form $\forall x \psi$, with a positive formula ψ.

$T_E(X) \models \forall x \psi [\rho]$ implies $T_E(X) \models \psi [\rho_{\overline{t}}^x]$, for all $\overline{t} \in T_E(X)$. If we choose $\overline{t} = \rho(x)$, we get $T_E(X) \models \psi [\rho_{\rho(x)}^x]$ or equivalently $T_E(X) \models \psi [\rho]$.

By the induction hypothesis, it follows that $A \models \psi [\alpha]$, for all models A of E and all variable assignments $\alpha : X \to A$. But this is equivalent to $A \models \forall x \psi [\alpha]$, for all models A of E and all variable assignments $\alpha : X \to A$. □

There are numerous results in the literature on the decidability and undecidability of these fragments, which are widely unknown in unification theory. We may profit by these results using the following consequence of Proposition 1.

Theorem 4 *Let E be a recursively enumerable set of conditional Σ-equations over a recursive signature Σ. Then the following are equivalent*

1. *The E-unification problem is decidable.*

2. *The diophantine theory of the free algebra $T_E(X)$ is decidable.*

3. *The diophantine theory of the model class $Mod_\Sigma(E)$ is decidable.*

If Σ is strict, then (1) to (3) are equivalent to

4. *The diophantine theory of the initial algebra T_E is decidable.*

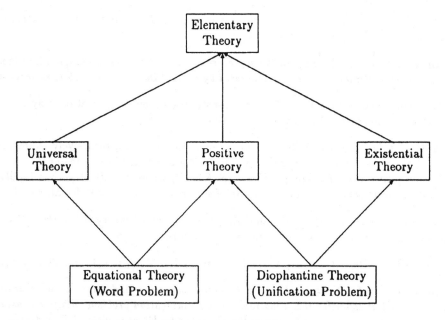

Figure 1: A Hierarchy of Decidability Problems

This result relates the unification problem to a hierarchy of decidability problems in mathematical logic (Figure 1).

The utility of this viewpoint is now illustrated by a number of examples.

4 Adding Free Constants

The first theorem, based on [Vaz74] and [Roz85], shows how adding a constant symbol to the signature may affect the decidability of the unification problem. The example is taken from the theory of inverse semigroups.

Theorem 5 *The unification problem for the theory*

$$
\begin{aligned}
IS \ = \ & \{(x * y) * z \doteq x * (y * z), \\
& (x^{-1})^{-1} \doteq x, \\
& x \doteq x * x^{-1} * x, \\
& x^{-1} * x * y^{-1} * y \doteq y^{-1} * y * x^{-1} * x\}
\end{aligned}
$$

of inverse semigroups *is decidable in the signature* $\Sigma = \{*, .^{-1}, a\}$ *and undecidable in the signature* $\Sigma' = \{*, .^{-1}, a, b\}$.

The theorem follows directly from Theorem 4 and the following results of Vazhenin [Vaz74] and Rozenblat [Roz85].

Theorem 6 (Vazhenin 74) *In the signature* $\Sigma = \{*, .^{-1}, a\}$ *the elementary theory of the free inverse semigroup generated by the element* a *is decidable. However, the elemen-*

tary theory of a free inverse semigroup that is generated by more than one element is undecidable.

Theorem 7 (Rozenblat 85) *In the signature $\Sigma' = \{*, .^{-1}, a, b\}$ the diophantine theory of a free inverse semigroup which is generated by at least two elements a, b is undecidable.*

Theorem 5 shows that by adding free constant symbols E-unification may become undecidable. This motivates the next definition.

Definition 8 Let $\Sigma = (S, F)$ be a recursive signature and E a recursively enumerable set of conditional Σ-equations. The *E-unification problem with free constants* is the E-unification problem in the signature $\Sigma' = (S, F \cup C)$, where $C = (C_s)_{s \in S}$ is a family of countably infinite sets C_s of new constants of sort s, i. e. $C_s \cap F = \emptyset$, for all $s \in S$.

Definition 9 Let $\mathcal{K}$ be a class of Σ-algebras. The *positive $\forall \exists$-theory* of $\mathcal{K}$ is the collection of those closed formulas valid in $\mathcal{K}$ that are of the form

$$\forall x_1 \ldots \forall x_k \exists y_1 \ldots \exists y_m ((t_{11} \doteq u_{11} \wedge \ldots \wedge t_{1j_1} \doteq u_{1j_1}) \vee \ldots \vee (t_{q1} \doteq u_{q1} \wedge \ldots \wedge t_{qj_q} \doteq u_{qj_q}))$$

Proposition 10 *Let $\Sigma = (S, F)$ be a recursive signature and let E be a recursively enumerable set of conditional Σ-equations. The E-unification problem with free constants is decidable if the positive $\forall \exists$-theory of the class $Mod_\Sigma(E)$ of models of E is decidable.*

Proof: Let

$$S' : s_1' \doteq t_1' \wedge \ldots \wedge s_n' \doteq t_n', n \geq 1,$$

be a system of equations over the extended signature $\Sigma' = (S, F \cup C)$. Let $y_1, \ldots, y_m, m \geq 0$, be the variables and $c_1, \ldots, c_k, k \geq 0$, be the constants from C occurring in S'. Let $x_1, \ldots, x_k$ be pairwise distinct new variables which do not occur in S' such that x_i and c_i are of the same sort, for $i = 1, \ldots, k$. We replace in S' the constant c_i with x_i, for $i = 1, \ldots, k$, and obtain a new system of equations

$$S : s_1 \doteq t_1 \wedge \ldots \wedge s_n \doteq t_n, n \geq 1$$

over the signature Σ.

Now we prove that

$$Mod_\Sigma(E) \models \forall x_1, \ldots, x_k \exists y_1, \ldots, y_m (s_1 \doteq t_1 \wedge \ldots \wedge s_n \doteq t_n),$$

if and only if

$$Mod_{\Sigma'}(E) \models \exists y_1, \ldots, y_m (s_1' \doteq t_1' \wedge \ldots \wedge s_n' \doteq t_n').$$

" $\Longrightarrow$: " Let A' be a Σ'-algebra which is a model of E. By forgetting the interpretation of the constant symbols from C, A' can be seen as a Σ-algebra A. It follows

$$A \models \forall x_1, \ldots, x_k \exists y_1, \ldots, y_m (s_1 \doteq t_1 \wedge \ldots \wedge s_n \doteq t_n),$$

and in particular

$$A \models \exists y_1, \ldots, y_m (s_1 \doteq t_1 \wedge \ldots \wedge s_n \doteq t_n)[\alpha]$$

for the variable assignment $\alpha : X \rightarrow A, \alpha(x_i) \stackrel{\text{def}}{=} c_i^{A'}$, for $i = 1, \ldots, k$. But this is equivalent to

$$A' \models \exists y_1, \ldots, y_m (s_1' \doteq t_1' \wedge \ldots \wedge s_n' \doteq t_n').$$

" $\Longleftarrow$: " Let A be a Σ–algebra that is a model of E and let $a_1, \ldots, a_k$ be elements of A such that a_i and c_i are of the same sort, for $i = 1, \ldots, k$.

We have to show that

$$A \models \exists y_1, \ldots, y_m (s_1 \doteq t_1 \wedge \ldots \wedge s_n \doteq t_n)[\beta],$$

for the variable assignment $\beta : X \to A, \beta(x_i) \overset{\text{def}}{=} a_i$, for $i = 1, \ldots, k$.

By choosing an interpretation of the constants in C such that c_i is interpreted by a_i, for $i = 1, \ldots, k$, A becomes a Σ'-algebra A'. By hypothesis,

$$A' \models \exists y_1, \ldots, y_m (s_1' \doteq t_1' \wedge \ldots \wedge s_n' \doteq t_n')$$

and the proposition is proved. $\qquad \qquad \square$

5 Locally Finite Varieties

The next result establishes a connection between the decidability of the word problem and the decidability of the unification problem. In general, the decidability of the equational theory does not imply the decidability of the diophantine theory, even in theories defined by a canonical term rewriting system [Boc87, HH87]. In locally finite varieties however, we can say more.

Definition 11 Let Σ be a signature. A class $\mathcal{V}$ of Σ-algebras is a *variety* iff there exists a set of Σ–equations E such that $\mathcal{V} = Mod_\Sigma(E)$. A variety $\mathcal{V}$ is *locally finite* iff every finitely generated Σ-algebra A in $\mathcal{V}$ is finite.

Theorem 12 (Vazhenin/Rozenblat 83b) *Let $\mathcal{V}$ be a locally finite variety over a finite signature. Then the positive theory of $\mathcal{V}$ is decidable if and only if the equational theory of $\mathcal{V}$ is decidable. If $\mathcal{V}$ is finitely axiomatizable, then both theories are decidable.*

Corollary 13 *Let E be a recursively enumerable set of equations over a finite signature Σ such that the variety $Mod_\Sigma(E)$ is locally finite. The unification problem for E is decidable if the word problem for E is decidable. If E is finite, both problems are decidable.*

Proof: By the Birkhoff's theorem, the word problem for E is decidable iff the equational theory of $Mod_\Sigma(E)$ is decidable. The unification problem for E is decidable iff the diophantine theory of $Mod_\Sigma(E)$ is decidable. If the equational theory is decidable, then by Theorem 12 the positive theory and consequently the diophantine theory of E are decidable. If E is finite, then again by Theorem 12 all three theories are decidable. $\quad \square$

Example 14 The varieties

$$AI = \{(x * y) * z \doteq x * (y * z), x * x \doteq x\}$$

of *idempotent semigroups* and

$$ACI = \{(x * y) * z \doteq x * (y * z), x * y \doteq y * x, x * x \doteq x\}$$

of *semilattices* over the signature $\Sigma = \{*\}$ are locally finite.

Therefore, these varieties have a decidable word and unification problem. By Theorem 12 they even have a decidable positive theory, whereas their elementary theory is undecidable [VR83b].

6 Unification in Groups: The Role of Commutativity

Next we give some results from group theory.

Consider the signature $\Sigma = \{*, .^{-1}, e, a_1, \ldots, a_n\}$, with n additional constant symbols $a_1, \ldots, a_n, n \geq 1$.

Makanin [Mak83] showed that unification is decidable in the Σ—theory

$$\{(x * y) * z \doteq x * (y * z), e * x \doteq x, x^{-1} * x \doteq e\}$$

of *groups*.

Later he proved the decidability of the positive theory and the universal theory of a free group [Mak85]. The decidability of the elementary theory [Tar53] of a free group is still open.

For the decidability of the elementary theory of a variety of groups commutativity plays an important role. It is well-known that the variety of abelian groups has a decidable elementary theory [Szm55]. Zamyatin [Zam78] showed that every variety of groups that contains at least one non-abelian group has an undecidable elementary theory.

What about the unification problem in non-abelian varieties of groups ?

The axiom of commutativity can be weakened in several ways. As an abbreviation we use the *commutator* $[x, y] \stackrel{\text{def}}{=} x^{-1} * y^{-1} * x * y$.

A group G is *abelian* iff $[x, y] = e$, for all $x, y \in G$.

A group G is *nilpotent of class 2* , iff $[[x, y], z] = e$, for all $x, y, z \in G$, and *metabelian* iff $[[x, y], [u, v]] = e$ for all $x, y, u, v \in G$.

Obviously, an abelian group is nilpotent of class 2, and a nilpotent group of class 2 is metabelian. The next results follow from [Rom79, Rep85].

Theorem 15 *1. In the Σ—theory of abelian groups the unification problem is decidable.*

2. In the Σ—theory of nilpotent groups of class 2 the unification problem for equations in one unknown is decidable.

3. In the Σ—theory of metabelian groups the unification problem is undecidable.

For nilpotent groups of class $c \geq 10^{20}$ the solvability of equations in one unknown is undecidable [Rep85]. For nilpotent groups of class 5 the solvability of equations in more than one unknown is undecidable [Rep88].

7 General Equational Problems

In [BSS89] H. J. Bürckert and M. Schmidt-Schauß formulated the following problem:

> Are arbitrary equational problems decidable, if E-unification with free constants is decidable ?

According to their definition, an *arbitrary equational problem* corresponds to a closed formula of first-order predicate logic with equality. The equational problem is called *solvable* iff this formula is valid in the free algebra $T_E(X)$. This means that arbitrary equational problems are decidable iff the elementary theory of $T_E(X)$ is decidable. Now we can give the following counterexample.

Theorem 16 *In the theory of* monoids

$$A1 = \{x \circ (y \circ z) \doteq (x \circ y) \circ z, 1 \circ x \doteq x \circ 1 \doteq x\}$$

over the signature $\Sigma = \{\circ, 1\}$ *the unification problem with free constants is decidable, whereas arbitrary equational problems are undecidable.*

Proof: By Proposition 10, $A1$-unification with free constants is decidable if the positive $\forall\exists$-theory of $Mod_\Sigma(A1)$, which by Proposition 3 is the same as the positive theory of the free monoid $T_{A1}(X)$, is decidable. Now the result is an immediate consequence of the following theorems of Quine [Qui46] and Vazhenin/Rozenblat [VR83a]. □

Theorem 17 (Quine 46) *The elementary theory of the free monoid* $T_{A1}(X)$ *over the signature* $\Sigma = \{\circ, 1\}$ *is undecidable.*

Theorem 18 (Vazhenin/Rozenblat 83a) *The positive theory of the free monoid* $T_{A1}(X)$ *over the signature* $\Sigma = \{\circ, 1\}$ *is decidable.*

The theorem of Vazhenin and Rozenblat is only true if we assume that $T_{A1}(X)$ is the free monoid with countably infinitely many generators. For a finitely generated free monoid with $n \geq 2$ generators the positive theory [Dur74] and even the positive $\forall\exists$-theory [Mar82] are undecidable. The proof of the result of Vazhenin/Rozenblat is based on Makanin's algorithm [Mak77] for deciding the solvability of equations over free semigroups.

References

[Boc87] A. Bockmayr. A note on a canonical theory with undecidable unification and matching problem. *Journal of Automated Reasoning*, 3:379 –381, 1987.

[Boc89] A. Bockmayr. On the decidability of the unification problem. In *UNIF 89, Extended Abstracts of the 3rd International Workshop on Unification, Lambrecht (Pfalz)*, SEKI SR-89-17. Univ. Kaiserslautern, 1989.

[BSS89] H. J. Bürckert and M. Schmidt-Schauß. On the solvability of equational problems. Technical Report SEKI SR-89-07, Univ. Kaiserslautern, June 1989.

[Dur74] V. G. Durnev. Positive formulas on free semigroups. *Sib. J. Math.*, 15:796–800, 1974.

[HH87] S. Heilbrunner and S. Hölldobler. The undecidability of the unification and matching problem for canonical theories. *Acta Inform.*, 24:157–171, 1987.

[HO80] G Huet and D. C. Oppen. Equations and rewrite rules, A survey. In R. V. Book, editor, *Formal Language Theory*. Academic Press, 1980.

[JK91] J. P. Jouannaud and C. Kirchner. Solving equations in abstract algebras: A rule-based survey of unification. In J.L. Lassez and G. Plotkin, editors, *Computational Logic: Essays in Honor of A. Robinson*. MIT Press, 1991.

[Mak77] G. S. Makanin. The problem of solvability of equations in a free semigroup. *Math. USSR Sbornik*, 32(2):129–198, 1977.

[Mak83] G. S. Makanin. Equations in a free group. *Math. USSR Izv.*, 21:483 – 546, 1983.

[Mak85] G. S. Makanin. Decidability of the universal and positive theories of a free group. *Math. USSR Izv.*, 25:75 – 88, 1985.

[Mar82] S. S. Marchenkov. Undecidability of the positive ∀∃-theory of a free semigroup (in Russian). *Sibirsk. Mat. Zh.*, 23:196–198, 1982. MR 83e: 03067.

[Qui46] W. V. Quine. Concatenation as a basis for arithmetic. *Journal of Symbolic Logic*, 11:105–114, 1946.

[Rep85] N.N. Repin. The solvability problem for equations in one unknown in nilpotent groups. *Math. USSR Izv.*, 25:601–618, 1985.

[Rep88] N. N. Repin. Some simply presented groups for which an algorithm recognizing solvability of equations is impossible (in Russian). *Vopr. Kibern., Mosk.*, 134:167 – 175, 1988. MR 89i: 20057.

[Rom79] V. A. Roman'kov. Equations in free metabelian groups. *Sib. J. Math.*, 20(1):469–471, 1979.

[Roz85] B. V. Rozenblat. Diophantine theories of free inverse semigroups. *Sib. J. Math.*, 26(6):860–864, 1985.

[Sie89] J. H. Siekmann. Unification theory. *Journal of Symbolic Computation*, 7:207–274, 1989.

[Szm55] W. Szmielew. Elementary properties of abelian groups. *Fund. Math.*, 41:203–271, 1955.

[Tar53] A. Tarski. *Undecidable Theories (in collaboration with A. Mostowski and R. M. Robinson)*. North Holland, 1953.

[Vaz74] Y. M. Vazhenin. On the elementary theory of free inverse semigroups. *Semigroup Forum*, 9:189–195, 1974.

[VR83a] Y. M. Vazhenin and B. V. Rozenblat. Decidability of the positive theory of a free countably generated semigroup. *Math. USSR Sbornik*, 44:109–116, 1983.

[VR83b] Y. M. Vazhenin and B. V. Rozenblat. On positive theories of free algebraic systems. *Sov. Math. (Izv. VUZ)*, 27(3):88–91, 1983.

[Zam78] A. P. Zamyatin. A non-abelian variety of groups has an undecidable elementary theory. *Algebra and Logic*, 17:13–17, 1978.

Model-Theoretic Aspects of Unification

Alexander Bockmayr
Max-Planck-Institut für Informatik
Im Stadtwald
D-6600 Saarbrücken
e-mail: bockmayr@cs.uni-sb.de

Unification is a fundamental operation in various areas of computer science, in particular in automated theorem proving and logic programming. In this paper we establish a relation between unification theory and classical model theory. We show how model-theoretic methods can be used to investigate a generalized form of unification, namely the problem whether, given an equational theory E and a system of equations S, there is an extension of the free algebra in E in which S is solvable.

1 Introduction

Unification is a fundamental operation in various areas of computer science, in particular in automated theorem proving and logic programming [Sie89, Kni89, JK91]. From an algebraic point of view unification can be seen as equation solving in the initial or free algebra of a given equational theory. In this paper we establish a relation between unification theory and classical model theory. We study a generalized form of unification. Inspired from work in classical mathematics we consider the following question: Given an arbitrary first-order theory T with some (standard) model A and a system of equations S, is there an extension B of A in $Mod(T)$ such that S is solvable in B ? We call this problem the solvability problem in T over A.

In the first part of the paper we give a necessary and sufficient condition for the solvability of a system of equations in an extension of the initial algebra of an equational theory and present a Knuth-Bendix procedure to check this property. In the second part we study the decidability of the solvability problem in T over A by model-theoretic methods. The main idea is to consider existentially closed extensions of the model A. A system of equations S is solvable in an extension of A if and only if it is solvable in an existentially closed extension. This leads us to consider the theory of existentially closed structures in T. We prove that the solvability problem is decidable for theories with a decidable model completion or with a decidable complete model companion. Necessary and sufficient conditions for the existence of decidable model companions have been given in [Bur89].

A typical example of a theory with decidable model companion is the theory of ordered fields. It is still an open problem whether the solvability of diophantine equations in the rational numbers is decidable. For the integer numbers this is Hilbert's Tenth

Problem. However, the solvability of diophantine equations in the real numbers is decidable, becauce the theory of real closed fields, which is a complete model companion of the theory of ordered fields, is decidable.

We believe that model-theoretic notions may give new insights into the difficulty of unification problems. For example we may classify theories according to the property whether they have a model completion (resp. model companion) or not.

Furthermore we get a sufficient condition for the unsolvability of unification problems: a system of equations is not solvable in an algebra A if it is not solvable in some extension B of A. Since unification is semi-decidable, but in general not decidable, we cannot always detect that a system of equations has no solution. In such cases decidable sufficient conditions for the unsolvability are particularly interesting.

2 Preliminaries

We recall briefly some basic notions that are needed in the sequel. More details can be found in [HO80, Kei77].

We will work in the framework of many-sorted first-order predicate logic with equality.

$\Sigma = (S, F)$ denotes a *signature* with a set S of sort symbols and a set F of function symbols together with an arity function. A sort $s \in S$ is *strict* iff there exists $f \in F$ such that either $f :\to s$ or $f : s_1 \times \ldots \times s_n \to s$ and s_i is strict for $i = 1, \ldots, n$. A signature Σ is *strict* iff each $s \in S$ is strict.

A Σ-*algebra* A consists of a family of non-empty sets $(A_s)_{s \in S}$ and a family of functions $(f^A)_{f \in F}$ such that if $f : s_1 \times \ldots \times s_n \to s$ then $f^A : A_{s_1} \times \ldots \times A_{s_n} \to A_s$. A Σ-*extension* of A is a Σ-algebra B such that $A_s \subseteq B_s$, for all $s \in S$, and $f^B \mid A = f^A$ for all $f \in F$.

X represents a family $(X_s)_{s \in S}$ of countably infinite sets X_s of *variables* of sort s. $T(F, X)$ is the Σ-algebra of *terms* with variables over Σ. If Σ is strict, then $T(F) = (T(F)_s)_{s \in S}$ is the Σ-algebra of *ground terms* over Σ.

For a term $t \in T(F, X)$, $Var(t)$ and $Occ(t)$ denote the set of *variables* and *occurrences* in t respectively. t/ω is the subterm of t at position $\omega \in Occ(t)$ and $t[\omega \leftarrow s]$ the term obtained from t by replacing the subterm t/ω with the term $s \in T(F, X)$.

A *substitution* is a mapping $\sigma : X \to T(F, X)$ which is different from the identity only for a finite subset $D(s)$ of X. We do not distinguish σ from its canonical extension to $T(F, X)$.

A binary relation $\to$ on a Σ-algebra A is Σ-*compatible* iff $t_1 \to u_1, \ldots, t_n \to u_n$ implies $f^A(t_1, \ldots, t_n) \to f^A(u_1, \ldots, u_n)$ for all $t_i, u_i \in A_{s_i}$ and all $f : s_1 \times \ldots \times s_n \to s$ in F. A *congruence* is a Σ-compatible equivalence relation.

A *system of equations* S is an expression of the form $t_1 \doteq u_1 \wedge \ldots \wedge t_n \doteq u_n$ where $n \geq 0$ and t_i, u_i are terms of $T(F, X)$ belonging pairwise to the same sorts. A *conditional equation* is an expression of the form $t_1 \doteq u_1 \wedge \ldots \wedge t_n \doteq u_n \Rightarrow l \doteq r$, with a system of equations $t_1 \doteq u_1 \wedge \ldots \wedge t_n \doteq u_n$ and an equation $l \doteq r$.

Let E be a set of conditional equations. The *conditional equational theory* $\equiv_E$ associated with E is the smallest congruence $\equiv$ on $T(F, X)$ with the property that $\sigma(t_1) \equiv \sigma(u_1) \wedge \ldots \wedge \sigma(t_n) \equiv \sigma(u_n)$ implies $\sigma(l) \equiv \sigma(r)$ for any conditional equation $t_1 \doteq u_1 \wedge \ldots \wedge t_n \doteq u_n \Rightarrow l \doteq r$ in E and any substitution $\sigma : X \to T(F, X)$.

The Σ-algebra $T_E(X) \stackrel{\text{def}}{=} T(F, X)/ \equiv_E)$ is the *free algebra* in E. If Σ is strict, then $T_E \stackrel{\text{def}}{=} T(F)/ \equiv_E$ is the *initial algebra* in E.

A system of equations $S : t_1 \doteq u_1 \wedge \ldots \wedge t_n \doteq u_n$ is *E-unifiable* iff there is a substitution $\sigma : X \to T(F, X)$ with $\sigma(t_i) \equiv_E \sigma(u_i)$ for all $i = 1, \ldots, n$.

A *term rewrite system* R is a set of directed equations $l \to r$ such that $Var(r) \subseteq Var(l)$. The *reduction relation* $\to_R$ associated with R is defined by $s \to_R t$ iff there is an occurrence $\omega \in Occ(s)$ and a rule $l \to r$ in R such that there exists a substitution $\sigma : X \to T(F, X)$ with $\sigma(l) = s/\omega$ and $t = s[\omega \leftarrow \sigma(r)]$. $\to_R^*$ denotes the reflexive-transitive closure of $\to_R$. The rewrite system R is *confluent* iff $s \to_R^* t_1$ and $s \to_R^* t_2$ implies the existence of a term u such that $t_1 \to_R^* u$ and $t_2 \to_R^* u$. R is *noetherian* iff there is no infinite sequence $t_1 \to_R t_2 \to_R \ldots \to_R t_n \to_R \ldots$. R is *canonical* iff it is confluent and noetherian. If R is canonical, then the *R-normal form* $NF_R(t)$ of a term t is the uniquely determined term t' such that $t \to_R^* t'$ and t' is irreducible.

A *reduction ordering* is a well-founded Σ-compatible ordering $>$ on $T(F, X)$ such that $s > t$ implies $\sigma(s) > \sigma(t)$ for all terms $s, t \in T(F, X)$ and all substitutions $\sigma : X \to T(F, X)$.

A *Σ-theory* T is a consistent set of closed formulas of first-order predicate logic with equality over Σ. By $Mod(T)$ we denote the model class of T. Two Σ-theories T_1, T_2 are *equivalent* iff $Mod(T_1) = Mod(T_2)$. If $\mathcal{K}$ is a class of Σ-algebras, then $Th(\mathcal{K})$ ist the set of all closed first-order Σ-formulas that are valid in all members of $\mathcal{K}$. The class $\mathcal{K}$ is *elementary* iff there is a Σ-theory T such that $\mathcal{K} = Mod(T)$. A Σ-theory T is *complete* iff there is a Σ-algebra A such that T is equivalent to $Th(A)$. Two Σ-algebras A_1, A_2 are *elementarily equivalent* iff $Th(A_1) = Th(A_2)$. A Σ-theory T over a recursive signature Σ is *decidable* iff $Th(Mod(T))$ is decidable.

For a first-order formula φ over Σ we denote by $FV(\varphi)$ the set of free variables in φ. If $FV(\varphi) = \{x_1, \ldots, x_n\}$ then $\underline{\forall}\varphi \stackrel{def}{=} \forall x_1 \ldots \forall x_n \varphi$ is the *universal closure* and $\underline{\exists}\varphi \stackrel{def}{=} \exists x_1 \ldots \exists x_n \varphi$ the *existential closure* of φ. A *universal formula* has the form $\forall x_1 \ldots \forall x_n \varphi$, where φ contains no quantifiers. A *universal theory* is a theory which is equivalent to a set of closed universal formulas. For a theory T we denote by $T_\forall$ the set of all closed universal formulas valid in T.

A mapping $h : A \to B$ of a Σ-algebra A in a Σ-algebra B is an *embedding* iff for all equations $t_1 \doteq t_2$ over Σ and all variable assignments $\mu : X \to A$ we have $A \models (t_1 \doteq t_2)[\mu]$ iff $B \models (t_1 \doteq t_2)[h \circ \mu]$.

3 Unification and Equation Solving

Our starting point is the following easy lemma stating that E-unification can be seen as equation solving in the free algebra $T_E(X)$ of E.

Lemma 1 *Let E be a set of conditional equations over the signature Σ. A system of Σ-equations S*

$$s_1 \doteq t_1 \wedge \ldots \wedge s_n \doteq t_n, n \geq 1,$$

is E-unifiable iff S is solvable in the free algebra $T_E(X)$, i.e. iff there is a variable assignment $\rho : X \to T_E(X)$ such that $\rho(s_i) = \rho(t_i)$, for $i = 1, \ldots, n$. If Σ is strict, this is equivalent to the condition that S is solvable in the initial algebra T_E, i.e. that there is a variable assignment $\gamma : X \to T_E$ such that $\gamma(s_i) = \gamma(t_i)$, for $i = 1, \ldots, n$.

Generalizing the notion of equation solving *in* an algebra A we now introduce the notion of equation solving *over* A.

Definition 2 Let Σ be a signature and A a Σ-algebra. A system of Σ-equations S

$$s_1 \doteq t_1 \wedge \ldots \wedge s_n \doteq t_n, n \geq 1,$$

is *solvable in* A iff there is a variable assignment $\alpha : X \to A$ such that $\alpha(s_i) = \alpha(t_i)$, for all $i = 1, \ldots, n$. If T is a Σ-theory and A a model of T, then S is *solvable in* T *over* A iff there exists an extension B of A in $Mod(T)$ such that S is solvable in B.

Now let E be a set of conditional equations over a strict signature Σ and let S

$$s_1 \doteq t_1 \wedge \ldots \wedge s_n \doteq t_n, n \geq 1,$$

be a system of equations. We want to give a criterion that S is solvable in E over the initial algebra T_E. For this purpose we substitute for the variables $x_1, \ldots, x_m$ in S new constant symbols $c_1, \ldots, c_m$ and obtain a system of ground equations S'

$$s'_1 \doteq t'_1 \wedge \ldots \wedge s'_n \doteq t'_n$$

over the signature Σ' obtained from Σ by adding the constant symbols $c_1, \ldots, c_m$. Now we consider the set E^S of conditional equations over Σ' defined by

$$E^S \stackrel{\text{def}}{=} E \cup \{s'_1 \doteq t'_1, \ldots, s'_n \doteq t'_n\}.$$

Theorem 3 *Let E be a set of conditional equations over a strict signature Σ. A system of equations S is solvable in E over the initial algebra T_E iff for all ground terms s and t over Σ we have:*

$$s \equiv_{E^S} t \text{ implies } s \equiv_E t.$$

Proof: First we repeat the above construction in a more formal way. Let $\Sigma = (Sorts, F)$ be the given signature and $Var(S) = \{x_1, \ldots, x_m\}$ the set of variables in the system of equations S. Let $C = \{c_1, \ldots, c_m\}$ be a set of new constant symbols, $C \cap F = \emptyset$, such that x_i and c_i are of the same sort, for all $i = 1, \ldots, m$. We define a substitution $\theta : Var(S) \to T(F \cup C, X)$ by $\theta(x_i) \stackrel{\text{def}}{=} c_i$, for $i = 1, \ldots, m$. Then $s'_j \stackrel{\text{def}}{=} \theta(s_j)$ and $t'_j \stackrel{\text{def}}{=} \theta(t_j)$, for $j = 1, \ldots, n$. The signature Σ' is given by $\Sigma' \stackrel{\text{def}}{=} (Sorts, F \cup C)$. Now we turn to the proof of "$\Longrightarrow$":

Let A be an extension of T_E in $Mod(E)$ in which S is solvable and let $\eta : Var(S) \to A$ be a solution, i. e. $A \models s_1 \doteq t_1 \wedge \ldots \wedge s_n \doteq t_n[\eta]$. Let $s, t \in T(F)$ with $s \equiv_{E^S} t$. Then there exist terms $u_0, \ldots, u_k \in T(F \cup C)$ with

$$s = u_0 \equiv_E u_1 \equiv_{S'} u_2 \equiv_E \ldots \equiv_{S'} u_{k-1} \equiv_E u_k = t,$$

where $\equiv_{S'}$ denotes the congruence on $T(F \cup C)$ generated by $\{s'_1 \doteq t'_1, \ldots, s'_n \doteq t'_n\}$.

By interpreting the constant symbols c_i by $\eta(x_i)$, for $i = 1, \ldots, m$, A can be seen as a Σ'-algebra. Then $A \models s'_1 \doteq t'_1 \wedge \ldots \wedge s'_n \doteq t'_n$.

For any terms $u, v \in T(F \cup C)$ with $u \equiv_{S'} v$ we get $A \models u \doteq v$. Since A is a model of E we know that for any terms $u', v' \in T(F \cup C)$ with $u' \equiv_E v'$ we must have $A \models u' \doteq v'$. Together this yields

$$A \models (s \doteq u_0 \doteq u_1 \doteq u_2 \doteq \ldots \doteq u_{k-1} \doteq u_k \doteq t).$$

But since $s, t \in T(F)$ and A is an extension of T_E, it follows that $T_E \models s \doteq t$ or $s \equiv_E t$.

To prove "$\Longleftarrow$" we suppose that $s \equiv_{E^S} t$ implies $s \equiv_E t$ for all $s, t \in T(F)$. We define the Σ-algebra A by

$$A \stackrel{\text{def}}{=} T(F \cup C)/ \equiv_{E^S} .$$

Then, by hypothesis, A is an extension of T_E. Since $E \subseteq E^S$, A is a model of E. The system of equations S is solvable in A, because

$$A \models s_1' \doteq t_1' \wedge \ldots \wedge s_n' \doteq t_n',$$

and consequently the variable assignment $\alpha : Var(S) \to A, \alpha(x_i) \stackrel{\text{def}}{=} c_i^A$, for $i = 1, \ldots, m$ is a solution. $\square$

In universal algebra similar theorems have been proved for the first time in [Dör51, Dör67].

Example 4 In the equational theory

$$E = \{0 + x \doteq x, s(x) + y \doteq s(x + y)\}$$

over the signature $\Sigma = \{0, s, +\}$ the equation

$$s(0) + x \doteq x, \quad (*)$$

is solvable in E over the initial algebra T_E. In order to obtain an extension of T_E where $(*)$ is solvable, we take the Σ-algebra A with the set of natural numbers $\mathcal{N} \cup \{\infty\}$ as carrier set and the usual operations, where $s(\infty) \stackrel{\text{def}}{=} \infty$ and $\infty + x = x + \infty \stackrel{\text{def}}{=} \infty$.

However, if we consider the equational theory

$$E' = \{ \quad s(p(x)) \doteq x, \qquad p(s(x)) \doteq x,$$
$$0 + x \doteq x, \qquad s(x) + y \doteq s(x + y), \quad p(x) + y \doteq p(x + y),$$
$$-0 \doteq 0, \qquad -s(x) \doteq p(-x), \qquad -p(x) \doteq s(-x),$$
$$(x + y) + z \doteq x + (y + z), \quad x + (-x) \doteq 0\}$$

over the signature $\Sigma' = \{0, s, p, +, -\}$, then the equation $(*)$ is not solvable in E' over the initial algebra $T_{E'}$. For if A were an extension of $T_{E'}$ in $Mod(E')$, where a solution $x \leftarrow c$ of $(*)$ exists, we would get the contradiction

$$0 = c + (-c) = (s(0) + c) + (-c) = s(0) + (c + (-c)) = s(0) + 0 = s(0).$$

Definition 5 Let T be a Σ-theory over a recursive signature Σ. For a model A of T the *solvability problem in T over A* is to decide for arbitrary systems of equations S over Σ, whether S is solvable in T over A.

In general, the solvability problem is undecidable. Consider for example a set of conditional equations E over a strict recursive signature with an undecidable ground word problem. Then the solvability problem in E over the initial algebra T_E is undecidable.

4 Solvability and Knuth-Bendix-Completion

Let R^0 be a finite canonical term rewriting system over the strict signature $\Sigma = (Sorts, F)$ and let S

$$s_1 \doteq t_1 \wedge \ldots \wedge s_n \doteq t_n, n \geq 1,$$

be a system of equations over Σ.

We want to decide, whether S is solvable in R^0 over the initial algebra T_{R^0}. As in the preceding theorem we substitute for all variables in S new constant symbols and consider the set of equations

$$S^0 = \{s_1' \doteq t_1', \ldots, s_n' \doteq t_n'\}$$

over the extended signature $\Sigma' = (Sorts, F \cup C)$.

Definition 6 A term $t \in T(F \cup C, X)$ is called *basic* iff $t \in T(F, X)$. A basic term $t \in T(F, X)$ is called *inductively reducible* by R^0 iff for any ground substitution $\gamma : X \rightarrow T(F)$ the term $\gamma(t)$ is reducible by R^0.

Inductive reducibility is decidable for finite term rewrite systems [Pla85].

The solvability of the system of equations S in R^0 over the initial algebra T_{R^0} can be checked by a Knuth-Bendix completion procedure which is similar to the inductive completion procedure of [JK89] (Figure 1). For the technical details, which are not interesting here, we refer to [Hue81]. The symbol $>$ denotes a reduction ordering on $T(F \cup C, X)$.

Theorem 7 *Let R^0 be a finite canonical term rewriting system over the strict signature Σ and let S be a system of equations. If the procedure Solvability($R^0, S^0, >$) does not fail ("Don't know"), then:*

- *If S is not solvable in R^0 over T_{R^0}, the procedure yields the answer "Unsolvable".*

- *If S is solvable in R^0 over T_{R^0}, the procedure yields the answer "Solvable" or it doesn't terminate.*

Proof: First suppose that the procedure yields the answer "Unsolvable". Then there exists a rule $l \rightarrow r$, which was generated from a pair in S, such that l and r are basic, but l is not inductively reducible by R^0. Consequently, there exists a ground substitution $\sigma : X \rightarrow T(F)$, such that the term $\sigma(l)$ is irreducible by R^0 and moreover

$$s \stackrel{\text{def}}{=} NF_{R^0}(\sigma(l)) = \sigma(l) \rightarrow_{\{l \rightarrow r\}} \sigma(r) \rightarrow^*_{R^0} NF_{R^0}(\sigma(r)) \stackrel{\text{def}}{=} t.$$

s and t are basic, because l and r are basic and because σ introduces only basic terms. Since $l > r$ and $>$ is a reduction ordering, it follows that $s > t$ and in particular $s \neq t$. Since s and t are irreducible by R^0, it is not possible that $s \equiv_{R^0} t$. However, by the correctness of the Knuth-Bendix-Procedure we have $s \equiv_{R^0 \cup S^0} t$. By Theorem 3 we deduce that S is not solvable in R^0 over T_{R^0}.

If the procedure does not answer "Unsolvable", then it gives either the answer "Solvable" or it does not terminate.

Let R^∞ denote the possibly infinite rewrite system that is generated. By the completeness of the Knuth-Bendix-Procedure [Hue81] we know that R^∞ is canonical and that it defines the same equational theory as $R^0 \cup S^0$.

INITIAL CALL: *Solvability(R^0, S^0, >)*, where all rules in R^0 are marked
PROCEDURE *Solvability(R, P, >)*
    CASE $P \neq \emptyset$ **THEN**
        Choose a pair (p,q) in P and remove it
        Compute the R-normal forms $p{\downarrow}$ and $q{\downarrow}$ of p and q
        CASE $p{\downarrow} = q{\downarrow}$ **THEN** *Solvability(R, P, >)*
                $p{\downarrow} > q{\downarrow}$ **THEN** $l := p{\downarrow}$ and $r := q{\downarrow}$
                $q{\downarrow} > p{\downarrow}$ **THEN** $l := q{\downarrow}$ and $r := p{\downarrow}$
                ELSE STOP and **RETURN** "Don't know"
        ENDCASE
        IF l and r are basic and l is not inductively reducible
        by R^0
            THEN STOP and **RETURN** "Unsolvable"
        ENDIF
        IF l is basic and r is not basic
            THEN STOP and **RETURN** "Don't know"
        ENDIF
        $(R, P) := Simplify(R, P, l \rightarrow r)$
        Solvability($R \cup \{l \rightarrow r\}$, P, >)
    All rules in R are marked THEN
        STOP and **RETURN** "Solvable"
    ELSE fairly choose an unmarked rule $l \rightarrow r$ in R
        $(R, P) := Critical\text{-}Pairs(l \rightarrow r, R, P)$
        Mark rule $l \rightarrow r$
        Solvability(R, P, >)
  ENDCASE
ENDPROCEDURE

Figure 1: Knuth-Bendix Procedure for Testing Solvability

Let $s, t \in T(F)$ be two basic ground terms such that $s \equiv_{R^0 \cup S^0} t$. By the canonicity of R^∞ it follows that $NF_{R^\infty}(s) = NF_{R^\infty}(t)$.

We want to show that $s \equiv_{R^0} t$ or equivalently $NF_{R^0}(s) = NF_{R^0}(t)$.

If $u \in T(F)$ is a basic ground term which is irreducible by R^0 then u is also irreducible by R^∞. Otherwise there would be a rule $l \rightarrow r$ in R^∞ that reduces u. $l \rightarrow r$ cannot belong to R^0, because this would imply that u is reducible by R^0. So the rule $l \rightarrow r$ must be generated during the completion process. Then l is inductively reducible by R^0 and the ground term u, which contains a ground instance of l, is reducible by R^0 in contradiction to our assumption.

We conclude that $NF_{R^0}(u) = NF_{R^\infty}(u)$ for all basic ground terms $u \in T(F)$.

In particular, we get $NF_{R^0}(s) = NF_{R^\infty}(s) = NF_{R^\infty}(t) = NF_{R^0}(t)$ and again by Theorem 3, S is solvable in R^0 over T_{R^0}. $\qquad\square$

Example 8 Consider a canonical term rewriting system for integer arithmetic

$$
\begin{aligned}
Int = \{ \quad & s(p(x)) \rightarrow x, & p(s(x)) \rightarrow x, \\
& 0 + x \rightarrow x, & s(x) + y \rightarrow s(x+y), & p(x) + y \rightarrow p(x+y), \\
& x + 0 \rightarrow x, & x + s(y) \rightarrow s(x+y), & x + p(y) \rightarrow p(x+y), \\
& -0 \rightarrow 0, & -s(x) \rightarrow p(-x), & -p(x) \rightarrow s(-x), \\
& 0 * x \rightarrow 0, & s(x) * y \rightarrow y + (x*y), & p(x) * y \rightarrow (-y) + (x*y), \\
& x * 0 \rightarrow 0, & x * s(y) \rightarrow (x*y) + x, & x * p(y) \rightarrow (x*y) + (-x), \\
& -(-x) \rightarrow x, & (-x) + x \rightarrow 0, & x + (-x) \rightarrow 0, \\
& x + ((-x) + z) \rightarrow z, & (-x) + (x+z) \rightarrow z, & -(x+y) \rightarrow (-y) + (-x), \\
& (x+y) + z \rightarrow x + (y+z) \}.
\end{aligned}
$$

Suppose that we want to check the solvability of the equation $s(0) + x \doteq x$.

After the initial call $Solvability(Int, \{s(c) \doteq c\}, >)$ the completion procedure first will orient the equation $s(c) \doteq c$ into a rule $s(c) \rightarrow c$ and superpose it on the left-hand side of the rule $s(x) + y \rightarrow s(x+y)$ from Int. This will lead to a critical pair $s(c+y) \doteq c+y$. Superposing the rule $x + (-x) \rightarrow 0$ from Int on the left-hand side of the corresponding rule $s(c \dotplus y) \rightarrow c + y$ will yield another critical pair $s(0) \doteq 0$. The terms $s(0)$ and 0 are both basic and $s(0)$ is not inductively reducible by Int. This shows that the original equation $s(0) + x \doteq x$ is unsolvable.

5 Existentially Closed Structures

Now we turn to the decidability of the solvability problem. Our aim is to give model-theoretic conditions that for a set of conditional equations E the solvability problem in E over the free algebra $T_E(X)$ is decidable. In this section, we will introduce the necessary model-theoretic background.

In order to decide whether a system of equations S is solvable over a given Σ-algebra A we have to study the extensions of A. The basic tool for that is the method of diagrams due to [Rob51].

Definition 9 Let $\Sigma = (S, F)$ be a signature and A a Σ-algebra. The *diagram signature* of A is the expansion $\Sigma_A = (S, F_A)$ of Σ formed by adding a new constant symbol c_a for each element $a \in A$. The *diagram expansion* of A is the Σ_A-algebra A_A with the same carrier sets and operations as A and such that each c_a is interpreted by a. The *diagram* of A is the set $Diag(A)$ of all ground and negated ground equations of Σ_A that are true in A_A.

The diagram can be used to axiomatize the extensions of A as is shown by the next lemma.

Lemma 10 A Σ-algebra A can be embedded into a Σ-algebra B iff B can be expanded to a Σ_A-algebra B_A which is a model of $Diag(A)$.

Proof: "$\Longrightarrow$": Let $h : A \rightarrow B$ be an embedding. Then for all Σ-equations $t_1 \doteq t_2$ and all variable assignments $\mu : X \rightarrow A$ we have $A \models (t_1 \doteq t_2)[\mu]$ iff $B \models (t_1 \doteq t_2)[h \circ \mu]$. We expand B to a Σ_A-algebra B_A by interpreting the constant symbol c_a by $h(a)$, for each $a \in A$. Let $u_1 \doteq u_2$ be a ground equation over the diagram signature Σ_A which contains the constant symbols $\{c_{a_1}, \ldots, c_{a_m}\}$ from F_A. By replacing the constants c_{a_i} with new variables x_i and by considering the variable assignment $\mu : X \to A$ that maps x_i to a_i, for $i = 1, \ldots, n$, we see that $A_A \models (u_1 \doteq u_2)$ iff $B_A \models (u_1 \doteq u_2)$. This implies that B_A is a model of $Diag(A)$.

"$\Longleftarrow$": Let the Σ_A-algebra B_A be an expansion of B which is a model of $Diag(A)$. Define a mapping $h : A \to B$ by $h(a) \stackrel{\text{def}}{=} c_a^{B_A}$, for all $a \in A$. Let $t_1 \doteq t_2$ be an equation over Σ_A and $\mu : X \to A$ a variable assignment such that $A \models (t_1 \doteq t_2)[\mu]$. Consider the ground substitution $\sigma : X \to T(F_A), \sigma(x) \stackrel{\text{def}}{=} c_{\mu(x)}$. Then $A \models (t_1 \doteq t_2)[\mu]$ iff $A_A \models (\sigma(t_1) \doteq \sigma(t_2))$ iff $B_A \models (\sigma(t_1) \doteq \sigma(t_2))$ iff $B \models (t_1 \doteq t_2)[h \circ \mu]$. This implies that h is an embedding of A in B. $\qquad \Box$

When we consider the theory of the extensions of a given Σ-algebra A in a theory T one of the most important concepts is the notion of model completeness [Rob51]. For an excellent survey on this subject see [Mac77].

Definition 11 A Σ-theory T is called *model complete* iff for each model A of T the Σ_A-theory $T \cup Diag(A)$ is complete.

The models of $T \cup Diag(A)$ are those models of T that are extensions of A. If $T \cup Diag(A)$ is complete, this means that any two extensions of A in $Mod(T)$ are elementarily equivalent with respect to Σ_A.

Definition 12 A Σ-algebra A is called *existentially closed* in a Σ-extension B iff any closed existential Σ_A-formula φ that is true in B is also true in A.

A close relationship between model completeness and existentially closed structures is indicated by Robinson's Test.

Theorem 13 (Robinson's Test) *For any theory T the following are equivalent:*

- *T is model complete,*
- *for any two models A, B of T such that $A \subseteq B$ we have that A is existentially closed in B,*
- *for any Σ-formula φ there is a universal Σ-formula ρ with $FV(\rho) \subseteq FV(\varphi)$, such that $T \models (\varphi \Leftrightarrow \rho)$.*

The aim of the next definition is to generalize the notion of algebraic closure known from field theory to arbitrary theories.

Definition 14 Let T and T^* be two Σ-theories. T^* is called a *model companion* of T iff

- T and T^* are *mutually model consistent*, i.e. every model of T is embeddable in a model of T^* and vice versa,
- T^* is *model complete*, i.e. $Diag(A^*) \cup T^*$ is complete for any model A^* of T^*.

T^* is called a *model completion* of T iff T^* is a model companion of T and moreover

- $Diag(A) \cup T^*$ is complete for any model A of T.

Not every theory T has a model companion. However, if a theory T has a model companion T^*, then T^* is unique up to equivalence [Rob63]. The classical example of a model completion is the theory of algebraically closed fields.

Example 15 Let T be the theory of fields over the signature $\Sigma = \{0, 1, +, *, -\}$ with the axiomatization

$$
\begin{aligned}
K_0: & & 0 \neq 1, \\
K_1: & \forall x, y, z & x + (y + z) \doteq (x + y) + z, \\
K_2: & \forall x & x + 0 \doteq x, \\
K_3: & \forall x & x + (-x) \doteq 0 \\
K_4: & \forall x, y, z & x * (y * z) \doteq (x * y) * z, \\
K_5: & \forall x & x * 1 \doteq x, \\
K_6: & \forall x \exists y & (x \doteq 0 \vee x * y \doteq 1), \\
K_7: & \forall x, y & x * y \doteq y * x, \\
K_8: & \forall x, y, z & (x + y) * z \doteq x * z + y * z.
\end{aligned}
$$

A model completion T^* of T is obtained by adding the axioms

$$AK_n : \forall x_0, \ldots, x_n \exists y : y^{n+1} + x_n * y^n + \ldots + x_0 \doteq 0, \text{ for } n \geq 1.$$

T^* is the theory of *algebraically closed fields*. The fact that T^* is a model completion of T means that any two algebraically closed extensions of a field K are elementarily equivalent in Σ_K. The theory T^* of algebraically closed fields is not complete. However, if we fix the characteristic of the field, that is if we add either an axiom

$$C_p : p * 1 \doteq 0, \text{ for some prime number } p$$

or the set of axioms

$$\{\neg C_q : q \text{ a prime number }\},$$

we get the theory of algebraically closed fields with a fixed characteristic, which is complete.

We have already defined, when a Σ-algebra A is existentially closed in a Σ-algebra B. Now we explain when A is existentially closed with respect to a theory T.

Definition 16 Let T be a Σ-theory. A Σ-algebra A is called *T-existentially closed* iff

- A is embeddable in a model B of T and
- A is existentially closed in any extension B of A in $Mod(T)$

$E(T)$ denotes the class of T-existentially closed structures and T^e denotes the theory of $E(T)$.

For any model A of $T_\forall$ there is a T-existentially closed structure B with $A \subseteq B$. Therefore, the theories T and T^e are mutually model consistent for any Σ-theory T. Thus T^e is a good candidate for a model companion of T. In fact, if T has a model companion T^*, then T^* is equivalent to the theory T^e of T-existentially closed structures. The point is that the class of T-existentially closed structures may not be elementary, that is in general it cannot be axiomatized by a set of first-order formulas. T has a model companion if and only if this is possible.

Theorem 17 (Eklof/Sabbagh 1971) *A theory T has a model companion T^* iff the class $E(T)$ of T-existentially closed structures is elementary. In that case*

$$T^* = Th(E(T)) \text{ and } E(T) = Mod(T^*).$$

We obtain the following classification of theories T with respect to the T-existentially closed structures.

1. The class of T-existentially closed structures is not elementary, T has no model companion. In this case the theory T^e of T-existentially closed structures has some models that are not T-existentially closed.

2. The class of T-existentially closed structures is elementary, T has a model companion T^*.

   (a) T^* is not a model completion of T, that is there exists a model A of T with two T-existentially closed extensions that are not elementarily equivalent with respect to Σ_A.

   (b) T^* is a model completion of T, that is for any model A of T the T-existentially closed extensions are elementarily equivalent with respect to Σ_A. In some sense A then has a unique existential closure.

Example 18 Here are some examples of theories with and without model-companion [HW75, Che76].

1. Theories with model completion: Field, ordered fields, ordered sets, Boolean algebras, Abelian groups [ES71], R-modules over a coherent ring R [ES71]

2. Theories with model companion, but without model completion: Formally real fields, commutative rings without nilpotent elements, trees [Par83]

3. Theories without model companion: Groups [Mac72], commutative rings [Che73], R-modules over a non-coherent ring R [ES71], lattices [Sch83], division rings, Peano arithmetic, number theory

6 Existentially Closed Structures and the Solvability Problem

After having exposed the model-theoretic background, we now investigate the decidability of the solvability problem.

Proposition 19 *Let T be a Σ-theory with model completion T^* and let A be a model of T. A system of equations S*

$$s_1 \doteq t_1 \wedge \ldots \wedge s_n \doteq t_n, n \geq 1,$$

with $s_i, t_i \in T(F, X), i = 1, \ldots, n$, is solvable in T over A iff

$$T^* \cup Diag(A) \models \exists (s_1 \doteq t_1 \wedge \ldots \wedge s_n \doteq t_n).$$

Proof: S is solvable in T over A iff there is an extension A' of A in $Mod(T)$ such that $A' \models \exists S$. Since T and T^* are mutually model consistent, this is equivalent to the existence of a T-existentially closed extension A^* of A in $Mod(T^*)$ such that $A^* \models \exists S$. By Lemma 10 the extensions of A are same as the models of $Diag(A)$. It follows that S is solvable in T over A iff there is a model A^* of $T^* \cup Diag(A)$ such that $A^* \models \exists S$. Since T^* is a model completion of T, the theory $T^* \cup Diag(A)$ is complete and we deduce that $A^* \models \exists S$ iff $T^* \cup Diag(A) \models \exists S$. $\qquad\square$

Theorem 20 *Let E be a recursively enumerable set of conditional equations over a recursive signature Σ. The solvability problem in E over the free algebra $T_E(X)$ is decidable if E has a decidable model completion E^*.*

Proof: We prove that the theory $E^* \cup Diag(T_E(X))$ over the signature $\Sigma_{T_E(X)}$ is decidable. The theorem then follows from the preceding proposition.

The decidability of E^* implies the decidability of the universal theory $E_\forall^*$. Since E and E^* are mutually model consistent and since universal theories are preserved under substructures, $E_\forall^*$ is identical with $E_\forall$. It follows that the universal theory $E_\forall$ and consequently the equational theory of $Mod(E)$ are decidable. This means that we can decide for any equation $s \doteq t$ over Σ whether $E \models s \doteq t$. By Birkhoff's theorem this is equivalent to $T_E(X) \models s \doteq t$.

Now we show that we can even decide for any equation $s' \doteq t'$ over the diagram signature $\Sigma_{T_E(X)}$ whether $T_E(X) \models s' \doteq t'$. For any element $[u]_E$ in $T_E(X)$ there exists a constant symbol $c_{[u]}$ in the diagram signature $\Sigma_{T_E(X)}$. If we replace any $c_{[u]}$ in $s' \doteq t'$ with a member of the congruence class $[u]$ then we obtain an equation $s \doteq t$ over the signature Σ. We have $T_E(X) \models s \doteq t$ iff $T_E(X) \models s' \doteq t'$. Consequently, the diagram of $T_E(X)$ is decidable. Since E^* and $Diag(T_E(X))$ are decidable, $E^* \cup Diag(T_E(X))$ is recursively enumerable. Since $E^* \cup Diag(T_E(X))$ is moreover complete, it is even decidable. $\qquad\square$

Next we consider the case where T has a model-companion, but no model-completion. This can happen, when T has not the amalgamation property.

Definition 21 A theory T has the *amalgamation property* iff whenever there are models A, B_1, B_2 of T and embeddings $\sigma_1 : A \to B_1, \sigma_2 : A \to B_2$ there exist a model C of T and embeddings $\tau_1 : B_1 \to C, \tau_2 : B_2 \to C$ such that $\tau_1 \circ \sigma_1 = \tau_2 \circ \sigma_2$.

A model companion T^* of a first-order theory T is a model completion iff T has the amalgamation property. A theory admits quantifier elimination iff it is model complete and has the amalgamation property [HW75].

We have seen that the solvability problem is decidable, if E has a decidable model completion. Now we prove that this result still holds in the case that E has no decidable model completion but a decidable complete model companion.

Proposition 22 *Let T be Σ-theory with complete model companion T^* and let A be a model of T. A system of equations S*

$$s_1 \doteq t_1 \wedge \ldots \wedge s_n \doteq t_n, n \geq 1,$$

with $s_i, t_i \in T(F, X), i = 1, \ldots, n$, is solvable in T over A, iff

$$T^* \models \exists(s_1 \doteq t_1 \wedge \ldots \wedge s_n \doteq t_n).$$

Proof: "⟹": Suppose S is solvable in T over A. Then there exists an extension A' of A in $Mod(T)$, such that S is solvable in A'. Since T and T^* are mutually model consistent, A' can be embedded in a T-existentially closed structure A^*. Then S is also solvable in A^*, that is $A^* \models \exists S$. Since T^* is the theory of T-existentially closed structures and T^* is complete, we get $T^* \models \exists S$.

"⟸": If $T^* \models \exists S$, then $A^* \models \exists S$, for any T-existentially closed structure A^*. Therefore this holds in particular for a T-existentially closed extension of A. Such an extension exists since T and T^* are mutually model consistent. □

Theorem 23 *Let E be a recursively enumerable set of conditional equations over a recursive signature Σ. The solvability problem in E over the free algebra $T_E(X)$ is decidable if E has a decidable and complete model companion E^*.*

Proof: This is an immediate consequence of the preceding proposition. □

A characterization of the universal theories with a decidable model companion has been given recently by [Bur89].

Example 24 Let T be the theory of ordered fields in the signature $\Sigma = \{0, 1, +, *, -, <\}$. Besides the field axioms $K_0 - K_8$ the theory T satisfies the ordering axioms

$$
\begin{aligned}
O_1 &: \quad \forall x \quad\quad \neg x < x, \\
O_2 &: \quad \forall x, y, z \quad (x < y \wedge y < z \Rightarrow x < z), \\
O_3 &: \quad \forall x, y \quad (x < y \vee x \doteq y \vee y < x),
\end{aligned}
$$

and the axioms

$$
\begin{aligned}
OK_1 &: \quad \forall x, y, z \quad (x < y \Rightarrow x + z < y + z), \\
OK_2 &: \quad \forall x, y \quad (0 < x \wedge 0 < y \Rightarrow 0 < x * y).
\end{aligned}
$$

The field of rational numbers Q is a prime model of T. As far as we know, it is still an open problem, whether the solvability of Σ-equations in the rational numbers is decidable [DMR76]. However, the solvability of Σ-equations in the real numbers is decidable. In addition to the axioms of ordered fields, the real numbers satisfy the axioms

$$
\begin{aligned}
RK_1 &: \quad \forall x \exists y \quad\quad (x < 0 \vee x \doteq y^2), \\
RK_2 &: \quad \forall x_0, \ldots, x_{2n} \exists y \quad y^{2n+1} + x_{2n} * y^{2n} + \ldots + x_1 * y + x_0 \doteq 0, \text{ for } n \geq 1.
\end{aligned}
$$

All these axioms together define the theory TRC of *real closed fields*. TRC is a model companion of T and is complete. It admits quantifier elimination and is decidable.

A model companion T^* of a theory T is complete iff T has the joint embedding property [HW75].

Definition 25 A Σ-theory T has the *joint embedding property* iff for any models A, B of T there exists a model C of T in which both A, B can be embedded.

If an equational theory E with the joint embedding property has a model companion E^*, then E^* is complete. By the theorem of Eklof/Sabbagh, E^* is the theory of E-existentially closed structures. The inverse, however, is false. The theory of E-existentially closed structures of an equational theory E with the joint embedding property can be complete without E having a model companion. This is shown by the next example.

Example 26 Let R be an associative ring with $1 \neq 0$. Consider the signature $\Sigma_R \overset{\text{def}}{=}$ $(\{M\}, \{0, +\} \cup \{g_r : M \rightarrow M \mid r \in R\}$. Instead of $g_r(x)$ we write $r * x$ and instead of $(-1) * x$ only $-x$. Then any R-module is in a natural way a Σ_R-algebra. The Σ_R-theory T_R of R-modules can be axiomatized as follows:

$$
\begin{array}{llll}
M_1 : & \forall x, y, z & (x + y) + z \doteq x + (y + z), & \\
M_2 : & \forall x & x + 0 \doteq 0, & \\
M_3 : & \forall x & x + (-x) \doteq 0, & \\
M_4 : & \forall x, y & x + y \doteq y + x, & \\
M_5 : & \forall x & 1 * x \doteq x, & \\
M_{6,r,s,t} : & \forall x & s * x + t * x \doteq r * x & \text{for all } r, s, t \in R \text{ with } r = s + t, \\
M_{7,r} : & \forall x, y & r * (x + y) \doteq r * x + r * y & \text{for all } r \in R, \\
M_{8,r,s,t} : & \forall x & r * (s * x) \doteq t * x & \text{for all } r, s, t \in R \text{ with } r * s = t.
\end{array}
$$

T_R has the joint embedding property, because any two R-modules A and B can be embedded into the direct product $A \times B$. A ring R is *coherent* iff any finitely generated ideal is finitely presented, that is for all $n > 0$, the kernel of any homomorphism $R^n \rightarrow R$ is finitely generated. The theory T_R of R-modules has a model companion iff the ring R is coherent [ES71]. The theory T_R^e of T_R-existentially closed R-modules is complete even if R is not coherent ([Che76], Theorem 27). This means that for non-coherent rings R the theory T_R^e is complete and mutually model-consistent with T_R, but T_R^e is not model-complete.

In general, the theory T^e of the T-existentially closed structures of a Σ-theory T is neither model-complete nor complete. However, if T has the joint embedding property, we get the following result.

Proposition 27 *Let T be a theory with the joint embedding property. Then T^e is $\forall\exists$-complete, that is for any closed $\forall\exists$-formula φ, either $T^e \models \varphi$ or $T^e \models \neg\varphi$.*

Proof: First we show that two T-existentially closed structures satisfy the same closed $\forall\exists$-formulas.

Let A be T-existentially closed and let φ be a closed $\forall\exists$-formula with $A \models \varphi$. Let B be T-existentially closed, too. By mutual model consistency, there exist models A' and B' of T such $A \subseteq A'$ and $B \subseteq B'$. Since T has the joint embedding property, there is a model C of T such $A' \subseteq C$ and $B' \subseteq C$. A is existentially closed in C. Therefore C can be extended to an elementary extension D of A ([HW75], Prop. 1.1) and so $D \models \varphi$. Since D is also an extension of B, B is existentially closed in D. In existentially closed extensions we may "pull down" not only $\exists$-formulas, but even $\forall\exists$-formulas ([HW75], Prop. 1.2). So we can conclude $B \models \varphi$.

Let us show now that for any closed $\forall\exists$-formula φ either $T^e \models \varphi$ or $T^e \models \neg\varphi$. If $T^e \not\models \varphi$, there is a T-existentially closed structure A with $A \not\models \varphi$, and so $A \models \neg\varphi$. But then it follows that $B \models \neg\varphi$ for all T-existentially closed structures B, und consequently $T^e \models \neg\varphi$. $\qquad\qquad\square$

References

[Boc87] A. Bockmayr. A note on a canonical theory with undecidable unification and matching problem. *Journal of Automated Reasoning*, 3:379–381, 1987.

[Bur89]   S. Burris. Decidable model companions. *Zeitschr. f. math. Logik und Grund-lagen d. Math.*, 35:225 – 227, 1989.

[Che73]   G. Cherlin. Algebraically closed commutative rings. *Journal of Symbolic Logic*, 38:493 – 499, 1973.

[Che76]   G. Cherlin. *Model-Theoretic Algebra - Selected Topics*, volume 521 of *Lecture Notes in Mathematics*. Springer-Verlag, 1976.

[DMR76] M. Davis, Y. Matiyasevich, and J. Robinson. Hilbert's tenth problem. Dio-phantine equations: Positive aspects of a negative solution. *Proc. Symp. Pure Math.*, 28:323–378, 1976.

[Dör51]   K. Dörge. Bemerkungen über Elimination in beliebigen Mengen mit Opera-tionen. *Mathematische Nachrichten*, 4:282 –297, 1951.

[Dör67]   K. Dörge. Über die Lösbarkeit allgemeiner algebraischer Gleichungssysteme und einige weitere Fragen. *Mathematische Annalen*, 171:1–21, 1967.

[ES71]   P. Eklof and G. Sabbagh. Model completions and modules. *Ann. Math. Logic*, 2:251 – 295, 1971.

[HO80]   G Huet and D. C. Oppen. Equations and rewrite rules, A survey. In R. V. Book, editor, *Formal Language Theory*. Academic Press, 1980.

[Hue81]   G. Huet. A complete proof of the Knuth-Bendix completion procedure. *J. Comp. Syst. Sc.*, 23:11–21, 1981.

[HW75]   J. Hirschfeld and W. H. Wheeler. *Forcing, Arithmetic, Division Rings*, volume 454 of *Lecture Notes in Mathematics*. Springer-Verlag, 1975.

[JK89]   J. P. Jouannaud and E. Kounalis. Automatic proofs by induction in theories without constructors. *Inform. and Comput.*, 82:1–33, 1989.

[JK91]   J. P. Jouannaud and C. Kirchner. Solving equations in abstract algebras: A rule-based survey of unification. In J.L. Lassez and G. Plotkin, editors, *Computational Logic: Essays in Honor of A. Robinson*. MIT Press, 1991.

[Kei77]   H. J. Keisler. Fundamentals of model theory. In J. Barwise, editor, *Handbook of Mathematical Logic*. North Holland, 1977.

[Kni89]   K. Knight. Unification, a multidisciplinary survey. *ACM Comp. Surveys*, 21:93–124, 1989.

[Mac72]   A. Macintyre. On algebraically closed groups. *Annals of Math.*, 96:53–97, 1972.

[Mac77]   A. Macintyre. Model completeness. In J. Barwise, editor, *Handbook of Math-ematical Logic*. North Holland, 1977.

[Par83]   M. Parigot. Le modèle compagnon de la théorie des arbres. *Zeitschr. f. math. Logik und Grundlagen d. Math.*, 29:137–150, 1983.

[Pla85]   D. Plaisted. Semantic confluence tests and completion methods. *Information and Control*, 65:233–264, 1985.

[Rob51]   A. Robinson. *On the Metamathematics of Algebra*. North Holland, 1951.

[Rob63]   A. Robinson. *Introduction to Model Theory and to the Metamathematics of Algebra*. North Holland, 1963.

[Sch83]   P. H. Schmitt. Algebraically complete lattices. *Algebra Universalis*, 17:135 – 142, 1983.

[Sie89]   J. H. Siekmann. Unification theory. *Journal of Symbolic Computation*, 7:207–274, 1989.

Complete Equational Unification Based on an Extension of the Knuth-Bendix Completion Procedure

Akihiko Ohsuga

Toshiba Corporation, 70 Yanagicho, Saiwai-ku, Kawasaki-shi 210, Japan

and

Kô Sakai

ICOT Research Center, 1-4-28 Mita, Minato-ku, Tokyo 108, Japan

ABSTRACT

A unifier is a substitution that makes two terms syntactically equal. In this paper, we discuss a more semantical unifier: an equational unifier, which is a substitution that makes two terms equal modulo a congruence relation. As a result we will give a general procedure that enumerates a complete set of equational unifiers for a given pair of terms under a given congruence.

1. Introduction

We assume the reader has elementary knowledge on universal algebra, in particular, on term rewriting systems (see [Huet 80], for example).

Let $\simeq$ be a congruence relation on terms and s, t be terms. A substitution θ is called a $\simeq$-**unifier** of s and t if $s\theta \simeq t\theta$. The set of all $\simeq$-unifiers of s and t are denoted $U(s,t)$. Let $V(s,t)$ be the set of all the variables occurring in s or t. Since we are only interested in substitutions for their effect on s and t in this paper, we regard two unifiers as identical if they differ only on the variables not occurring in s or t, so that we can avoid the subtle treatment of the domains of substitutions. Accordingly, we relativize all notions on unifiers to $V(s,t)$. For example, a unifier θ is said to be **more general** than another unifier ϕ under $\simeq$ (denoted $\theta \leq \phi$) if there is a substitution ψ such that $(v\theta)\psi \simeq v\phi$ for any variable $v \in V(s,t)$. A subset C of $U(s,t)$ is said to be **complete** if, for any $\theta \in U(s,t)$, there exists a unifier $\theta' \in C$ such that $\theta' \leq \theta$. Moreover, a complete subset C is called the **minimum** if $\theta = \theta'$ for any $\theta, \theta' \in C$ such that $\theta \leq \theta'$. We write $\theta \approx \theta'$ if $\theta \leq \theta'$ and $\theta' \leq \theta$. Then, relation $\approx$ is an equivalence on unifiers and, if the minimum complete set exists, it is unique up to $\approx$ [Fages 86].

In the case of ordinary unification (or, more precisely, in the case that $\simeq$ is the identity relation), unifiability is decidable and the most general unifier always exists for any unifiable pair of terms [Robinson 65]. For a general congruence $\simeq$, however, the existence of the most general unifier is not guaranteed. In this situation, a complete set of $\simeq$-unifiers plays the role that the most general unifier plays in ordinary unification: a representative of all unifiers.

We call a pair of terms an **equation**. As well-known, a set of equations presents a congruence relation. Let $\Re$ be a set of equations. First of all, we give a definition of the congruence presented by $\Re$ in terms of reduction. A term u is **reduced** to another term u' by $\Re$ (denoted by $u \Rightarrow u'$) if there is an equation $\langle l, r \rangle \in \Re$, a context $c[]$, and a substitution θ such that $c[l\theta] = u$ and $c[r\theta] = u'$. In other words, if a term has a subterm matched with the left hand side of an equation, then it is reduced to the term obtained by replacing the subterm with the right hand side. We write $u \Leftrightarrow u'$ if $u \Rightarrow u'$ or $u' \Rightarrow u$. Then, the **congruence** presented by $\Re$ is defined as the reflexive transitive closure of relation $\Leftrightarrow$.

In what follows, we assume that a finite presentation of congruence $\simeq$ is given. We call problems concerning $\simeq$-unifiers for such an congruence $\simeq$ equational unification problems. The main equational unification problems are the following.

(1) Is $\simeq$-unifiability decidable?

(2) Does the minimum complete set of $\simeq$-unifiers exist? Can it be enumerated?

(3) Is there a finite complete set of $\simeq$-unifiers?

(4) Is there an efficient procedure to enumerate a complete set of $\simeq$-unifiers?

It is undecidable in general whether two given terms have a $\simeq$-unifier. As for the answers to these problems on specific sets of equations, there is a wide-ranged survey by Siekman [Siekman 89]. The result on AC-unification, in which the equation set consists only of the associative and the commutative laws, seems to be the most important from a practical point of view. That is, the minimum complete set of AC-unifiers always exists and it is finite and computable [Stickel 81][Fages 84][Huet 78].

In this paper, we address problem (4) for a general set of equations. A procedure is said to be a **complete equational unification procedure** if it enumerates a complete set of $\simeq$-unifiers of given terms s and t. It is clear from the definition that, for any terms s and t, the set $U(s,t)$ is recursively enumerable and complete. Therefore, enumeration of $U(s,t)$ is a complete (but not interesting) equational unification procedure. What is interesting is a more efficient procedure than simple enumeration of all unifiers.

From a theoretical point of view, the minimum complete set may be the most interesting since it is unique and not redundant. However, there is no reality of computation of the minimum complete set for the following reasons. First of all, the minimum set may not exist. That is, there may be a complete set C of $\simeq$-unifiers with the following property: for any $\theta \in C$, there is a unifier $\theta' \in C$ such that $\theta' < \theta$ (that is, $\theta' \leq \theta$, but $\theta \not\leq \theta'$) [Fages 86]. Even if the minimum set exists, there may be no procedures to enumerate its elements. Even if it is enumerable, it may need more cost to compute than other (redundant) complete sets.

Several researchers proposed equational unification procedures based on narrowing [Fay 79] and basic narrowing [Hullot 80][Bosco 87], under the assumption that the given set of equations (viewed as left-to-right rewrite rules) is confluent and terminating. These are efficient, but the assumption is seldom satisfied in actual cases. Gallier and Snyder proposed a universal equational unification procedure [Gallier 87], but it does not seem efficient enough for actual applications.

We propose another general procedure and prove its completeness. We confirmed that it has little redundancy (and, therefore, hopefully efficient) in many cases by actual implementation and experiment using simple but real mathematical problems. We do not discuss the implementation details in this paper, but Examples 4.1 and 4.2 show some material of experiment.

The procedure is based on a combination of the Knuth-Bendix completion [Knuth 70][Huet 81] (or, more precisely, completion without failure [Bachmair 87]) and narrowing. The procedure applies narrowing to s and t, while constructing a (possibly infinite) confluent and terminating set of equations (viewed as rewrite rules). Since, as shown in [Huet 81] and [Bachmair 87], a confluent and terminating set can be obtained virtually even if the completion process does not terminate, the narrowing process eventually enumerates a complete set of unifiers. Moreover, since the procedure is an extension of the Knuth-Bendix completion, it may obtain a finite confluent and terminating set on the way of equational unification. Once such a set is obtained, the subsequent process becomes ordinary narrowing. Therefore, Fay's result [Fay 79] is viewed as a special case.

The essential idea is common with the refutational theorem proving in first-order logic with equality proposed by Hsiang and Rusinowitch [Hsiang 87][Rusinowitch 88]. The purpose of this paper is not to claim originality of the idea but to claim its naturality and effectiveness and to give a proof of its completeness from the viewpoint of equational unification.

2. Inference rules for equational unification

In the following discussion, let $\preceq$ be a fixed strong simplification order on terms, namely, a simplification order [Dershowitz 82] which is total on ground terms. We use the lexicographic subterm ordering [Sakai 84] in the examples as such an order. In the following definitions, we assume that given terms and equations do not have common variables for simplicity of discussion.

First we change the concept of reduction by set of equations. Usually, as defined in the previous section, when an equation is viewed as a rewrite rule, it is assumed to be used from left to right only. However, we do not assume this any longer, that is, an equation is used as a rewrite rule in both directions. Instead, we control the direction of rewriting by order $\preceq$. For this definition of reduction, it is simpler to consider an equation as an unordered pairs of terms. Therefore, from now on, we regard equations $\langle l, r \rangle$ and $\langle r, l \rangle$ as the same.

To be precise, the definition of reduction is the following: a term u is **reduced** to another term u' (denoted $u \Rightarrow u'$) if $u' \prec u$ and there is an equation $\langle l, r \rangle$ (or its equivalent $\langle r, l \rangle$), a context $c[]$, and a substitution σ such that $c[l\sigma] = u$ and $c[r\sigma] = u'$. Let us denote the reflexive transitive symmetric closure of $\Rightarrow$ by $\simeq'$. It is a routine to verify that $\simeq'$ is a congruence relation.

This above reduction has somewhat different properties from the ordinary left-to-right reduction. First, since $\preceq$ is well-founded [Dershowitz 82][Sakai 84], it is always terminating. Second, congruence $\simeq'$ may be weaker than the congruence relation $\simeq$ presented by the given set of equation. For example, let us consider equation set $\{\langle x+y, y+x \rangle\}$. Then, reduction $\Rightarrow$ is not terminating in the ordinary sense (defined in the previous section) since $u + u' \Rightarrow u' + u \Rightarrow u + u' \Rightarrow u' + u \Rightarrow \cdots$, and $v + w \not\simeq' w + v$ for any variables v and w since neither $v + w \prec w + v$ nor $w + v \prec v + w$.

The word problem involves the decision of not $\simeq'$ but $\simeq$. However, in many cases, we can assume that terms are ground without loss of generality by substituting fresh constants to variables in the terms. For ground terms, the symmetric closures $\Leftrightarrow$ coincide in both definitions of reduction since $\preceq$ is total and, therefore, so do congruences $\simeq$ and $\simeq'$.

Next, we define narrowing, which is somewhat modified of that by Fay [Fay 79] as well. A term u is said to be **narrowed** to another term u' (denoted $u \hookrightarrow u'$) if there are a non-variable subterm u_0 of u, an equation $\langle l, r \rangle$ such that $u_0 \theta = l\theta$ and $u\theta \not\preceq u' = c[r]\theta$, where $u = c[u_0]$ and θ is the most general unifier of u_0 and l. If necessary, we suffix the most general unifier, for example, as $u \hookrightarrow_\theta u'$. In what follows, we discuss narrowing of a pair of terms. A pair is narrowed to another pair if one of the terms is narrowed. To be precise, notation $\langle u_1, u_2 \rangle \hookrightarrow_\theta \langle u_1', u_2' \rangle$ means that either $u_1 \hookrightarrow_\theta u_1'$ and $u_2\theta = u_2'$, or $u_1\theta = u_1'$ and $u_2 \hookrightarrow_\theta u_2'$.

Let us extend the definition of critical pairs [Knuth 70] as well. Let $\langle l_1, r_1 \rangle$ and $\langle l_2, r_2 \rangle$ be equations and u be a non-variable subterm of l_2 unifiable with l_1. If $l_1\theta \not\preceq r_1\theta$ and $l_2\theta \not\preceq r_2\theta$, then pair $\langle c[r_1]\theta, r_2\theta \rangle$ is called a **critical pair**, where $l_2 = c[u]$ and θ is the most general unifier of l_1 and u.

Now, we give a universal equational unification procedure, which inputs a set $\Re$ of equations and terms s, t and outputs $\simeq$-unifier of s and t for the congruence $\simeq$ presented by $\Re$. It is given below in the form of inference rules.

$E\text{-generation:} \quad \dfrac{(E, R, G, U)}{(E \cup \{\langle u_1, u_2 \rangle\}, R, G, U)} \quad \langle u_1, u_2 \rangle$ is a critical pair between equations in R

$E\text{-reduction:} \quad \dfrac{(E \cup \{\langle u_1, u_2 \rangle\}, R, G, U)}{(E \cup \{\langle u_1, u_2' \rangle\}, R, G, U)} \quad u_2 \Rightarrow u_2'$ by an equation in R

$E\text{-deletion:} \quad \dfrac{(E \cup \{\langle u, u \rangle\}, R, G, U)}{(E, R, G, U)}$

$R\text{-generation:} \quad \dfrac{(E \cup \{\langle u_1, u_2 \rangle\}, R, G, U)}{(E, R \cup \{\langle u_1, u_2 \rangle\}, G, U)}$

$G\text{-generation:} \quad \dfrac{(E, R, G \cup \{\langle u_1, u_2, \theta \rangle\}, U)}{(E, R, G \cup \{\langle u_1, u_2, \theta \rangle, \langle u_1', u_2', \theta \circ \theta' \rangle\}, U)}$

$$\langle u_1, u_2 \rangle \hookrightarrow_{\theta'} \langle u_1', u_2' \rangle \text{ by an equation in } R$$

U-generation:
$$\frac{(E, R, G \cup \{\langle u_1, u_2, \theta \rangle\}, U)}{(E, R, G \cup \{\langle u_1, u_2, \theta \rangle\}, U \cup \{\theta \circ \theta'\})}$$
$$\theta' \text{ is the most general unifier of } u_1 \text{ and } u_2$$

Each inference rule expresses operation to transform the quadruple above the horizental line to the quadruple below. Both E and R are sets of equations. G is a set of triples $\langle u_1, u_2, \theta \rangle$ (called **goals**) where u_1 and u_2 are terms and θ is a substitution. We regare triples $\langle u_1, u_2, \theta \rangle$ and $\langle u_2, u_1, \theta \rangle$ as the same similarly to equations. U is a set of substitutions. At the beginning of the procedure, these are set as follows:

$$E = \Re, \quad R = \emptyset, \quad G = \{\langle s, t, \epsilon \rangle\}, \quad U = \emptyset,$$

where ϵ denotes the identity substitution. The procedure enumerates $\simeq$-unifiers of s and t as elements of U.

When one of the inference rules is applied, a quadruple (E, R, G, U) is transformed to another quadruple (E', R', G', U'), denoted by $(E, R, G, U) \vdash (E', R', G', U')$. If necessary, the name of the applied inference rule are suffixed to symbol $\vdash$. Let

$$(E_0, R_0, G_0, U_0) \vdash (E_1, R_1, G_1, U_1) \vdash (E_2, R_2, G_2, U_2) \vdash \cdots$$

be a sequence of applications of the inference rules. We denote $\bigcup_{i=0}^{\infty} E_i$ by E_∞, $\bigcup_{i=0}^{\infty} R_i$ by R_∞, $\bigcup_{i=0}^{\infty} G_i$ by G_∞, and $\bigcup_{i=0}^{\infty} U_i$ by U_∞. An inference sequence is called **fair**, if it satisfies the following conditions.

(1) Any critical pair between equations in R_∞ is contained in E_∞.

(2) $\bigcup_{i=0}^{\infty} \bigcap_{j=i}^{\infty} E_j = \emptyset$. In other words, any equation in E_∞ is removed from E eventually by E-reduction, E-deletion, or R-generation.

(3) Any goal obtained from a goal in G_∞ and an equation in R_∞ by G-generation is contained in G_∞.

(4) Any substitution obtained from a goal in G_∞ by U-generation is contained in U_∞.

We claim that any fair inference sequence can enumerate a complete set of $\simeq$-unifiers as U_∞.

Example 2.1

Consider equation set $\Re = \{\langle x \times x, x^2 \rangle, \langle 1 \times y, y \rangle\}$ of equations and terms $s = z^2$ and $t = 1$, where $\preceq$ is the lexicographic subterm ordering based on total order $1 <^2 < \times$

on the function symbols. Then, the following is a fair inference sequence.

$$(E_0 = \{\langle x \times x, x^2 \rangle, \langle 1 \times y, y \rangle\}, R_0 = \emptyset, G_0 = \{\langle z^2, 1, \epsilon \rangle\}, U_0 = \emptyset)$$
$$\vdash_{R\text{-generation}} (E_1 = \{\langle 1 \times y, y \rangle\}, R_1 = \{\langle x \times x, x^2 \rangle\}, G_1 = G_0, U_1 = \emptyset)$$
$$\vdash_{R\text{-generation}} (E_2 = \emptyset, R_2 = R_1 \cup \{\langle 1 \times y, y \rangle\}, G_2 = G_1, U_2 = \emptyset)$$
$$\vdash_{E\text{-generation}} (E_3 = \{\langle 1^2, 1 \rangle\}, R_3 = R_2, G_3 = G_2, U_3 = \emptyset)$$
$$\vdash_{R\text{-generation}} (E_4 = \emptyset, R_4 = R_3 \cup \{\langle 1^2, 1 \rangle\}, G_4 = G_3, U_4 = \emptyset)$$
$$\vdash_{G\text{-generation}} (E_5 = \emptyset, R_5 = R_4, G_5 = G_4 \cup \{\langle 1, 1, [1/z] \rangle\}, U_5 = \emptyset)$$
$$\vdash_{U\text{-generation}} (E_6 = \emptyset, R_6 = R_5, G_6 = G_5, U_6 = \{[1/z]\})$$

Thus, $\simeq$-unifier $[1/z]$ of $s = z^2$ and $t = 1$ is obtained as an element of U_6 in the above sequence, where notation $[1/z]$ expresses the substitution θ such that $z\theta = 1$ and $v\theta = v$ for any variables v other than z.

3. Completeness of the unification procedure

First of all, we prove the soundness of the procedure given in the previous section.

Theorem 3.1

Let

$$(\Re, \emptyset, \{\langle s, t, \epsilon \rangle\}, \emptyset) = (E_0, R_0, G_0, U_0) \vdash (E_1, R_1, G_1, U_1) \vdash (E_2, R_2, G_2, U_2) \vdash \cdots$$

be an inference sequence. Then, any element of U_∞ is an $\simeq$-unifier of s and t.

Proof: Let $\simeq_i$ be the congruence relation presented by $E_i \cup R_i$. Then it is easy to prove the following by induction on i.

(1) $\simeq_i = \simeq$.

(2) For any $\langle u_1, u_2, \theta \rangle \in G_i$, $s\theta \simeq u_1$ and $t\theta \simeq u_2$.

(3) For any $\theta \in U_i$, $s\theta \simeq t\theta$. ∎

The proof of completeness of the procedure consists of two parts. First, R_∞ is proved to be confluent by evidence transformation method [Bachmair 86]. Second, narrowing is proved to be able to trace any rewriting by R_∞.

Hereafter, we use symbols $\Rightarrow$ to denote reduction in the new sense defined in Section 2. On the other hand, we use symbols $\Leftrightarrow$ to denote the symmetric closure of reduction in the old sense defined in Section 1.

Let $\Re$ be a set of equations, and g and g' ground terms such that $g \simeq g'$. Then, from the definition, there is a finite sequence of terms

$$g = g_0 \Leftrightarrow g_1 \Leftrightarrow \cdots \Leftrightarrow g_m = g'.$$

Let us define sequences of this form in a more general framework. A sequence $g = g_0 \Xi_1 g_1 \Xi_2 \cdots \Xi_m g_m = g'$ is called an **evidence** of $g \simeq g'$ by E and R, if each g_i is a ground term and Ξ_i is one of the following symbols:

(1) $\Leftrightarrow$, which indicates that $g_{i-1} \Leftrightarrow g_i$ by E.

(2) $\Leftarrow$, which indicates that $g_i \Rightarrow g_{i-1}$ by R.

(3) $\Rightarrow$, which indicates that $g_{i-1} \Rightarrow g_i$ by R.

An evidence is said to be **normal** if it has the following form

$$g = g_0 \Rightarrow g_1 \Rightarrow \cdots \Rightarrow g_{m-1} \Rightarrow h \Leftarrow g'_{n-1} \Leftarrow \cdots \Leftarrow g'_1 \Leftarrow g'_0 = g' \quad (m \geq 0, n \geq 0)$$

Now, we will define the weight of an evidence. First, the weight $w(g \Xi g')$ of each step $g \Xi g'$ of an evidence is defined as follows:

$$w(g \Leftrightarrow g') = \{g, g'\}, \quad w(g \Leftarrow g') = \{g'\}, \quad w(g \Rightarrow g') = \{g\}$$

where $\{g, g'\}$, $\{g'\}$, and $\{g\}$ are not sets but multi-sets, and are compared by the multi-set ordering [Dershowitz 79]. The weight of an evidence is defined as the multi-set consisting of the weights of all the steps of the evidence. Note that, since the weight of a step is a multi-set, the weight of an evidence is a doubly-multi-set (a multi-set of multi-sets of terms). The set of the weights of evidences is well-founded since the base order is well-founded. Let us denote the order also by $\preceq$.

Theorem 3.2

Let
$$(E_0 = \Re, R_0 = \emptyset, G_0, U_0) \vdash (E_1, R_1, G_1, U_1) \vdash (E_2, R_2, G_2, U_2) \vdash \cdots$$

be a fair inference sequence. Then, R_∞ is a confluent set of equations for $\simeq$ w.r.t. ground terms.

Proof: It is sufficient to prove that, for any ground terms g and g' such that $g \simeq g'$, there exist a normal evidence of $g \simeq g'$ by E_∞ and R_∞. (Since the evidence is normal, it expresses reduction of g and g' to the same term by R_∞.) Let g and g' be arbitrary ground terms such that $g \simeq g'$. Then, there is an evidence of $g \simeq g'$ by E_0 and R_0, which is also an evidence by E_∞ and R_∞ of course. Let ¶ be an evidence by E_∞ and R_∞ with minimal weight. We prove that ¶ is normal. First we prove that ¶ contains no steps of the form

$$c[u_1\theta] \Leftrightarrow c[u_2\theta] \tag{A}$$

where $\langle u_1, u_2 \rangle \in E_i$ for some i. Suppose that such a step exists. From fairness condition (2), for some j such that $i < j$, equation $\langle u_1, u_2 \rangle$ must be deleted from E_j; that is, inference rule E-reduction, E-deletion, or R-generation must be applied to $\langle u_1, u_2 \rangle$. If it is E-reduction, $\langle u_1, u \rangle \in E_j$ (or $\langle u, u_2 \rangle \in E_j$) for some u such that $u_2 \Rightarrow u$ (or $u_1 \Rightarrow u$ by R_j). Therefore, by replacing the step of form (A) with two steps

$$c[u_1\theta] \Leftrightarrow c[u\theta] \Leftarrow c[u_2\theta] \quad (\text{or} \quad c[u_1\theta] \Rightarrow c[u\theta] \Leftrightarrow c[u_2\theta]),$$

we can obtain a new evidence ¶'. Comparing the weight of the steps, that is, $\{c[u_1\theta], c[u_2\theta]\}$ in ¶ and $\{c[u_1\theta], c[u\theta]\}, \{c[u_2\theta]\}$ (or $\{c[u_1\theta]\}, \{c[u\theta], c[u_2\theta]\}$) in ¶', we can easily see that $w(¶') \preceq w(¶)$, which contradicts that ¶ has minimal weight. If the inference step is E-deletion, u_1 must be equal to u_2. Therefore, by simply removing the step of form (A), we can obtain a new evidence, which again contradicts that ¶ has minimal weight. If the inference step is R-generation, R_j contains equation $\langle u_1, u_2 \rangle$. In this case, the step of form (A) can be replaced with

$$c[u_1 \theta] \Rightarrow c[u_2 \theta] \quad \text{or} \quad c[u_1 \theta] \Leftarrow c[u_2 \theta]$$

since $\preceq$ is total for ground terms, and a contradiction follows. Next, we prove that ¶ contains no steps of the form

$$h_1 \Leftarrow h \Rightarrow h_2 \tag{B}$$

Suppose that there are steps of form (B), in which term h is reduced in two ways, say, to h_1 by equation $\langle l_1, r_1 \rangle \in R_i$ and to h_2 by equation $\langle l_2, r_2 \rangle \in R_j$. There are several cases. First assume that the reduced parts do not overlap, that is h, h_1, and h_2 have forms $c[l_1\theta_1, l_2\theta_2]$, $c[r_1\theta_1, l_2\theta_2]$, and $c[l_1\theta_1, r_2\theta_2]$. In this case, by replacing the steps of form (B) with

$$h_1 \Rightarrow c[r_1\theta_1, r_2\theta_2] \Leftarrow h_2$$

we can obtain a new evidence, which contradicts that ¶ has minimal weight. Next assume that the reduced parts overlap. Since the discussion is symmetrical, we can assume that $h = d[c[l_1\theta_1]] = d[l_2\theta_2]$, $h_1 = d[c[r_1\theta_1]]$, and $h_2 = d[r_2\theta_2]$ without loss of generality. If $l_1\theta_1$ occurs at a variable position in l_2, we can easily arrive at a contradiction similarly to the non-overlapping case. Otherwise, $\langle c[r_1\theta_1], r_2\theta_2 \rangle$ is an instance of a critical pair of equations $\langle l_1, r_1 \rangle$ and $\langle l_2, r_2 \rangle$. From fairness condition (1), the critical pair must be in some E_k. Then, by replacing the steps of form (B) with

$$h_1 = d[c[r_1\theta_1]] \Leftrightarrow d[r_2\theta_2] = h_2,$$

we arrive at a contradiction again. Thus, we have proved that ¶ contains no steps of form (A) or (B). Such an evidence is clearly normal. ∎

If there is a normal evidence

$$g_0 \Rightarrow g_1 \Rightarrow \cdots \Rightarrow g_{m-1} \Rightarrow h \Leftarrow g'_{n-1} \Leftarrow \cdots \Leftarrow g'_1 \Leftarrow g'_0 \tag{C}$$

we can always convert it to a one-way reduction sequence of pairs of terms of the following form:

$$\langle g_0, g'_0 \rangle = p_0 \Rightarrow p_1 \Rightarrow \cdots \Rightarrow p_{m+n} = \langle h, h \rangle. \tag{C'}$$

In each step, either the left or the right element of pairs is reduced. In what follows, sequences of form (C') are called normal evidences instead of those of form (C) for simplicity of discussion.

A substitution σ is said to be **irreducible** if $v\sigma$ is irreducible for any variable v.

Theorem 3.3 [Hullot 80]

Let u be a term (or pair of terms) and θ be an irreducible substitution. Then, for any sequence of reduction

$$u\theta = g_0 \Rightarrow g_1 \Rightarrow \cdots \Rightarrow g_n,$$

there is a sequence of narrowing

$$u = u_0 \hookrightarrow_{\theta_0} u_1 \hookrightarrow_{\theta_1} \cdots \hookrightarrow_{\theta_{n-1}} u_n$$

and a sequence of irreducible substitutions $\psi_0, \psi_1, \ldots, \psi_n$ *such that*

$$g_i = u_i \psi_i \ (i = 0, 1, \ldots, n)$$

and

$$\theta = \psi_0 = \theta_0 \circ \psi_1 = \cdots = \theta_0 \circ \cdots \circ \theta_{n-1} \circ \psi_n.$$

In the original form of the above theorem, the concepts of reduction and narrowing are the conventional left-to-right-only ones, the set of equations is assumed to be confluent and terminating, and substitution θ is assumed to be normal. However, the above form of the theorem can also be proved in the same way as the original.

Now, we are ready to prove the completeness of the $\simeq$-unification procedure.

Theorem 3.4

Let $\Re$ *be a set of equations, s and t be terms, and*

$$(\Re, \emptyset, \langle s, t, \epsilon \rangle, \emptyset) = (E_0, R_0, G_0, U_0) \vdash (E_1, R_1, G_1, U_1) \vdash (E_2, R_2, G_2, U_2) \vdash \cdots$$

be a fair inference sequence. Then, U_∞ is a complete set of $\simeq$-unifiers of s and t. That is, for any $\simeq$-unifier θ of s and t, there is a substitution $\theta' \in U_\infty$ more general than θ.

Proof: By replacing variables in $s\theta$ and $t\theta$ with fresh constants, we can assume that $s\theta$ and $t\theta$ are ground terms without loss of generality. Moreover, by replacing the value of θ at each variable with its normal form w.r.t. R_∞, we can assume that θ is irreducible. Since $s\theta \simeq t\theta$, there is a term h and a normal evidence

$$\langle s, t \rangle \theta = p_0 \Rightarrow p_1 \Rightarrow \cdots \Rightarrow p_n = \langle h, h \rangle$$

by R_∞. Then, from Theorem 3.3, there is a sequence of narrowing by R_∞

$$\langle s, t \rangle = \langle s_0, t_0 \rangle \hookrightarrow_{\theta_0} \langle s_1, t_1 \rangle \hookrightarrow_{\theta_1} \cdots \hookrightarrow_{\theta_{n-1}} \langle s_n, t_n \rangle$$

and a sequence of irreducible substitutions $\psi_0, \psi_1, \ldots, \psi_n$ such that $p_i = \langle s_i, t_i \rangle \psi_i$ $(i = 0, 1, \ldots, n)$ and

$$\theta = \psi_0 = \theta_0 \circ \psi_1 = \cdots = \theta_0 \circ \cdots \circ \theta_{n-1} \circ \psi_n.$$

From fairness condition (3), we can easily prove by induction that, for each i, $\langle s_i, t_i, \theta_0 \circ \cdots \circ \theta_{i-1} \rangle \in G_\infty$, in particular, $\langle s_n, t_n, \theta_0 \circ \cdots \circ \theta_{n-1} \rangle \in G_\infty$. Since $s_n \psi_n = h = t_n \psi_n$, s_n and t_n are unifiable. Let ψ be the most general unifier of s_n and t_n. Then, from fairness condition (4), $\theta' = \theta_0 \circ \cdots \circ \theta_{n-1} \circ \psi \in U_\infty$, which is more general than $\theta = \theta_0 \circ \cdots \circ \theta_{n-1} \circ \psi_n$. $\blacksquare$

As shown in Theorem 3.2, the $\simeq$-unification is an extension of the Knuth-Bendix completion procedure. In particular, if $R_i = R_\infty$ for some i, a finite confluent and terminating set of equations is obtained after a finite number of steps of inference. Then, the subsequent process can be assumed to consist only of G-generations and U-generations since the other rules cause no essential change in R_i, G_i, and U_i. Therefore, the procedure can be viewed as an extension of Fay's procedure. Moreover, if $G_j = G_\infty$ for some j (in fact, Example 2.1 is this case), we can obtain a finite complete set U_∞ of $\simeq$-unifiers of s and t. Note that, even in this case, U_∞ is not necessarily the minimum complete set.

4. Implementation issues and examples

There are a lot of things to be considered for efficiency in actual implementation of the procedure discussed in the previous section.

If the proof of Theorem 3.2 is examined, it can be easily seen that the inference rules E-reduction and E-deletion do not contribute to the completeness of the procedure. In fact, these rules are introduced for efficiency. To improve efficiency further, the following inference rules should be taken into consideration. If these rules are given priority over the generation rules, they will save a lot of time by not applying useless inferences.

R-reduction: $\dfrac{(E, R \cup \{\langle u_1, u_2 \rangle\}, G, U)}{(E \cup \{\langle u_1, u_2' \rangle\}, R, G, U)}$ $u_2 \Rightarrow u_2'$ by an equation in R

G-reduction: $\dfrac{(E, R, G \cup \{\langle u_1, u_2, \theta \rangle\}, U)}{(E, R, G \cup \{\langle u_1, u_2', \theta \rangle\}, U)}$ $u_2 \Rightarrow u_2'$ by an equation in R

G-deletion: $\dfrac{(E, R, G \cup \{\langle u_1, u_2, \theta \rangle\}, U)}{(E, R, G, U)}$
θ is reducible by R or an element of U is more general than θ

U-deletion: $\dfrac{(E, R, G, U \cup \{\theta\})}{(E, R, G, U)}$
θ is reducible by R or an element of U is more general than θ

The reader can clearly see the role of R-reduction and G-reduction. Rules G-deletion and U-deletion play a similar role to that the basic narrowing plays in Hullot's procedure [Hullot 80]

Even if the above inference rules are also employed, the procedure is still complete. To prove its completeness, however, the evidence order and the limits need more subtle treatment, and this would introduce a simple but long discussion, which we have avoided in the proof of Theorem 3.2. For example, if R-reduction is employed, R_∞ must not be defined as $\bigcup_{i=1}^{\infty} R_i$ but as $\bigcup_{i=1}^{\infty} \bigcap_{j=i}^{\infty} R_j$, since R_i is no longer increasing.

We will show several examples of $\simeq$-unifications in combinatory logic. In the examples, we use the strong simplification order $\preceq$ based on lexicographic subterm ordering. Terms of the form $*(\cdots(*(x, y), \cdots), z)$ are abbreviated to the form $xy \cdots z$ in the following inference sequence.

Example 4.1

An identity combinator $\mathbf{i}$ is defined as a combinator with property $\forall x\ \mathbf{i}x = x$. Here, we show the example of automatic construction of $\mathbf{i}$ from $\mathbf{s}$ and $\mathbf{k}$ by $\simeq$-unification. Let $\Re$ be $\{\langle sxyz, xz(yz) \rangle, \langle kxy, x \rangle\}$ (that is, consist of the defining equation for $\mathbf{s}$ and $\mathbf{k}$), and let us try to $\simeq$-unify $s = vc$ and $t = c$. Function symbols are ordered as

$c < k < s < *.$

$$
\begin{aligned}
&(E_0 = \{\langle sxyz, xz(yz)\rangle, \langle kxy, x\rangle\}, R_0 = \emptyset, G_0 = \{\langle vc, c, \epsilon\rangle\}, U_0 = \emptyset)\\
&\vdash_{R\text{-generation}}(E_1 = \{\langle sxyz, xz(yz)\rangle\}, R_1 = \{\langle kxy, x\rangle\}, G_1 = G_0, U_1 = \emptyset)\\
&\vdash_{R\text{-generation}}(E_2 = \epsilon, R_2 = R_1 \cup \{\langle sxyz, xz(yz)\rangle\}, G_2 = G_1, U_2 = \emptyset)\\
&\vdash_{E\text{-generation}}(E_3 = \{\langle skxy, y\rangle\}, R_3 = R_2, G_3 = G_2, U_3 = \emptyset)\\
&\vdash_{R\text{-generation}}(E_4 = \epsilon, R_4 = R_3 \cup \{\langle skxy, y\rangle\}, G_4 = G_3, U_4 = \emptyset)\\
&\vdash_{G\text{-generation}}(E_5 = \epsilon, R_5 = R_4, G_5 = G_4 \cup \{\langle c, c, [skx/v]\rangle\}, U_5 = \emptyset)\\
&\vdash_{U\text{-generation}}(E_6 = \epsilon, R_6 = R_5, G_6 = G_5, U_6 = \{[skx/v]\})
\end{aligned}
$$

Thus, identity combinator skx is obtained as the term substituted to v.

Remark: Strictly speaking, an $\simeq$-unifier of vc and c is not necessarily an identity combinator, since it may depend on c (that is, the term substituted to v may contain c as its subterm). If we want to restrain such a unifier from being generated, we should $\simeq$-unify $vc(v)$ and $c(v)$. Note that disequation $vc(v) \neq c(v)$ is the Skolem form of the negation of formula $\forall x \, vx = x$.

Example 4.2

Next let us try the mockingbird problem [Smullyan 85]. A mockingbird is a combinator m with property $\forall x \, mx = xx$. The problem is to construct a fixed point of a given combinator c from m, b, and c itself, where b is a composition combinator, which has property $\forall x \, \forall y \, \forall z \, bxyz = x(yz)$. A fixed point of c is defined as a combinator f with property $cf = f$.

We set E_0 to be $\{\langle mx, xx\rangle, \langle byzw, y(zw)\rangle\}$, R_0 to be $\emptyset$, G_0 to be $\{\langle cv, v, \epsilon\rangle\}$, and U_0 to be $\emptyset$, and execute the $\simeq$-unification procedure. We do not trace the details, but a fixed point of c is obtained through the following process.

(1) New equation $\langle m(byz), y(z(byz))\rangle$ is obtained as a critical pair of equations $\langle mx, xx\rangle$ and $\langle byzw, y(zw)\rangle$ by E-generation.

(2) New goal $\langle m(bcz), z(bcz), [z(bcz)/v]\rangle$ is obtained from goal $\langle cv, v, \epsilon\rangle$ and equation $\langle m(byz), y(z(byz))\rangle$ by G-generation.

(3) Finally, we can generate $\simeq$-unifier $[m(bcm)/v]$ of cv and v from the above goal $\langle m(bcz), z(bcz), [z(bcz)/v]\rangle$ by U-generation, and $m(bcm)$ is a fixed point of c in fact.

REFERENCES

[Bachmair 86] Bachmair, L., Dershowitz, N., and Hsiang, J.: *Ordering for equational proof*, Proc. Symp. Logic in Computer Science, Cambridge, Massachusetts (June 1986)

[Bachmair 87] Bachmair, L., Dershowitz, N., and Plaisted, D. A.: *Completion without failure*, Proc. Colloquium on Resolution of Equations in Algebraic Structures (1987)

[Bosco 87]   Bosco, P. G., Giovannetti, F., and Moiso, C.: *Refined strategies for equational proofs*, TAPSOFT '87, LNCS 250, pp. 276–290 (1987)

[Dershowitz 79]   Dershowitz, N. and Manna, Z.: *Proving termination with multiset orderings*, Comm. ACM 22, pp. 465–467 (1979)

[Dershowitz 82]   Dershowitz, N.: *Orderings for term-rewriting systems*, Theoretical Computer Science 17, pp. 279–301 (1982)

[Fages 84]   Fages, F.: *Associative-commutative unification*, 7th International Conference on Automated Deduction, LNCS 170, pp. 194–208 (1984)

[Fages 86]   Fages, F. and Huet, G.: *Complete sets of unifiers and matchers in equational theories*, Theoretical Computer Science 43, pp. 189–200 (1986)

[Fay 79]   Fay, M.: *First order unification in an equational theory*, 4th workshop on Automated Deduction, Austin, Texas, pp. 161–167 (1979)

[Gallier 87]   Gallier, J. H. and Snyder, W.: *A general complete E-unification procedure*, Rewriting Techniques and Applications, LNCS 256, pp. 216–227 (1987)

[Hsiang 87]   Hsiang, J.: *Rewrite method for theorem proving in first order theory with equality*, J. Symbolic Computation, 3, 133–151 (1987)

[Huet 78]   Huet, G.: *An algorithm to generate the basis of solutions to homogeneous linear Diophantine equations*, Inform. Process. Lett. 7, pp. 144–147 (1978)

[Huet 80]   Huet, G. and Oppen, D. C.: *Equations and Rewrite Rules: a survey*, Formal Language: Perspectives and Open Problems Academic Press, pp. 349–405 (1980)

[Huet 81]   Huet, G.: *A complete proof of correctness of the Knuth-Bendix completion algorithm*, J. Computer and System Science 23, pp. 11–21 (1981)

[Hullot 80]   Hullot, J. M.: *Canonical forms and unification*, 5th Workshop on Automated Deduction, LNCS 87, pp. 318–334 (1980)

[Knuth 70]   Knuth, D. E. and Bendix, P. B.:   *Simple word problems in universal algebras*, Computational problems in abstract algebra, Pergamon Press, Oxford (1970)

[Makanin 77]   Makanin, G. S.: *The problem of solvability of equations in a free semigroup*, Soviet Akad. Nauk SSSR, Tom 233, No. 2 (1977)

[Robinson 65]   Robinson, J. A.: *A machine-oriented logic based on the resolution principle*, J. ACM 12, pp. 23–41 (1965)

[Rusinowitch 88]   Rusinowitch, M.: *Theorem-proving with resolution and superposition: an extension of the Knuth and Bendix procedure to acomplete set of inference rules,* International Conference on Fifth Generation Computer Systems, 1988, pp. 524–531 (1988)

[Sakai 84] Sakai, K.: *An ordering method for term rewriting systems,* Technical Report 062, ICOT (1984)

[Siekmann 89] Siekmann, J. H.: *Unification theory,* J. Symbolic Computation 7, pp. 207–274 (1989)

[Smullyan 85] Smullyan, R. M.: *To Mock a Mockingbird,* Alfred A. Knopf, Inc. (1985)

[Stickel 81] Stickel, M.E.: *A unification algorithm for associative-commutative functions,* J. ACM, 28, 3, pp. 423–434 (1981)

Unification in
Varieties of Completely Regular Semigroups

Franz Baader

German Research Center for AI (DFKI)

Postfach 2080

W-6750 Kaiserslautern, Germany

e-mail: baader@dfki.uni-kl.de

Abstract

All varieties of idempotent semigroups have been classified with respect to the unification types of their defining sets of identities. With the exception of eight finitary unifying theories, they are all of unification type zero. This yields countably many examples of theories of this type which are more "natural" than the first example constructed by Fages and Huet.

The lattice of all varieties of idempotent semigroups is a sublattice of the lattice of all varieties of orthodox bands of groups, and this lattice is a sublattice of the lattice of all varieties of completely regular semigroups. The proof which was used to establish the result for the varieties of idempotent semigroups of type zero can—with some modifications—also be applied to the larger lattice of all varieties of completely regular semigroups. This shows that type zero is not an exception, but rather common for varieties of semigroups.

To establish the results for the eight exceptional finitary varieties of idempotent semigroups we have developed a method which under certain conditions allows to deduce the unification type of a join of varieties from the types of the varieties participating in this join. This method can also be employed for varieties of orthodox bands of abelian groups. Any variety of orthodox bands of abelian groups is the join of a variety of idempotent semigroups and a variety of abelian groups. It turns out that the unification type of such a join is just the type of the variety of idempotent semigroups taking part in this join.

The emphasis of the paper is on describing the tools necessary for proving all the mentioned results.

1   Introduction

$\mathcal{E}$-unification is concerned with solving term equations modulo an equational theory $\mathcal{E}$. The theory is called *unitary* (*finitary*) if the solutions of an equation can always be represented by one (finitely many) *most general* solutions. Otherwise the theory is of type *infinitary* or *zero*. For infinitary theories, the solutions of an equation can be represented by possibly infinitely many most general solutions; whereas theories of type zero are characterized by the fact that there exists an equation whose set of solutions cannot be represented by its most general solutions. This distinction into four different classes of theories yields the so-called *unification hierarchy* of equational theories.

Equational theories which are of unification type unitary or finitary play an important rôle in automated theorem provers with built in theories (see e.g., [Pl72,Ne74,Sl74,St85]), in generalizations of the Knuth-Bendix algorithm (see e.g., [Hu80,PS81,JK86,Bc87]), and in logic programming with equality (see e.g., [JL84]). Investigating the unification hierarchy and determining unification types of equational theories is thus not only interesting for unification theory, but also has consequences for automated reasoning. In this context, theories of type zero constitute the worst possible case.

Unfortunately, but not at all surprisingly, there cannot be a general algorithm which computes the unification type of a given equational theory from its defining axioms (see [Nu90b]). For this reason, it is important to develop diverse methods which can sometimes be used to determine the unification types of equational theories, and to illustrate the utility of these tools by various examples.

In the present paper, we shall consider equational theories defining varieties of so-called completely regular semigroups under this aspect. This class of semigroups has thoroughly been investigated by semigroup theoreticians (see e.g., [Pe74,Cl79,Pe82,Ge83,Po85]), and several sublattices of the lattice of all varieties of completely regular semigroups have been described completely (see e.g., [Bi70,Ge70,Fe71,Pe75]). In the following, we shall not only determine the unification types of some specific theories defining varieties of completely regular semigroups, but shall characterize two of the sublattices—namely, the lattice of all varieties of idempotent semigroups and the lattice of all varieties of orthodox bands of abelian groups—with respect to the unification types of the defining equational theories. However, the present paper will not contain the complete proofs of the correctness of these classifications. The emphasis will be on describing the tools which have been developed to make these proofs feasible. For the first lattice, the proof has already been published in [Ba87]. The proof for the other lattice uses the same ideas, but is rather involved and requires more knowledge about completely regular semigroups. A complete proof for the lattice of all varieties of orthodox bands of abelian groups can be found in [Ba89b].

It has turned out that the worst case, i.e., type zero, is the normal case for this kind of equational theories. In particular, we thus get a countably infinite number of "natural" examples of equational theories of type zero. Such examples are interesting because, until in 1983 F. Fages and G. Huet [FH83] described the first equational theory of type zero, the existence of such a theory was an open problem for more than a decade. But their example is not natural in the sense that the theory was just constructed for that purpose, and has no mathematical meaning of its own. In 1986, M. Schmidt-Schauß [Sc86] and the author [Ba86] gave the first natural example by showing that the theory of idempotent semigroups is of type zero.

The fact that "most" varieties of completely regular semigroups are of type zero can be established by a single proof. It employs a condition which is sufficient for showing that a theory is of type zero, but need not hold for all theories of type zero (see Section 5.2 below, and [Ba89a]). The proof for the finitary theories splits into two different parts.

1. Some of the finitary theories are treated directly. For two very simple theories, this is done in an "ad hoc" manner; but for the other theories one can use the fact that they are closely connected to the class of so-called commutative theories (see [Ba89c]), and thus can employ the tools developed for this class (see Section 5.4 below).

2. The varieties defined by the remaining finitary theories can be obtained as finite joins of varieties already treated in the above mentioned direct way. For these joins one can employ a method which under certain conditions allows one to deduce the unification type for a join of varieties from the types for the varieties participating in this join (see Section 5.3 below).

The method developed for the join of varieties should also be seen under the following aspect. In unification theory, the so-called *combination problem* has attracted considerable attention. This problem is concerned with the question of how to deduce unification properties of a theory $E = E_1 \cup E_2$ obtained as a union of equational theories from unification properties of the single theories E_1, E_2. Until now the attention was mostly restricted to the case where the theories E_1 and E_2 are built over disjoint signature (see e.g. [Ba91b] for more details and references). For the defined varieties, union of the equational theories means meet of the varieties. In this sense, the problem considered in Section 5.3 is dual to the combination problem.

2 Definitions and Notations

In the following we assume that the reader is familiar with the basic notions of universal algebra [Co65,Gr68]. For more information on unification theory see e.g. [Si89,JK90, Ba91b]. The composition of mappings is written from left to right, that is, $\phi \circ \psi$ or simply $\phi\psi$ means first ϕ and then ψ. Consequently, we use suffix notation for mappings.

2.1 Equational Theories

We assume that two disjoint infinite sets of symbols are given, a set of function symbols and a set of variables. A *signature* Σ is a finite set of function symbols each of which is associated with its arity. Every signature Σ determines a class of Σ-algebras and Σ-homomorphisms. We define Σ-terms and Σ-substitutions as usual. By $\{x_1 \mapsto t_1, \ldots, x_n \mapsto t_n\}$ we denote the substitution which replaces the variables x_i by the terms t_i.

An *equational theory* $\mathcal{E} = (\Sigma, E)$ is a pair consisting of a signature Σ and a set of identities E. The equality of Σ-terms induced by $\mathcal{E}$ will be denoted by $=_{\mathcal{E}}$. Every equational theory $\mathcal{E}$ determines a variety $\mathcal{V}(\mathcal{E})$, the class of all Σ-algebras satisfying each identity of E. For any set of generators X, the variety $\mathcal{V}(\mathcal{E})$ contains a free algebra over $\mathcal{V}(\mathcal{E})$ with generators X, which will be denoted by $\mathcal{F}_{\mathcal{E}}(X)$. Thus any mapping of X into a Σ-algebra A can be uniquely extended to a Σ-homomorphism of $\mathcal{F}_{\mathcal{E}}(X)$ into A.

The set of all subvarieties of a given variety $\mathcal{V}(\mathcal{E})$ forms a complete lattice with respect to set inclusion. This lattice will be denoted by $\mathcal{L}(\mathcal{E})$. The greatest lower bound (meet) of $\mathcal{V}(\mathcal{F})$ and $\mathcal{V}(\mathcal{G})$ in such a lattice is $\mathcal{V}(\mathcal{F}) \cap \mathcal{V}(\mathcal{G})$, and the least upper bound (join) is denoted by $\mathcal{V}(\mathcal{F}) \vee \mathcal{V}(\mathcal{G})$. Please note that in general, the union of $\mathcal{V}(\mathcal{F})$ and $\mathcal{V}(\mathcal{G})$ is a proper subset of $\mathcal{V}(\mathcal{F}) \vee \mathcal{V}(\mathcal{G})$. The following correspondence between the equalities induced by equational theory and the varieties defined by these theories will be important later on.

$$\mathcal{V}(\mathcal{E}) \subseteq \mathcal{V}(\mathcal{F}) \quad \text{iff} \quad =_{\mathcal{F}} \subseteq =_{\mathcal{E}}$$
$$\mathcal{V}(\mathcal{G}) = \mathcal{V}(\mathcal{E}) \vee \mathcal{V}(\mathcal{F}) \quad \text{iff} \quad =_{\mathcal{G}} = =_{\mathcal{E}} \cap =_{\mathcal{F}}$$

2.2 Unification

Let $\mathcal{E} = (\Sigma, E)$ be an equational theory. An $\mathcal{E}$-unification problem is a finite set of equations $\Gamma = \{s_i \doteq t_i \mid 1 \le i \le n\}$, where s_i and t_i are Σ-terms. A substitution θ is called an $\mathcal{E}$-unifier of Γ if $s_i\theta =_{\mathcal{E}} t_i\theta$ for each i. The set of all $\mathcal{E}$-unifiers of Γ is denoted by $U_{\mathcal{E}}(\Gamma)$. In general, one does not need the set of all $\mathcal{E}$-unifiers. A complete set of $\mathcal{E}$-unifiers, i.e., a set of $\mathcal{E}$-unifiers from which all unifiers may be generated by $\mathcal{E}$-instantiation, is usually sufficient. More precisely, for every set of variables X we extend $=_{\mathcal{E}}$ to a relation $=_{\mathcal{E},X}$ between substitutions, and introduce the $\mathcal{E}$-instantiation quasi-ordering $\le_{\mathcal{E},X}$ as follows:

- $\sigma =_{\mathcal{E},X} \theta$ iff $x\sigma =_{\mathcal{E}} x\theta$ for all $x \in X$,

- $\sigma \le_{\mathcal{E},X} \theta$ iff   there exists a substitution λ such that $\theta =_{\mathcal{E},X} \sigma \circ \lambda$.

If $\sigma \le_{\mathcal{E},X} \theta$ and $\theta \not\le_{\mathcal{E},X} \sigma$ then we shall write $\sigma <_{\mathcal{E},X} \theta$. A set $cC_{\mathcal{E}}(\Gamma) \subseteq U_{\mathcal{E}}(\Gamma)$ is a complete set of $\mathcal{E}$-unifiers of Γ iff for every $\mathcal{E}$-unifier θ of Γ there exists $\sigma \in cC_{\mathcal{E}}(\Gamma)$ such that $\sigma \le_{\mathcal{E},X} \theta$, where X is the set of all variables occurring in Γ. For reasons of efficiency, this set should be as small as possible. Thus one is interested in minimal complete sets of $\mathcal{E}$-unifiers, i.e., in complete sets where two different elements are not comparable with respect to $\mathcal{E}$-instantiation.

The unification type of a theory $\mathcal{E}$ is defined with reference to the existence and cardinality of minimal complete sets. The theory $\mathcal{E}$ is unitary (finitary, infinitary, respectively) iff minimal complete sets of $\mathcal{E}$-unifiers always exist, and their cardinality is at most one (always finite, at least once infinite, respectively). The theory $\mathcal{E}$ is of unification type zero iff there exists an $\mathcal{E}$-unification problem without a minimal complete set of $\mathcal{E}$-unifiers.

The distinction between unification type zero and the other three types can also be rephrased as follows. Let us call an $\mathcal{E}$-unifier σ of Γ most general iff $U_{\mathcal{E}}(\Gamma)$ does not contain another unifier θ such that $\theta <_{\mathcal{E},X} \sigma$. It is easy to show that Γ does not have a minimal complete set of $\mathcal{E}$-unifiers iff the set of all most general elements of $U_\mathcal{E}(\Gamma)$ is not complete (see [Ba89a,Ba88] for more information on type zero).

3 Varieties of completely regular semigroups

Let $\Sigma = \{\cdot\}$ be a signature consisting of one binary function symbol "$\cdot$", and let $A = \{x \cdot (y \cdot z) = (x \cdot y) \cdot z\}$ be the set of identities expressing associativity of this symbol. Then $\mathcal{A} = (\Sigma, A)$ defines the variety of all semigroups. As usual, we shall often omit the multiplication sign "$\cdot$", and simply write xy in place of $x \cdot y$. In addition, since we are only dealing with semigroups, we shall often omit all parentheses.

The theory of semigroups has different origins, of which one was the attempt to generalize group theory. For this reason, semigroups which are—in a certain way—close to groups have been extensively studied. Completely regular semigroups are close to groups in the following sense.

A subsemigroup G of a given semigroup S is called a group in S iff G is a group with respect to the multiplication in S; G is maximal iff it is not contained in another group in S. A semigroup is *completely regular* iff it is the disjoint union of its maximal groups (see e.g. [Pe74,Cl79,Pe82,Ge83,Po85,Ba89b] for information on completely regular semigroups). The class of all completely regular semigroups is not a semigroup variety, that means, it cannot be defined by an equational theory over $\Sigma = \{\cdot\}$. But if we augment the signature by a unary symbol "$^{-1}$", the class of all completely regular semigroups is defined by the equational theory $\mathcal{CR} = (\{\cdot,^{-1}\}, CR)$ where the set of identities is

$$CR := \{x \cdot (y \cdot z) = (x \cdot y) \cdot z, x \cdot (x^{-1} \cdot x) = x, x \cdot x^{-1} = x^{-1} \cdot x\}$$

For an element s of a completely regular semigroup S the expression s^{-1} is interpreted by the group inverse of s in the maximal group in S containing s. Thus $e := ss^{-1} = s^{-1}s$ is the neutral element of this group, which explains why $s \cdot (s^{-1} \cdot s) = s$ holds. In particular, e satisfies $ee = e$, i.e., e is an *idempotent* of S. Each of the maximal groups contains exactly one idempotent. We may thus index the groups by the corresponding idempotent, i.e., write G_e for the maximal group in S containing e. In the following, we shall abbreviate the expression ss^{-1} by s^0.

In general, the product of two idempotents in a completely regular semigroup S need not be idempotent. But if this is always the case, i.e., if the idempotents form a subsemigroup of S, then S is called *orthodox*. The class of all orthodox completely regular semigroups (see e.g. [GP81]) is a subvariety of $\mathcal{V}(\mathcal{CR})$ which can be defined by the set of identities $CR \cup \{xy = xyy^{-1}x^{-1}xy\}$.

If S is an orthodox completely regular semigroup, and e, f are idempotents in S then ef is also idempotent. If we consider the corresponding groups G_e, G_f, G_{ef} then the product of an element of G_e with an element of G_f need not be in G_{ef}. However, if $G_e \cdot G_f \subseteq G_{ef}$ for all idempotents e, f in S, then S is called *orthodox band of groups*. The class of all orthodox bands of groups (see e.g. [Pe75]) is also a subvariety of $\mathcal{V}(\mathcal{CR})$, defined by the identities $OBG := CR \cup \{x^0 y^0 = (xy)^0\}$. If in addition all the maximal groups in an orthodox band of groups are abelian then we have an *orthodox band of abelian groups*.

If an orthodox band of groups contains only one idempotent then it is a group. On the other hand, if all its maximal groups are trivial then we have an *idempotent semigroup* (*band*), i.e., a semigroup having only idempotent elements. As a subvariety of $\mathcal{V}(\mathcal{CR})$, the class of all groups can be defined by the identities $G := CR \cup \{x^0 = y^0\}$, and the class of all idempotent semigroups by $B := CR \cup \{xx = x\}$. Since in idempotent semigroups we have $x^{-1} = x$, the variety of idempotent semigroups can also be defined by identities over the signature $\Sigma = \{\cdot\}$, which means that it can be seen as a subvariety of the variety of all semigroups.

Now we are ready to introduce the lattices whose elements will later on be characterized with respect to the unification types of their defining equational theories.

3.1 The lattice of all varieties of idempotent semigroups

Complete descriptions of the lattice $\mathcal{L}(B)$ of all varieties of idempotent semigroups have been given independently by Gerhard [Ge70], Birjukov [Bi70], and Fennemore [Fe71] (see Figure 1).

The lattice has countably many elements, and each of these varieties can be defined by $xx = x$ and one additional identity. Furthermore, the proofs of Gerhard and Fennemore show that the word problem in these varieties is decidable.

3.2 The lattice of all varieties of orthodox bands of abelian groups

Petrich [Pe75] has shown that the *lattice of all varieties of orthodox bands of groups* can be obtained as the direct product of the lattice of all varieties of idempotent semigroups and the lattice of all varieties of groups. More precisely, Petrich proves that the mapping

$$\mathcal{L}(\mathcal{G}) \times \mathcal{L}(\mathcal{B}) \rightarrow \mathcal{L}(\mathcal{OBG}) : (U, V) \mapsto U \vee V$$

is a lattice isomorphism.

Of course, this result determines the lattices of all varieties of orthodox band of groups only modulo the lattice of all varieties of groups, which itself is very complex. However, the situations improves significantly if we restrict our attention to orthodox bands of abelian groups. In fact, the lattice of all varieties of abelian groups has a very simple structure.

Obviously, the variety of all abelian groups is defined by the set of identities $AB := G \cup \{xy = yx\}$. For each nonnegative integer n, we define the set of identities $AB_n := AB \cup \{x^n = x^0\}$.[1] It is easy to see that the varieties $\mathcal{V}(AB_n)$ are all the subvarieties of

---

[1] Recall that x^0 is the unit element of the group.

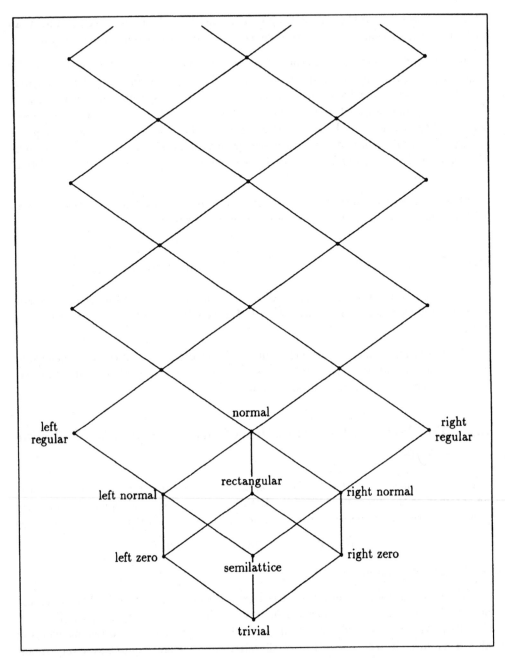

Figure 1: The lattice of all varieties of idempotent semigroups

$\mathcal{V}(\mathcal{AB})$. In addition, one has $\mathcal{V}(\mathcal{AB}_m) \subseteq \mathcal{V}(\mathcal{AB}_n)$ iff m devides n. This means that the lattice $\mathcal{L}(\mathcal{AB})$ is isomorphic to the lattice induced by divisability on nonnegative integers.

Together with Petrich's result this provides us with a complete description of the *lattices of all varieties of orthodox band of groups*: it is the direct product of the lattice depicted in Figure 1 and the lattice of nonnegative integers with respect to divisability.

4 The results on unification in varieties of completely regular semigroups

In [Ba87] the lattice of all varieties of idempotent semigroups has completely been characterized with respect to the unification types of the sets of identities defining the varieties. This result is summarized in the following theorem.

Theorem 4.1 *Let $\mathcal{E}$ be an equational theory defining a variety of idempotent semigroups.*

1. *$\mathcal{E}$ is unitary iff it defines one of the following four varieties: the variety of all trivial bands, the variety of all left zero bands, the variety of all right zero bands, or the variety of all rectangular bands.*

2. *$\mathcal{E}$ is finitary (but not unitary) iff it defines one of the following four varieties: the variety of all semilattices, the variety of all left normal bands, the variety of all right normal bands, or the variety of all normal bands.*

3. *Otherwise, $\mathcal{E}$ is of type zero.*

A look at the lattice shown in Figure 1 makes clear that the third part of the theorem can also be rephrased as follows: If a variety of idempotent semigroups contains the variety of all left regular bands or the variety of all right regular bands, then its defining equational theory is of unification type zero. In [Ba89b] I have shown that this result can be generalized to all varieties of completely regular semigroups.

Theorem 4.2 *Let $\mathcal{E}$ be an equational theory defining a variety of completely regular semigroups. If $\mathcal{V}(\mathcal{E})$ contains the variety of all left regular bands or the variety of all right regular bands, then $\mathcal{E}$ is of type zero.*

In the previous section we have seen that any variety of orthodox bands of abelian groups can be obtained as the join of a variety of idempotent semigroups and a variety of abelian groups. As an easy consequence of the above two theorems we get that such a variety is defined by a theory of type zero if the variety of idempotent semigroups participating in this join is defined by a theory of type zero. In [Ba89b] I have shown that the opposite direction is also true.

Theorem 4.3 *Let $\mathcal{E}$ be an equational theory defining a variety of orthodox bands of abelian groups, and let $\mathcal{V}(\mathcal{E}) = \mathcal{V}(\mathcal{AB}_n) \vee \mathcal{V}(\mathcal{F})$ where n is a nonnegative integer and $\mathcal{F}$ defines a variety of idempotent semigroups. The unification type of $\mathcal{E}$ is the same as the unification type of $\mathcal{F}$.*

In the remainder of this paper we shall first describe the tools which have been developed for proving all these results. Then the utility of the tools will be demonstrated by showing how they can be used in the proofs of the above theorems. However, in the present paper we can only give some parts of the proofs. The complete proof for the first theorem can be found in [Ba87], and for the other two theorems in [Ba89b].

5 The employed tools

In Section 2, unification problems have been introduced as finite systems of term equations. In the first subsection, we shall point out that for the purpose of the present paper it is enough to restrict the attention to single equations, i.e., unification problems of cardinality one. We shall then introduce a condition which can be used to prove the results for the theories of type zero. In the third subsection, we shall describe a method which under certain conditions allows one to deduce the unification type for a join of varieties from the types for the varieties participating in this join. Subsequently, we shall describe conditions under which commutative theories are unitary.

5.1 Single equations versus systems of equations

In order to show that a theory $\mathcal{E}$ is of type unitary or finitary, it is enough to demonstrate this for $\mathcal{E}$-unification problems of cardinality one, i.e., problems of the form $\{s \doteq t\}$. This fact is e.g. helpful for theories which are treated in a direct, ad hoc manner. To be more precise, we have the following proposition.

Proposition 5.1 *If for all $\mathcal{E}$-unification problems of the form $\{s \doteq t\}$ there exist complete sets of $\mathcal{E}$-unifiers of cardinality at most one (finite cardinality) then $\mathcal{E}$ is unitary (finitary).*

Proof. If a finite complete set of unifiers exists then there also exists a finite minimal complete set. This set can be obtained by simply removing redundant elements. The remaining part of the proof is an immediate consequence of the following fact: Let Γ be a finite unification problem, and let s, t be terms. If $cU_{\mathcal{E}}(\Gamma)$ is a finite complete set of $\mathcal{E}$-unifiers of Γ, and if for all substitutions $\sigma \in cU_{\mathcal{E}}(\Gamma)$, the sets $cU_{\mathcal{E}}(s\sigma, t\sigma)$ are finite complete sets of $\mathcal{E}$-unifiers of $\{s\sigma \doteq t\sigma\}$ then

$$\bigcup_{\sigma \in cU_{\mathcal{E}}(\Gamma)} \sigma \circ cU_{\mathcal{E}}(s\sigma, t\sigma) := \{\sigma \circ \tau \mid \sigma \in cU_{\mathcal{E}}(\Gamma) \text{ and } \tau \in cU_{\mathcal{E}}(s\sigma, t\sigma)\}$$

is a finite complete set of $\mathcal{E}$-unifiers of $\Gamma \cup \{s \doteq t\}$ (see e.g., [He87]). Obviously, if the complete sets $cU_\mathcal{E}(\Gamma)$ and $cU_\mathcal{E}(s\sigma, t\sigma)$ are of cardinality one then this new set is also a singleton. □

On the other hand, if we are able to show that an $\mathcal{E}$-unification problem of the form $\{s \doteq t\}$ does not have a minimal complete set of $\mathcal{E}$-unifiers then we obviously have shown that $\mathcal{E}$ is of type zero.

But note that in other cases there may be a difference between consider unification problems of cardinality one (single equations) or arbitrary finite unification problems (systems of equations). For example, there exists a theory $\mathcal{E}$ such that unification is infinitary for single equations, but of type zero for systems of equations (see [Ba91b] for references).

5.2 A sufficient condition for unification type zero

Let $\mathcal{E}$ be an equational theory, Γ be an $\mathcal{E}$-unification problem, and X be the set of all variables occurring in Γ.

Condition 5.2 *There is a decreasing chain $\theta_1 \geq_{\mathcal{E},X} \theta_2 \geq_{\mathcal{E},X} \theta_3 \geq_{\mathcal{E},X} \ldots$ in $U_\mathcal{E}(\Gamma)$ without lower bound in $U_\mathcal{E}(\Gamma)$ which has the property: for all $n \geq 1$ and all $\theta \in U_\mathcal{E}(\Gamma)$ with $\theta_n \geq_{\mathcal{E},X} \theta$ there exists $\hat{\theta} \in U_\mathcal{E}(\Gamma)$ such that $\theta \geq_{\mathcal{E},X} \hat{\theta}$ and $\theta_{n+1} \geq_{\mathcal{E},X} \hat{\theta}$.*

The following picture illustrates this condition.

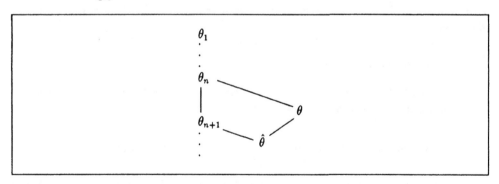

Proposition 5.3 *The equational theory $\mathcal{E}$ is of type zero if there exists an $\mathcal{E}$-unification problem satisfying Condition 5.2.*

Proof. Assume that the $\mathcal{E}$-unification problem Γ has a minimal complete set $\mu U_\mathcal{E}(\Gamma)$, although it satisfies the condition. Since θ_1 is an $\mathcal{E}$-unifier of Γ, there exists an $\mathcal{E}$-unifier θ in $\mu U_\mathcal{E}(\Gamma)$ such that $\theta_1 \geq_{\mathcal{E},X} \theta$. The fact that the decreasing chain has no lower bound in $U_\mathcal{E}(\Gamma)$ implies that there exists an $n \geq 1$ with $\theta_n \geq_{\mathcal{E},X} \theta$ but $\theta_{n+1} \not\geq_{\mathcal{E},X} \theta$. The condition yields an $\mathcal{E}$-unifier $\hat{\theta}$ such that $\theta \geq_{\mathcal{E},X} \hat{\theta}$ and $\theta_{n+1} \geq_{\mathcal{E},X} \hat{\theta}$. Now completeness of the set $\mu U_{\mathcal{E}}(\Gamma)$ yields an $\mathcal{E}$-unifier $\tau \in \mu U_{\mathcal{E}}(\Gamma)$ satisfying $\hat{\theta} \geq_{\mathcal{E},X} \tau$. We thus have $\theta \geq_{\mathcal{E},X} \tau$, which by minimality of $\mu U_{\mathcal{E}}(\Gamma)$ implies $\theta = \tau$. But this gives us $\theta_{n+1} \geq_{\mathcal{E},X} \hat{\theta} \geq_{\mathcal{E},X} \tau = \theta$, which finally is a contradiction to our choice of n. $\qquad\square$

The proposition shows that it is sufficient to establish Condition 5.2 for some $\mathcal{E}$-unification problem, if one wants to prove that $\mathcal{E}$ is of type zero. But note that not all unification problems without minimal complete set of unifiers satisfy this condition (see [Ba89a] for an example).

5.3   Unification and the join of varieties

Let $\mathcal{E}, \mathcal{F}, \mathcal{G}$ be equational theories (over the same signature) such that $\mathcal{V}(\mathcal{G}) = \mathcal{V}(\mathcal{E}) \vee \mathcal{V}(\mathcal{F})$. Please recall that this means $=_{\mathcal{G}} = \,=_{\mathcal{E}} \cap =_{\mathcal{F}}$. We are interested in the following question: Provided that $\mathcal{E}, \mathcal{F}$ are finitary, is $\mathcal{G}$ also finitary? In general this need not be the case. In fact, there exist unitary theories $\mathcal{E}, \mathcal{F}$ and a theory $\mathcal{G}$ such that $\mathcal{V}(\mathcal{G}) = \mathcal{V}(\mathcal{E}) \vee \mathcal{V}(\mathcal{F})$, but $\mathcal{G}$ is not even finitary (see [Ba89b]).

However, in this subsection we shall give sufficient conditions on which the answer to the above question is "yes." But note that these conditions are only sufficient but not necessary for getting a positive answer (see [Ba89b] for an example).

Let X be a finite set of variables, and let $\mathcal{E}, \mathcal{G}$ be equational theories satisfying $=_{\mathcal{G}} \subseteq =_{\mathcal{E}}$.

Condition 5.4 *For any substitution ν there exists a finite set $\Sigma_{\mathcal{E}}(\nu)$ of substitutions with the properties:*

1. *Any ν' in $\Sigma_{\mathcal{E}}(\nu)$ satisfies $\nu' \geq_{\mathcal{E},X} \nu$.*

2. *For any σ, λ such that $\sigma =_{\mathcal{E},X} \nu \circ \lambda$ there exist a substitution ν' in $\Sigma_{\mathcal{E}}(\nu)$ and a substitution λ' such that $\sigma =_{\mathcal{G},X} \nu' \circ \lambda'$.*

The next proposition states that this condition has the desired property.

Proposition 5.5 *Let $\mathcal{E}, \mathcal{F}, \mathcal{G}$ be equational theories over the same signature such that $\mathcal{E}, \mathcal{F}$ are finitary and $\mathcal{V}(\mathcal{G}) = \mathcal{V}(\mathcal{E}) \vee \mathcal{V}(\mathcal{F})$. If for any finite set X of variables, Condition 5.4 holds for $\mathcal{E}, \mathcal{G}$ and $\mathcal{F}, \mathcal{G}$, then $\mathcal{G}$ is also finitary.*

Proof. By Proposition 5.1 we know that it is sufficient to consider unification problems of the form $\Gamma = \{s \doteq t\}$, and to show that they have finite complete sets of $\mathcal{G}$-unifiers.

(1) First we show that there exists a finite set T of $\mathcal{F}$-unifiers of Γ such that any $\mathcal{G}$-unifier θ of Γ may be written as $\theta =_{\mathcal{G},X_0} \tau \circ \sigma$ for some τ in T and some substitution σ

(where X_0 denotes the set of variables occurring in s, t).

Let θ be a $\mathcal{G}$-unifier of Γ, and let $\mu U_{\mathcal{F}}(\Gamma)$ be a finite minimal complete set of $\mathcal{F}$-unifiers of Γ. Since $=_{\mathcal{G}} \subseteq =_{\mathcal{F}}$, the substitution θ is also an $\mathcal{F}$-unifier of Γ, and thus there exists $\delta \in \mu U_{\mathcal{F}}(\Gamma)$ and a substitution λ with $\theta =_{\mathcal{F}, X_0} \delta \circ \lambda$. Condition 5.4 (applied for $\mathcal{F}, \mathcal{G}$ and $X = X_0$) provides us with a finite set $\Sigma_{\mathcal{F}}(\delta)$. Now there exists some $\delta' \in \Sigma_{\mathcal{F}}(\delta)$ and a substitution λ' such that $\theta =_{\mathcal{G}, X_0} \delta' \circ \lambda'$. Thus we may define

$$T := \bigcup_{\delta \in \mu U_{\mathcal{F}}(\Gamma)} \Sigma_{\mathcal{F}}(\delta).$$

Obviously, T is finite, and since any $\delta' \in \Sigma_{\mathcal{F}}(\delta)$ is an $\mathcal{F}$-instance of the $\mathcal{F}$-unifier δ of Γ, the set T contains only $\mathcal{F}$-unifiers of Γ.

(2) Let $T = \{\tau_1, \ldots, \tau_n\}$. For $i = 1, \ldots, n$ we consider the unification problems $\Gamma_i := \{s_i \doteq t_i\}$ where $s_i := s\tau_i$ and $t_i := t\tau_i$. By X_i we denote the set of variables occurring in s_i, t_i. Since τ_i is an $\mathcal{F}$-unifier of Γ, we have $s_i =_{\mathcal{F}} t_i$. Let N_i be a finite complete set of $\mathcal{E}$-unifiers of Γ_i. We apply Condition 5.4 (for $\mathcal{E}, \mathcal{G}$ and $X = X_i$), and define

$$\Sigma_{\mathcal{E}}(N_i) := \bigcup_{\nu \in N_i} \Sigma_{\mathcal{E}}(\nu) \quad \text{and} \quad \Sigma := \bigcup_{i=1}^{n} \{\tau_i \circ \nu' \mid \nu' \in \Sigma_{\mathcal{E}}(N_i)\}.$$

Obviously, Σ is a finite set of substitutions, and we shall show that it is a complete set of $\mathcal{G}$-unifiers of Γ.

(3) Let σ be an element of Σ, i.e., $\sigma = \tau_i \circ \nu'$ for some i, $1 \le i \le n$, and some $\nu' \in \Sigma_{\mathcal{E}}(N_i)$. Since $\nu' \ge_{\mathcal{E}, X_i} \nu$ for some ν in N_i, we know that ν' is an $\mathcal{E}$-unifier of Γ_i, i.e., $s\tau_i\nu' =_{\mathcal{E}} t\tau_i\nu'$. On the other hand, $s\tau_i =_{\mathcal{F}} t\tau_i$ implies $s\tau_i\nu' =_{\mathcal{F}} t\tau_i\nu'$. Thus $\sigma = \tau_i \circ \nu'$ is a $\mathcal{G}$-unifier of $\Gamma = \{s \doteq t\}$.

(4) Let θ be a $\mathcal{G}$-unifier of Γ. By (1), there exist a substitution τ_i in T and a substitution σ such that $\theta =_{\mathcal{G}, X_0} \tau_i \circ \sigma$. Now $(s\tau_i)\sigma =_{\mathcal{G}} s\theta =_{\mathcal{G}} t\theta =_{\mathcal{G}} (t\tau_i)\sigma$ implies that σ is a $\mathcal{G}$-unifier, and thus an $\mathcal{E}$-unifier of $\Gamma_i = \{s_i \doteq t_i\}$. Hence there exist some $\nu \in N_i$ and some substitution λ satisfying $\sigma =_{\mathcal{E}, X_i} \nu \circ \lambda$. Condition 5.4 yields a substitution λ' and an element ν' of $\Sigma_{\mathcal{E}}(N_i)$ with the property $\sigma =_{\mathcal{E}, X_i} \nu' \circ \lambda'$. But then we have $\theta =_{\mathcal{G}, X_0} \tau_i \circ \nu' \circ \lambda'$, which means that $\theta \ge_{\mathcal{G}, X_0} \tau_i \circ \nu'$ for some $\tau_i \circ \nu'$ in Σ. $\qquad \square$

The following corollary states an analogous result for unitary theories. It is an immediate consequence of the definition of Σ in the above proof.

Corollary 5.6 *Let $\mathcal{E}, \mathcal{F}, \mathcal{G}$ be equational theories satisfying the hypothesis of the proposition. If in addition, $\mathcal{E}, \mathcal{F}$ are unitary and the sets $\Sigma_{\mathcal{E}}(\nu)$ and $\Sigma_{\mathcal{F}}(\nu)$ of the condition are always singletons, then $\mathcal{G}$ is unitary.*

When proving that Condition 5.4 holds for given theories $\mathcal{E}, \mathcal{G}$, it is often advantageous to use a slightly generalized condition which can be proved by induction. Let $X = \{x_1, \ldots, x_n\}$. Then Condition 5.4 is the case $k = n$ of the following condition.

Condition 5.7 *For any substitution ν and any $k \le n$ there exists a finite set $\Sigma_k(\nu)$ of substitutions with the properties:*

1. *Any ν' in $\Sigma_k(\nu)$ satisfies $\nu' \geq_{\mathcal{E},X} \nu$.*

2. *For any σ, λ such that $\sigma =_{\mathcal{E},X} \nu \circ \lambda$ there exist a substitution ν' in $\Sigma_k(\nu)$ and a substitution λ' such that*

$$x_1\sigma =_{\mathcal{G}} x_1\nu'\lambda', \ldots, x_k\sigma =_{\mathcal{G}} x_k\nu'\lambda', \quad \text{and}$$
$$x_{k+1}\sigma =_{\mathcal{E}} x_{k+1}\nu'\lambda', \ldots, x_n\sigma =_{\mathcal{E}} x_n\nu'\lambda'.$$

In order to prove this condition by induction on k, it is sufficient to establish:

Condition 5.8 *For all $k < n$ and any substitution δ there exists a finite set $\Gamma_k(\delta)$ of substitutions with the properties:*

1. *Any δ' in $\Gamma_k(\delta)$ satisfies $\delta' \geq_{\mathcal{E},X} \delta$.*

2. *If $x_1\sigma =_{\mathcal{G}} x_1\delta\lambda, \ldots, x_k\sigma =_{\mathcal{G}} x_k\delta\lambda$ and $x_{k+1}\sigma =_{\mathcal{E}} x_{k+1}\delta\lambda, \ldots, x_n\sigma =_{\mathcal{E}} x_n\delta\lambda$ then there exist a substitution δ' in $\Gamma_k(\delta)$ and a substitution λ' such that*

$$x_1\sigma =_{\mathcal{G}} x_1\delta'\lambda', \ldots, x_{k+1}\sigma =_{\mathcal{G}} x_{k+1}\delta'\lambda', \quad \text{and}$$
$$x_{k+2}\sigma =_{\mathcal{E}} x_{k+2}\delta'\lambda', \ldots, x_n\sigma =_{\mathcal{E}} x_n\delta'\lambda'.$$

In fact, if this condition holds one can define

$$\Sigma_{k+1}(\nu) := \bigcup_{\nu' \in \Sigma_k(\nu)} \Gamma_k(\nu')$$

during the inductive proof of Condition 5.7.

5.4 Unification in commutative theories

Motivated by a categorical reformulation of $\mathcal{E}$-unification, the class of commutative theories was defined in [Ba89c] by properties of the category $\mathcal{C}(\mathcal{E})$ of finitely generated $\mathcal{E}$-free objects as follows: an equational theory $\mathcal{E}$ is commutative iff the corresponding category $\mathcal{C}(\mathcal{E})$ is semiadditive (see e.g., [HS73,Ba89c] for the definition and for properties of semiadditive categories). In order to give a more algebraic definition, we need some more notation from universal algebra. Let $\mathcal{E} = (\Sigma, E)$ be an equational theory.

A constant symbol e of the signature Σ is called *idempotent in $\mathcal{E}$* iff for all symbols $f \in \Sigma$ we have $f(e, \ldots, e) =_{\mathcal{E}} e$. Note that for nullary f this means $f =_{\mathcal{E}} e$.

Let $\mathcal{K}$ be a class of Σ-algebras. An *n-ary implicit operation* in $\mathcal{K}$ is a family $o = \{o_A \mid A \in \mathcal{K}\}$ of mappings $o_A : A^n \to A$ which is compatible with all homomorphisms, i.e., for all homomorphisms $\omega : A \to B$ with $A, B \in \mathcal{K}$ and all $a_1, \ldots, a_n \in A$, $o_A(a_1, \ldots, a_n)\omega = o_B(a_1\omega, \ldots, a_n\omega)$ holds. We shall usually omit the index and just write o in place of o_A.

In the following algebraic definition of commutative theories, $\mathcal{F}(\mathcal{E})$ denotes the class of all free algebra over $\mathcal{V}(\mathcal{E})$ with finite sets of generators.

An equational theory $\mathcal{E} = (\Sigma, E)$ is called *commutative* iff the following holds:

1. The signature Σ contains a constant symbol e which is idempotent in $\mathcal{E}$.

2. There is a binary implicit operation "$*$" in $\mathcal{F}(\mathcal{E})$ such that

   (a) The constant e is a neutral element for "$*$" in any algebra $A \in \mathcal{F}(\mathcal{E})$.

   (b) For any n-ary function symbol $f \in \Sigma$, any algebra $A \in \mathcal{F}(\mathcal{E})$, and any $s_1, \ldots, s_n, t_1, \ldots, t_n \in A$ we have

   $$f(s_1 * t_1, \ldots, s_n * t_n) = f(s_1, \ldots, s_n) * f(t_1, \ldots, t_n).$$

Though it is not explicitly required by the definition, the implicit operation "$*$" turns out to be associative and commutative (see [Ba89c], Corollary 5.4). This justifies the name "commutative theory."

Well-known examples of commutative theories are the theory $\mathcal{AM}$ of abelian monoids, the theory $\mathcal{AIM}$ of idempotent abelian monoids, and the theory $\mathcal{AB}$ of abelian groups (see [Ba89c]). In these theories, the implicit operation "$*$" is given by the explicit binary operation in the signature. An example for a commutative theory where "$*$" is really implicit can also be found in [Ba89c] (Example 5.1).

W. Nutt observed [Nu90a] that commutative theories are—modulo a translation of the signature—what he calls monoidal theories (see [BN91] for a proof), and that unification in a monoidal theory $\mathcal{E}$ may be reduced to solving linear equations in a certain semiring $\mathcal{S}(\mathcal{E})$. Recall that a semiring is similar to a ring, with the only difference that it need not be a group, but simply a monoid with respect to addition.

Roughly speaking, the *semiring* $\mathcal{S}(\mathcal{E})$ can be obtained from a commutative theory $\mathcal{E}$ as follows (see [Nu90a,Ba91a] for details). Let x be an arbitrary generator.

- The elements of $\mathcal{S}(\mathcal{E})$ are the endomorphisms of $\mathcal{F}_{\mathcal{E}}(\{x\})$.

- The addition and the zero of $\mathcal{S}(\mathcal{E})$ are "induced" by the idempotent constant and the implicit operation.

- The multiplication and the unit of $\mathcal{S}(\mathcal{E})$ are given by the composition of the endomorphisms and the identity mapping.

For the theory $\mathcal{AM}$ of abelian monoids, $\mathcal{S}(\mathcal{AM})$ is the semiring of non-negative integers; the theory $\mathcal{AIM}$ of idempotent abelian monoids has the two-element Boolean algebra as corresponding semiring; and the semiring corresponding to the theory $\mathcal{AB}$ of abelian groups is the ring of integers.

It can be shown that commutative theories are either unitary or of type zero (see [Ba89c,Nu90a]), and the type unitary can be characterized by algebraic properties of the corresponding semiring as follows (see [Nu90a,Ba91a]).

Proposition 5.9 *A commutative theory $\mathcal{E}$ is unitary iff the corresponding semiring satisfies the following condition: For any $n, m \geq 1$, and any pair M_1, M_2 of $m \times n$-matrices over $S(\mathcal{E})$ the set*

$$\mathcal{U}(M_1, M_2) := \{\underline{x} \in S(\mathcal{E})^n \mid M_1 \cdot \underline{x} = M_2 \cdot \underline{x}\}$$

is a finitely generated right $S(\mathcal{E})$-semimodule.

Since the ring of integers is noetherian, one gets as an immediate consequence of this proposition that the theory $\mathcal{AB}$ of abelian groups is unitary. If the $\mathcal{E}$-free algebra in one generator, $\mathcal{F}_\mathcal{E}(\{x\})$, is finite then $S(\mathcal{E})$ is also finite. Consequently, the condition of the proposition is trivially satisfied.

Corollary 5.10 *Let $\mathcal{E}$ be a commutative theory. If the $\mathcal{E}$-free algebra in one generator is finite then $\mathcal{E}$ is of unification type unitary.*

An example of a theory to which this corollary can be applied is the theory $\mathcal{AIM}$ of idempotent abelian monoids.

6 Some examples of how the tools can be used

In order to demonstrate the utility of the tools developed above we shall now describe some of the proofs of our results on unification in varieties of completely regular semigroups. As mentioned before, complete proofs can be found in [Ba87,Ba89b].

6.1 Left zero, right zero, and rectangular bands

The variety of all left zero bands can be defined by the equational theory $\mathcal{LZ} = (\Sigma, LZ)$ where $\Sigma := \{\cdot\}$ and $LZ := \{x \cdot y = x\}$. For a Σ-term $s = x_1 \cdot x_2 \cdot \ldots \cdot x_n$ we denote by $head(s)$ the first symbol of s, i.e., x_1. Obviously $s =_{\mathcal{LZ}} head(s)$, and it is easy to see that $s =_{\mathcal{LZ}} t$ iff $head(s) = head(t)$.

By Proposition 5.1 it is enough to consider $\mathcal{LZ}$-unification problems of cardinality one if one wants to show that this theory is unitary. Let $\Gamma = \{s \doteq t\}$ be such a problem, and let z be a variable not occurring in s or t. Evidently, the substitution $\theta := \{head(s) \mapsto z, head(t) \mapsto z\}$ is an $\mathcal{LZ}$-unifier of Γ. Now let τ be an arbitrary unifier. Then we have $head(s)\tau =_{\mathcal{LZ}} s\tau =_{\mathcal{LZ}} t\tau =_{\mathcal{LZ}} head(t)\tau$. This shows that, if we define

$$\lambda := \{z \mapsto head(s)\tau\} \cup \{x \mapsto x\tau \mid x \text{ is a variable different from } z, head(s), head(t)\},$$

we have $\theta\lambda =_{\mathcal{LZ},X} \tau$ where X is the set of variables occurring in Γ. This proves that $\{\theta\}$ is a complete set of $\mathcal{LZ}$-unifiers of Γ, which means that we have shown that $\mathcal{LZ}$ is unitary.

An analogous proof can be used to establish this result for the theory $\mathcal{RZ}$ of right zero bands. Since the variety of all rectangular bands is the join of the varieties $\mathcal{V}(\mathcal{LZ})$ and

$V(\mathcal{RZ})$, one can try to apply Corollary 5.6 to show that the theory $\mathcal{RB}$ of rectangular bands is unitary. The set of identities for $\mathcal{RB}$ is $RB = B \cup \{(x \cdot y) \cdot x = x\}$, and it is well-known that $s =_{\mathcal{RB}} t$ iff $head(s) = head(t)$ and $tail(s) = tail(t)$ (where $tail(s), tail(t)$ denote the last symbols of s, t).

Because of the symmetry between $\mathcal{LZ}$ and $\mathcal{RZ}$ it is enough to *establish Condition 5.4 for $\mathcal{LZ}$ and $\mathcal{RB}$*:

For any finite set of variables X, and any substitution ν we define $\Sigma_{\mathcal{LZ}}(\nu) := \{\nu'\}$, where $x\nu' := (x\nu) \cdot z_x$ for all variables $x \in X$. The variables z_x in this definition are meant to be new variables.

Obviously, we have $x\nu' =_{\mathcal{LZ}} x\nu$ for all $x \in X$, which shows $\nu' \geq_{\mathcal{LZ},X} \nu$. Now assume that σ, λ are substitutions such that $\sigma =_{\mathcal{LZ},X} \nu \circ \lambda$, i.e., $head(x\nu\lambda) = head(x\sigma)$ for all $x \in X$. For $\lambda' := \lambda \cup \{z_x \mapsto tail(x\sigma) \mid x \in X\}$ we thus have $head(x\sigma) = head(x\nu\lambda) = head(x\nu\lambda \cdot z_x\lambda') = head(x\nu'\lambda')$, and $tail(x\sigma) = z_x\lambda' = tail(x\nu\lambda \cdot z_x\lambda') = tail(x\nu'\lambda')$ for all $x \in X$. But that means that $\sigma =_{\mathcal{RB},X} \nu' \circ \lambda'$ as required by Condition 5.4.

6.2 Abelian groups and semilattices

It is easy to see that for all $m \geq 0$, the theory $\mathcal{AB}_m$ is a commutative theory (in the sense of Section 5.4). Since for $m > 0$, the $\mathcal{AB}_m$-free algebra in one generator is the cyclic group of cardinality m, Corollary 5.10 applies, and we immediately obtain the result that the theories $\mathcal{AB}_m$ for $m > 0$ are unitary. For the case $m = 0$, we have already mentioned above that the semiring $\mathcal{S}(\mathcal{AB})$ is the ring of integers. Since this ring is noetherian, one gets as an immediate consequence of Proposition 5.9 that the theory $\mathcal{AB}$ of abelian groups is unitary.

The theory $\mathcal{SL}$ of semilattices has $SL := B \cup \{x \cdot y = y \cdot x\}$ as its set of identities. This theory is not commutative because there need not exist a unit element for the the multiplication. However, this theory is closely connected to the theory $\mathcal{AIM}$ of idempotent abelian monoids, which is a commutative theory. The theory $\mathcal{AIM}$ is obtained from SL by augmenting SL by a unit element. We have already mentioned above that $\mathcal{AIM}$ is unitary, and that this fact is an immediate consequence of Corollary 5.10.

The theory $\mathcal{SL}$ is not unitary but finitary. A finite complete set of $\mathcal{SL}$-unifiers for a unification problem Γ can easily be obtained from the most general $\mathcal{AIM}$-unifier σ of Γ by erasing variables in the range of σ in all possible ways (see [Ba87] for details and for the proof of correctness). For example, if $\sigma := \{x_1 \mapsto y_1 \cdot y_2, x_2 \mapsto y_1 \cdot y_2 \cdot y_3\}$ is the most general $\mathcal{AIM}$-unifier of Γ, we get the $\mathcal{SL}$-unifiers $\{x_1 \mapsto y_2, x_2 \mapsto y_2 \cdot y_3\}$ by erasing y_1, $\{x_1 \mapsto y_1, x_2 \mapsto y_1 \cdot y_3\}$ by erasing y_2, $\{x_1 \mapsto y_1 \cdot y_2, x_2 \mapsto y_1 \cdot y_2\}$ by erasing y_3, $\{x_1 \mapsto y_2, x_2 \mapsto y_2\}$ by erasing y_1 and y_3, and $\{x_1 \mapsto y_1, x_2 \mapsto y_1\}$ by erasing y_2 and y_3. But note that we cannot erase both y_1 and y_2 because then the immage of x_1 would be empty. This is not allowed for $\mathcal{SL}$ since there is no unit element.

6.3 Left normal bands

Since we already know that the theories $\mathcal{LZ}$ and $\mathcal{SL}$ are finitary, and since the variety of all left normal bands is the join of $\mathcal{V}(\mathcal{LZ})$ and $\mathcal{V}(\mathcal{SL})$, we can try to apply Proposition 5.5 when showing that the theory $\mathcal{LN}$ of left normal bands is finitary. Condition 5.4 can easily be established for $\mathcal{LZ}$ and $\mathcal{LN}$ by a proof which is very similar to the one given above for $\mathcal{LZ}$ and $\mathcal{RB}$.

Instead of proving Condition 5.4 for $\mathcal{SL}$ and $\mathcal{LN}$ directly, we establish Condition 5.8. We shall use the following characterizations of the equalities on terms induced by $\mathcal{SL}$ and $\mathcal{LN}$ in the proof. Recall that $head(s)$ denotes the first variable occurring in the term s, and let $contents(s)$ denote the set of all variables occurring in s. Then it is easy to show that

$$s =_{\mathcal{SL}} t \quad \text{iff} \quad contents(s) = contents(t)$$
$$s =_{\mathcal{LN}} t \quad \text{iff} \quad s =_{\mathcal{SL}} t \text{ and } s =_{\mathcal{LZ}} t, \text{ i.e.,}$$
$$\text{iff} \quad contents(s) = contents(t) \text{ and } head(s) = head(t)$$

Now we show how the set $\Gamma_k(\delta)$ required by Condition 5.8 can be defined. Let $contents(x_{k+1}\delta) := \{y_1, \ldots, y_r\}$ be the variables occurring in $x_{k+1}\delta$. For each $j, 1 \leq j \leq r$, we define a substitution δ_j in the following way. Let z be a new variable, and let ν_j be defined as $\nu_j := \{y_j \mapsto y_j \cdot z\}$. Then

$$x_i\delta_j := x_i\delta\nu_j \text{ for } i \neq k+1 \text{ and } x_{k+1}\delta_j := z \cdot y_j \cdot (x_{k+1}\delta_j\nu_j).$$

The set $\Gamma_k(\delta)$ consists of these substitutions δ_j for all $j, 1 \leq j \leq r$.

Since y_j occurs in $x_{k+1}\delta$, we know that y_j and z occur in $x_{k+1}\delta\nu_j$, and this obviously means that $x_{k+1}\delta_j =_{\mathcal{SL}} x_{k+1}\delta\nu_j$. Consequently, we have $\delta_j \geq_{\mathcal{SL},X} \delta$, which shows that the first part of Condition 5.8 is satisfied.

To establish the second part of the condition, assume that σ, λ are substitutions such that

$$x_1\sigma =_{\mathcal{LN}} x_1\delta\lambda, \ldots, x_k\sigma =_{\mathcal{LN}} x_k\delta\lambda, \text{ and } x_{k+1}\sigma =_{\mathcal{SL}} x_{k+1}\delta\lambda, \ldots, x_n\sigma =_{\mathcal{SL}} x_n\delta\lambda.$$

We denote $head(x_{k+1}\sigma)$ by y. Since y occurs in $x_{k+1}\sigma$ and $x_{k+1}\sigma =_{\mathcal{SL}} x_{k+1}\delta\lambda$, y also occurs in $x_{k+1}\delta\lambda$. Thus there exists a variable v in $x_{k+1}\delta$ such that y occurs in $v\lambda$, i.e., there exists $j, 1 \leq j \leq r$, such that y occurs in $y_j\lambda$.

We take $\delta_j \in \Gamma_k(\delta)$, and define $\lambda' := \lambda \cup \{z \mapsto y\}$. Since y occurs in $y_j\lambda$, we have $(y_j \cdot z)\lambda' =_{\mathcal{LN}} y_j\lambda$ and $(z \cdot y_j)\lambda' =_{\mathcal{SL}} y_j\lambda$. As a consequence, we get $x_i\delta_j\lambda' =_{\mathcal{LN}} x_i\sigma$ for $i = 1, \ldots, k$ and $x_i\delta_j\lambda' =_{\mathcal{SL}} x_i\sigma$ for $i = k+1, \ldots, n$. The fact that $head(x_{k+1}\delta_j\lambda') = z\lambda' = y = head(x_{k+1}\sigma)$ finally yields $x_{k+1}\delta_j\lambda' =_{\mathcal{LN}} x_{k+1}\sigma$.

6.4 The other unitary and finitary theories

The result for the variety of all right normal bands can be proved analogously, and the proof for the variety of all normal band is very similar. The remaining varieties are of the form $\mathcal{AB}_m \vee \mathcal{V}(\mathcal{F})$ where $\mathcal{F}$ is one of the eight unitary or finitary theories defining a variety of idempotent semigroups. For reasons of symmetry one does not have to consider all these cases, but the proofs for the remaining cases are rather complex (see [Ba89b]). Nevertheless they demonstrate the strength of the tools developed in Section 5 because it is not all clear how these results could be proved directly, without employing the join conditions.

6.5 The theories of type zero

In this subsection we shall give a sketch of the proof of Theorem 4.2. The complete proof can be found in [Ba89b]; and a similar proof is given in [Ba87] for the special case of varieties of idempotent semigroups.

Now assume that $\mathcal{E}$ is an equational theory defining a variety of completely regular semigroups such that $\mathcal{V}(\mathcal{E})$ contains the variety of all left regular bands (the case where $\mathcal{V}(\mathcal{E})$ contains the variety of all right regular bands can be treated analogously). Since $\mathcal{V}(\mathcal{E})$ lies between the variety $\mathcal{V}(\mathcal{LR})$ of all left regular bands and the variety $\mathcal{V}(\mathcal{CR})$ of all completely regular semigroups, we know that $=_{\mathcal{CR}} \subseteq =_{\mathcal{E}} \subseteq =_{\mathcal{LR}}$. Thus, whenever we want to prove that two terms are equal w.r.t. $=_{\mathcal{E}}$ it is enough to show that they are equal w.r.t. $=_{\mathcal{CR}}$; and if we want to prove that two terms are not equal w.r.t. $=_{\mathcal{E}}$ it is enough to show that they are not equal w.r.t. $=_{\mathcal{LR}}$.

In order to prove the theorem, we shall consider the $\mathcal{E}$-unification problem $\Gamma = \{(xzuzx)^0 = (xzx)^0\}$, and show that this problem satisfies Condition 5.2. Recall that w^0 is an abbreviation for ww^{-1}. Let X denote the set $\{u, x, z\}$ of all variables occurring in Γ. In the following, all substitutions will be compared with respect to the instantiation ordering $\geq_{\mathcal{E},X}$.

The unifiers θ_n whose existence is required by the condition are defined by

$$
\begin{aligned}
x\theta_n &:= x_1^0 \ldots x_n^0, \\
z\theta_n &:= z_1^0 \ldots z_n^0, \\
u\theta_n &:= (z_n x_n)^0 \ldots (z_1 x_1)^0,
\end{aligned}
$$

where $x_1, \ldots, x_n, z_1, \ldots, z_n$ are variables. Using well-known properties of $=_{\mathcal{CR}}$ it can now be shown (see [Ba89b]) that the sequence $(\theta_n)_{n\geq 1}$ is in fact a decreasing chain $\theta_1 \geq_{\mathcal{E},X} \theta_2 \geq_{\mathcal{E},X} \theta_3 \geq_{\mathcal{E},X} \ldots$ of $\mathcal{E}$-unifiers of Γ, which does not have a lower bound in $U_{\mathcal{E}}(\Gamma)$.

Now assume that θ is an $\mathcal{E}$-unifier of Γ such that $\theta_n \geq_{\mathcal{E},X} \theta$. We define the substitution $\hat{\theta}$ as follows:

$$
\begin{aligned}
x\hat{\theta} &:= (x\theta)(x_{n+1}^0)q, \\
z\hat{\theta} &:= (z\theta)(z_{n+1}^0)q', \\
u\hat{\theta} &:= (z_{n+1}x_{n+1})^0(u\theta),
\end{aligned}
$$

where x_{n+1}, z_{n+1}, q, q' are new variables. In [Ba89b] it is shown that this substitution satisfies the properties required by Condition 5.2, i.e., that $\hat{\theta}$ is an $\mathcal{E}$-unifier of Γ such that $\theta \geq_{\mathcal{E},X} \hat{\theta}$ and $\theta_{n+1} \geq_{\mathcal{E},X} \hat{\theta}$. By Proposition 5.3 we have thus proved that $\mathcal{E}$ is of type zero.

References

[Ba86]   F. Baader, "The Theory of Idempotent Semigroups is of Unification Type Zero," *J. Automated Reasoning* **2**, 1986.

[Ba87]   F. Baader, "Unification in Varieties of Idempotent Semigroups," *Semigroup Forum* **36**, 1987.

[Ba88]   F. Baader, "A Note on Unification Type Zero," *Information Processing Letters* **27**, 1988.

[Ba89a] F. Baader, "Characterizations of Unification Type Zero," *Proceedings of the Third International Conference on Rewriting Techniques and Applications,* Springer LNCS 355, 1989.

[Ba89b] F. Baader, *Unifikation und Reduktionssysteme für Halbgruppenvarietäten,* Dissertation, Institut für Mathematische Maschinen und Datenverarbeitung, Universität Erlangen-Nürnberg , 1989.

[Ba89c] F. Baader, "Unification in Commutative Theories," in C. Kirchner (ed.), *Special Issue on Unification, J. Symbolic Computation* **8**, 1989.

[Ba91a] F. Baader, "Unification in Commutative Theories, Hilbert's Basis Theorem, and Gröbner Bases," to appear in *J. ACM.*

[Ba91b] F. Baader, "Unification Theory," these proceedings.

[BN91]   F. Baader, W. Nutt, "Adding Homomorphisms to Commutative/Monoidal Theories, or: How Algebra Can Help in Equational Unification," *Proceedings of the 4th International Conference on Rewriting Techniques and Applications,* Springer LNCS 488, 1991.

[Bc87]   L. Bachmair, *Proof Methods for Equational Theories,* Ph.D. Thesis, Dep. of Comp. Sci., University of Illinois at Urbana-Champaign, 1987.

[Bi70]   A.P. Birjukov, "Varieties of Idempotent Semigroups," *Algebra i Logica* **9**, 1970; English Translation in *Algebra and Logic* **9**, 1970.

[Cl79]   A.H. Clifford, "The Free Completely Regular Semigroup on a Set," *J. Algebra* **59**, 1979.

[Co65]   P.M. Cohn, *Universal Algebra,* Harper and Row, New York, 1965.

[FH83]   F. Fages, G.P. Huet, "Complete Sets of Unifiers and Matchers in Equational Theories," *Proceedings of the CAAP'83,* Springer LNCS 170, 1983.

[Fe71]   C.F. Fennemore, "All Varieties of Bands I, II," *Math. Nachr.* **48**, 1971.

[Gr68]   G. Grätzer, *Universal Algebra*, Van Nostrand, Princeton, 1968.

[Ge70]   J.A. Gerhard, "The Lattice of Equational Classes of Idempotent Semigroups," *J. Algebra* **15**, 1970.

[Ge83]   J.A. Gerhard, "Free Completely Regular Semigroups I, II," *J. Algebra* **82**, 1983.

[GP81]   J.A. Gerhard, M. Petrich "The Word Problem for Orthogroups," *Can. J. Math.* **4**, 1981.

[He87]   A. Herold, *Combination of Unification Algorithms in Equational Theories*, Dissertation, Fachbereich Informatik, Universität Kaiserslautern , 1987.

[HS73]   H. Herrlich, G.E. Strecker, *Category Theory*, Allyn and Bacon, Boston, 1973.

[Hu80]   G.P. Huet, "Confluent Reductions: Abstract Properties and Applications to Term Rewriting Systems," *J. ACM* **27**, 1980.

[JL84]   J. Jaffar, J.L. Lassez, M. Maher, "A Theory of Complete Logic Programs with Equality," *J. Logic Programming* **1**, 1984.

[JK86]   J.P. Jouannaud, H. Kirchner, "Completion of a Set of Rules Modulo a Set of Equations," *SIAM J. Computing* **15**, 1986.

[JK90]   J.P. Jouannaud, C. Kirchner, "Solving Equations in Abstract Algebras: A Rule-Based Survey of Unification," Preprint, 1990. To appear in the Festschrift to Alan Robinson's birthday.

[Ne74]   A.J. Nevins, "A Human Oriented Logic for Automated Theorem Proving," *J. ACM* **21**, 1974.

[Nu90a] W. Nutt, "Unification in Monoidal Theories," *Proceedings 10th International Conference on Automated Deduction*, Springer LNCS 449, 1990.

[Nu90b] W. Nutt, "The Unification Hierarchy is Undecidable," to appear in *J. Automated Reasoning*. Also SEKI-Report SR-89-06, Universität Kaiserslautern, West Germany, 1989.

[PS81]   G. Peterson, M. Stickel, "Complete Sets of Reductions for Some Equational Theories," *J. ACM* **28**, 1981.

[Pe74]   M. Petrich, "The Structure of Completely Regular Semigroups," *T. AMS* **189**, 1974.

[Pe75]   M. Petrich, "Varieties of Orthodox Bands of Groups," *Pacific J. Math.* **56**, 1975.

[Pe82]   M. Petrich, "On Varieties of Completely Regular Semigroups," *Semigroup Forum* **25**, 1982.

[Pl72]   G. Plotkin, "Building in Equational Theories," *Machine Intelligence* **7**, 1972.

[Po85]   L. Polak, "On Varieties of Completely Regular Semigroups," *Semigroup Forum* **32**, 1985.

[Sc86]   M. Schmidt-Schauß, "Unification under Associativity and Idempotence is of Type Nullary," *J. Automated Reasoning* **2**, 1986.

[Si89]   J.H. Siekmann, "Unification Theory: A Survey," in C. Kirchner (ed.), *Special Issue on Unification, Journal of Symbolic Computation* **7**, 1989.

[Sl74]   J.R. Slagle, "Automated Theorem Proving for Theories with Simplifiers, Commutativity and Associativity," *J. ACM* **21**, 1974.

[St85]   M.E. Stickel, "Automated Deduction by Theory Resolution," *J. Automated Reasoning* **1**, 1985.

A NOTE ON CONFLUENT THUE SYSTEMS *

Ronald V. Book
Department of Mathematics
University of California
Santa Barbara, CA 93106, USA

1 Introduction

A rewriting system is a set of "rewrite rules" that may be applied to objects in a specific domain. (See Dershowitz and Jouannaud [DJ90] for an introduction and survey.) For any rewriting system, there is the "word problem":

INSTANCE: two objects in the domain;
QUESTION: are the two objects "equivalent" modulo the set of rewrite rules?

If one is to use a rewriting system, then it is important to know whether the word problem is decidable. If a finite rewriting system is both confluent and has the property that there are no infinite reduction chains, then the word problem is decidable (see [Hu80] for a discussion). In this note string rewriting is considered and Thue systems are viewed as string rewriting systems, that is, a Thue system is considered to be a rewriting system where the domain consists of the set of all strings over some fixed finite alphabet. Here it is shown that for a finite Thue system T such that (i) T is confluent and (ii) there is a total recursive function R that bounds the lengths of reduction chains of T, there is an algorithm to solve the word problem for T and the algorithm has running time $O(R(n))$, where n is the length of the input.

If a finite Thue system is viewed as a rewriting system and reduction of the length of strings is the basis for defining the notion of reduction for rewriting rules, then for any string w, no sequence of consecutive reductions applied to w can have length (i.e., the number of reductions performed) greater than $|w|$, where $|w|$ denotes the length of w. This fact leads to a linear-time algorithm for the word problem for confluent Thue systems [Bo82] where length is the basis for defining reduction. In this paper a more general situation is studied.

Let $R : N \to N$ be a function that is strictly increasing. Supppose that T is a Thue system with the property that no sequence of consecutive reductions that begins with an application of a reduction to a string w can have length that is greater than $R(|w|)$; such a system is called "R-noetherian." Suppose that T is a finite Thue system that is both

*This paper was written while the author visited the Institut für Informatik, Technische Universität München, Germany. The preparation of this paper was supported in part by the Alexander von Humboldt Stiftung and by the National Science Foundation under Grant CCR-8913584.

confluent and R-noetherian. Then it is intuitively clear that there is an algorithm to solve the word problem for T that has running time $2^{O(R(n))}$, where n is the size of the input. It is not difficult to improve this bound to a polynomial in $R(n)$ by using a nondeterministic procedure (use elementary results about formal languages and Turing machines). The main result of the present note is that one can do better. It is shown that for any finite R-noetherian Thue system T, if T is confluent, then there is an algorithm whose running time is $O(R(n))$ for solving the word problem for T. This is still true when each step in a reduction chain is required to be a leftmost application of a reduction rule.

It should be noted that if T is a finite Thue system that is noetherian, then there is a total recursive function R such that T is R-noetherian. Thus, the result yields an upper bound on the time complexity of the word problem for any finite Thue system that is confluent and noetherian. Madlener and Otto [MO85] have shown that the complexity may be arbitrarily high in the sense that for any total recursive function f, there is a finite Thue system that is confluent and noetherian and whose word problem has time complexity that exceeds f.

2 Preliminaries

Let Σ be a finite alphabet. If w is a string in Σ^*, then w has length denoted by $|w|$.

A Thue system T on Σ is a subset of $\Sigma^* \times \Sigma^*$. Each pair in T is a *relation* or *rewriting rule*. The *Thue congruence generated by* T is the reflexive transitive closure $\overset{*}{\longleftrightarrow}$ of the relation $\leftrightarrow$ defined as follows: for every $x, y \in \Sigma^*$, if $(u, v) \in T$ or $(v, u) \in T$, then $xuy \leftrightarrow xvy$. For each $w \in \Sigma^*$ the *congruence class of* w, denoted $[w]$, is the set $\{z \mid z \overset{*}{\longleftrightarrow} w\}$.

Notice that two strings x and y are congruent if and only if $[x] \cap [y] \neq \emptyset$.

The *word problem* for T is stated as follows:

INSTANCE: two strings x and y;

QUESTION: are x and y congruent? (i.e., $x \overset{*}{\longleftrightarrow} y$?)

We assume that there is a partial order on Σ^*, and that T and $>$ have the properties that (i) if $(u, v) \in T$, then u and v are comparable (i.e., $u > v$ or $u < v$), and (ii) the rewriting rules of T are compatible with this ordering, that is, if $(u, v) \in T$ and $u > v$, then for all $x, y \in \Sigma^*$, $xuy > xvy$. (Notice that ordering by length has properties (i) and (ii).) Based on this, define the binary relation $\overset{*}{\longrightarrow}$ as the reflexive, transitive closure of the relation $\rightarrow$ defined as follows: for every $x, y \in \Sigma^*$, if $(u, v) \in T$ and $u > v$, then $xuy \rightarrow xvy$. Each application of a rewriting rule that results in $w \rightarrow z$ for any $w, z \in \Sigma^*$, is a *reduction*. Each sequence $w_0 \rightarrow w_1 \rightarrow \ldots \rightarrow w_k$, $k \geq 1$, is a *reduction chain of length* k.

A string w is $\rightarrow$-*irreducible* with respect to $(\Sigma^*, \rightarrow)$ (or, just "irreducible" when the context allows) if there is no string z such that $w \rightarrow z$. If $x \overset{*}{\longrightarrow} y$ and y is irreducible, then y is an *irreducible descendant* of x.

It is of interest to consider those Thue systems with the property that all reduction chains have finite length. In this case it is clear that every $w \in \Sigma^*$ has at least one irreducible descendant: either w is irreducible or there exists a string y_1 such that $w \rightarrow y_1$; if y_1 is irreducible, then it is an irreducible descendant of w, and if y_1 can be reduced, then there is a string y_2 such that $y_1 \rightarrow y_2$. This process can be repeated only finitely often if all reduction chains have finite length.

We will make the following convention: if $(u, v) \in T$, then $u > v$ (so that $u \to v$) and $|u| \neq 0$. With this convention it will be convenient to write $(\Sigma^*, \to)$ instead of T in certain places; this is due to the fact that a restriction $\to^L$ of $\to$ will be considered.

If $T = (\Sigma^*, \to)$ has the property that for every $w \in \Sigma^*$, every reduction chain that begins with w is finite in length, then $(\Sigma^*, \to)$ is *noetherian*; if $R : \mathbf{N} \to \mathbf{N}$ is a function such that for every $w \in \Sigma^*$, every reduction chain that begins with w has length at most $R(|w|)$, then $(\Sigma^*, \to)$ is *R-noetherian*. Clearly, every R-noetherian Thue system is noetherian.

If a finite Thue system $T = (\Sigma^*, \to)$ is noetherian, then there is a total recursive function R such that T is R-noetherian. To see this, first notice that for a finite Thue system, every reduction is finitely branching. Hence, if $(\Sigma^*, \to)$ is noetherian, then for any string w, there are only finitely many reduction chains that begin with w. This means that for every n, there are only finitely many reduction chains that begin with a string of length n. If one lets $R(n)$ be the maximum length of these finitely many reduction chains, then R becomes a total recursive function.

If $T = (\Sigma^*, \to)$ has the property that for every $w, x, y \in \Sigma^*$, $w \xrightarrow{*} x$ and $w \xrightarrow{*} y$ imply that there exists a $z \in \Sigma^*$ such that $x \xrightarrow{*} z$ and $y \xrightarrow{*} z$, then $(\Sigma^*, \to)$ is *confluent*. If $(\Sigma^*, \to)$ has the property that for every $w, x, y \in \Sigma^*$, $w \to x$ and $w \to y$ imply that there exists a $z \in \Sigma^*$ such that $x \xrightarrow{*} z$ and $y \xrightarrow{*} z$, then $(\Sigma^*, \to)$ is *locally confluent*.

If a Thue system $T = (\Sigma^*, \to)$ is confluent, then for every string w, the congruence class of w has at most one irreducible element. Thus, irreducible elements are considered to be "normal forms," and so for confluent systems, normal forms are unique when they exist.

The property of local confluence is important since a system that is both locally confluent and noetherian is confluent.

A Thue system $T = (\Sigma^*, \to)$ is *complete* (or *canonical*) if it is both confluent and noetherian.

Notice that if a Thue system $T = (\Sigma^*, \to)$ is complete, then for every string w, the congruence class of w has a unique normal form.

A Thue system T is *Church-Rosser* if for all $x, y \in \Sigma^*$, $x \xleftrightarrow{*} y$ implies that there exists a $z \in \Sigma^*$ such that $x \xrightarrow{*} z$ and $y \xrightarrow{*} z$.

It is known that a Thue system is Church-Rosser if and only if it is confluent. (The reader should note that the notion of reduction used here is not based on the length of strings so that the definition of confluent is different from that in [Bo82]; that is what allows the statement to be true here while it was false in [Bo82].)

To solve the word problem, it is useful to consider the Church-Rosser property. Given two strings x and y, to determine whether x is congruent to y it is sufficient to determine whether there exists a string z such that $x \xrightarrow{*} z$ and $y \xrightarrow{*} z$. If $T = (\Sigma^*, \to)$ is complete, then both $[x]$ and $[y]$ have unique normal forms. Thus, to determine whether x is congruent to y, it is sufficient to compute the unique normal forms of x and y and compare them; they are identical if and only if x is congruent to y. Hence, to solve the word problem efficiently, it is sufficient to have an efficient algorithm for the computation of normal forms (since the comparison of strings to determine whether they are identical can be done efficiently).

For more background on Thue systems, see [Ja88].

3 Results

In this paper we are concerned with a specific restriction of the reduction relation $\rightarrow$, that is obtained by considering only "leftmost" reductions.

If $\alpha = xu_1y$, $\beta = xv_1y$, $(u_1, v_1) \in T$, every proper prefix of xu_1 is irreducible, and x is the shortest prefix of α with these properties, then the reduction $\alpha \rightarrow \beta$ is *leftmost*. We write $\alpha \rightarrow^L \beta$ if $\alpha \rightarrow \beta$ is leftmost. Let $\xrightarrow{*}^{L}$ denote the reflexive transitive closure of $\rightarrow^L$. Each sequence $w_0 \rightarrow^L w_1 \rightarrow^L \ldots \rightarrow^L w_k$, $k \geq 1$, is a *leftmost reduction chain of length k*.

If T has the property that every leftmost reduction chain is finite, then T is *left-noetherian*; if $R : \mathbb{N} \rightarrow \mathbb{N}$ is a function such that for every $w \in \Sigma^*$, every leftmost reduction chain that begins with w has length at most $R(|w|)$, then T is *R-left-noetherian*.

Notice that if T is noetherian, then T is left-noetherian. However, the converse does not hold.

A string w is $\rightarrow^L$-*irreducible* if there is no z such that $w \rightarrow^L z$.

Clearly, a string is $\rightarrow^L$-irreducible if and only if it is $\rightarrow$-irreducible, so that we need only say "irreducible." This fact will be useful.

It is easy to see that the following is true. (Similar facts are established in [Bo82].)

Lemma *Let T be a Thue system.*

(a) *Let $xuy \rightarrow^L xvy$ where $(u, v) \in T$. Then x is irreducible.*

(b) *If T is left-noetherian, then for every $w \in \Sigma^*$ there is an irreducible z such that $w \xrightarrow{*}^{L} z$.*

(c) *Suppose that T is confluent and left-noetherian. For every $x, y \in \Sigma^*$, if y is irreducible, $x \xleftrightarrow{*} y$ if and only if $x \xrightarrow{*}^{L} y$. Thus, for every $w \in \Sigma^*$ there is a unique irreducible z such that $w \xrightarrow{*}^{L} z$.*

The following result is the main technical tool. It allows for simple proofs of the principal results.

Theorem 1 *Let $R : \mathbb{N} \rightarrow \mathbb{N}$ be a strictly increasing function. Suppose that T is a finite Thue system that is R-left-noetherian. Then there is an algorithm with running time $O(R(n))$, where n is the length of the input, that on input string x (on the alphabet of T) computes a string y such that $x \xrightarrow{*}^{L} y$ and y is irreducible (mod T).*

Proof Since we are concerned only with the case that T is R-left-noetherian, no generality is lost if we assume that if (u, v_1) and (u, v_2) are rewriting rules in T, then $v_1 = v_2$. Let $k = max\{|u|, |v| \mid (u, v) \in T\}$.

Construct a deterministic Turing machine M with two pushdown stores that operates as follows. Initially, Store 1 is empty and Store 2 contains the input string x with the leftmost symbol of x on the top of Store 2. The step-by-step computation of M is described in terms of three operations.

(i) READ. M attempts to read a new symbol from Store 2, popping that symbol from Store 2 and pushing that symbol onto the top of Store 1. If M is able to read such a symbol, then it performs the SEARCH operation; otherwise, M halts.

(ii) SEARCH. M reads the top k symbols from Store 1 and determines whether there exists a string u stored on the top $|u|$ squares of Store 1 such that there exists $v, (u, v) \in T$. If such a u exists, then SEARCH "succeeds": in this case, M remembers (u, v) and performs the REWRITE operation. Otherwise, SEARCH "fails"; in this case, M restores the top $|u|$ symbols of Store 1 and performs the READ operation.

(iii) REWRITE. Having remembered the rewrite rule (u, v), M pops the string u from the top of Store 1 and pushes the string v onto the top $|v|$ squares of Store 2 so that the leftmost symbol of v is on the top of Store 2. Then M performs the READ operation.

By the Lemma and the assumption about T, if an application of SEARCH succeeds, then both the strings u and v are unique. Thus, M is deterministic. By the assumption made about T and the ordering $>$, if $(u, v) \in T$, then $u \rightarrow v$. Thus, it is easy to see that the computation of M on input x is a leftmost reduction using the rewrite rules of T. By hypothesis, T is R-noetherian, so that M's computation on x must halt; let y be the string contained on Store 1 when the computation halts, with the leftmost symbol of y being at the bottom of Store 1. Thus, $x \xrightarrow{*}_{L} y$. From the Lemma, it is clear that y is irreducible.

Consider M's running time. The process of reading, writing, and matching up to k symbols takes an amount of time independent of the input string x (although it does depend on the Thue system T), so M's running time on x is proportional to the total number $r(x)$ of READ steps performed. The number $r(x)$, in turn, is equal to the number of symbols originally appearing on Store 2 plus the number of symbols written onto Store 2 when REWRITE is performed, so $r(x)$ is $|x|$ plus the sum of the lengths of the right-hand sides v of rules (u, v) found in SEARCH. Since T is R-noetherian, there are at most $R(|x|)$ successful applications of SEARCH; also, each right-hand side v has length at most k, so $r(x) \leq |x| + k \cdot R(|x|)$. Thus, the length of m's computation on x is bounded above by $c(|x| + k \cdot R(|x|))$ for some constant c. Since R is strictly increasing, M's running time is $O(R(n))$, as claimed. □

The reader should note that the proof of Theorem 1 depends heavily on the fact that the binary operation of concatenation of strings in Σ^* is an associative operation and that each string in Σ^* has a unique factorization as a concatenation of elements of Σ.

Theorem 2 *Let $R : N \rightarrow N$ be a strictly increasing function. Suppose that T is a finite Thue system that is R-left-noetherian (R-noetherian) and that is confluent. Then there is an algorithm with running time $O(R(n))$, where n is the length of the input, to decide the word problem for T.*

Proof On input x and y, the algorithm given in the proof of Theorem 1 computes irreducible strings $\overline{x}$ and $\overline{y}$ such that $x \xrightarrow{*}_{L} \overline{x}$ and $y \xrightarrow{*}_{L} \overline{y}$. This can be accomplished in time $O(R(|xy|))$. Since T is confluent, $\overline{x}$ and $\overline{y}$ are identical if and only if x is congruent to y. Since it takes time at most $O(|\overline{x}|) + O(|\overline{y}|) = O(R(|x|)) + O(R(|y|)) = O(R(|xy|))$ to determine whether the two strings $\overline{x}$ and $\overline{y}$ are identical, and since R is strictly increasing, the running time is $O(R(|xy|))$. □

It should be noted that in Theorems 1 and 2, the size of the Thue system is treated as a constant. The running time of the algorithm depends on the Thue system that is given. However, the scheme of the proof is uniform in the Thue system.

Theorem 2 provides an algorithm for the word problem for finite Thue systems that are confluent and left-noetherian. As noted above, if a Thue system is noetherian, then it is left-noetherian so the bounds on the running time given in Theorem 2 apply to finite Thue systems that are complete. It is sometimes the case that a proof that a given system is noetherian (generally, an undecidable question) yields an upper bound on the length of reduction chains; hence, such a proof gives an upper bound on the complexity of the word problem.

Theorem 1 provides a basis for showing that there is an algorithm to determine whether an R-noetherian Thue system is confluent. Recall that since the Thue system is noetherian, it is sufficient to determine if the system is locally confluent. When considering the situation where the definition of reduction is based on length, Book and O'Dunlaing [BO81] described a polynomial time algorithm that on input a finite Thue system T will determine whether T is locally confluent. That same algorithm can be used in the case that T is R-noetherian but now the bound on the running time depends on R; an upper bound is $O(R(|T|)|T|^4)$, where $|T|$ denotes the size of T.

References

[Bo82] R. Book. Confluent and other types of Thue systems. *J. Assoc. Comput. Mach.*, vol. 29 (1982), 171–182.

[BO81] R. Book and C. O'Dunlaing. Testing for the Church-Rosser property. *Theoret. Comput. Sci.*, vol. 16 (1981), 216–223.

[DJ90] N. Dershowitz and J.-P. Jouannaud. *Rewrite Systems*, Chapter 6 in J. Van Leeuwen (ED.), *Formal Models and Semantics*, Handbook of Theoretical Computer Science, vol. B, Elsevier, (1990) 243–320.

[Hu80] G. Huet. Confluent Reductions: Abstract properties and applications to term rewriting systems. *J. Assoc. Comput. Mach.*, vol. 27, (1980) 797–821.

[Ja88] M. Jantzen. *Confluent Thue Systems*. Springer-Verlag (1988).

[MO85] K. Madlener and F. Otto. Pseudo-natural algorithms for decision problems in certain types of string-rewriting systems. *J. Symbolic Computation*, vol. 1 (1985), 383–418.

Confluence of One-Rule Thue Systems *

C. Wrathall
Department of Mathematics
University of California
Santa Barbara, CA 93106, USA

1 Introduction

A Thue system is a set of pairs of strings (each pair called a "rule") to be used for rewriting strings over some given alphabet. The way the rules are applied gives rise to a congruence relation on strings that is specific to the Thue system. When the rewriting is restricted so that each rule can be applied in only one direction, the resulting semi-Thue system determines a reduction relation on strings. A rewriting system is called "confluent" if the reduction relation has the "diamond property": different sequences of reductions applied to a single string can eventually be joined to produce a common descendant. In general, the question of whether a rewriting system is confluent is undecidable: there is no algorithm that will decide, for an arbitrary (finite) Thue system, whether it is confluent. However, for Thue systems with just one rule, the confluence property can be decided easily. This is shown in the present paper as a consequence of a characterization of confluent one-rule Thue systems.

Book [3] gave a characterization of the structure of the strings for a specific class of one-rule Thue systems that are confluent. Otto and Wrathall [15] extended that initial work, and the final piece of the characterization was recently provided by Kurth [13]. This paper is intended to present a simplified proof of the characterization in a uniform notation. The characterization is such that its conditions can be decided in linear time.

This characterization of confluence for one-rule Thue systems is based on comparison of the "(self-)overlaps" of the two strings making up the rule, where an overlap of a string is another string that is both a prefix and a suffix of the given string. For example, the proper overlaps of the string *ababa* are the three strings *e* (the empty string), *a* and *aba*, with *aba* the longest overlap of *ababa*. The notion of the overlaps of a string is encountered frequently in the design and analysis of algorithms for manipulating or matching strings.

The characterization has extensions to a restricted class of Thue systems with more than one rule [4], and also to some systems in which all the rules but one express commutation of letters [6].

A rewriting system is called "Noetherian" if its reduction relation is terminating in the sense that every sequence of reductions reaches (after a finite number of steps) an

---

*Preparation of this report was facilitated by the Lehrstuhl für Theoretische Informatik, Institut für Informatik, Universität Würzburg.

irreducible descendant, that is, a descendant to which no reduction rule can be applied. For example, any length-reducing Thue system (that is, one in which application of each rule decreases the length of the string) is Noetherian. A system that is both confluent and Noetherian has the useful computational property that every equivalence class has a unique irreducible representative that can be found by applying reductions until no further reduction is possible. For Noetherian systems, the confluence property reduces to a question of "local confluence," which is decidable for finite systems (using the notion of "critical pair"). It is not known whether the Noetherian property is decidable for one-rule Thue systems, although it is undecidable for arbitrary finite systems. The characterization of confluence for one-rule Thue systems given in the present paper is independent of the Noetherian property.

When a given one-rule Thue system proves not to be confluent, is it possible that there is an equivalent, nicely-behaved system? Some examples are collected here, but no general answer to this question is proposed.

2   Preliminary Definitions and Results

An *alphabet* is a finite set of letters or symbols, and A^* denotes the free monoid (semigroup with identity) generated by alphabet A. The elements of A^* are strings, the operation is concatenation of strings, and the identity element is the empty string, denoted by e. For a string x and integer $k \geq 1$, x^k is the concatenation of x with itself k times, and $x^0 = e$. Also, $x^+ = \{x^k \mid k \geq 1\}$ and $x^* = \{x^k \mid k \geq 0\}$.

A string x is *imprimitive* if there is some string y and some integer $k \geq 2$ such that $x = y^k$; otherwise, x is *primitive*. For a nonempty string x, there is a unique primitive string r, called the (primitive) *root* of x, such that x is a power of r.

The basic properties of strings described in the first chapter of Lothaire [14] are sufficient for the work here. Part (ii) of the following simple lemma is proved there, and part (i) in [15].

Lemma 1   (i) *If y and s are nonempty strings, xy is primitive, and $xy = rs$ and $yx = sr$, then $x = r$ and $y = s$.*

(ii) *If $xy = yz$ and x and z are nonempty, then there exist r, s such that s is nonempty, $x = rs$, $z = sr$ and $y = x^k r$ for some $k \geq 0$.*

Definition   *For a string $w \in A^*$, let $OVL(w)$ denote the set of (proper self-)overlaps of w:*

$$OVL(w) = \{u \in A^* \mid w = ur = su \text{ for some nonempty strings } r, s\}.$$

Note that the empty word, but not w itself, is included as an overlap of w. Overlaps have also been termed "borders" of strings. Pairs of strings in $OVL(w)$ have the property that the shorter is an overlap of the longer: if $u, v \in OVL(w)$ and u is longer than v, then $v \in OVL(u)$. In particular, the set $OVL(w)$ contains a longest element, which will be denoted by $ov(w)$. It is clear that $OVL(w) = \{ov(w)\} \cup OVL(ov(w)) = \{ov(w), ov^2(w), \ldots, e\}$.

Let $\pi(w)$ denote the *period* of string w: $\pi(w) = |w| - |ov(w)|$. (This is one of several equivalent definitions of the period of a string [14, chapter 8].)

Lemma 2 *Suppose u is the longest overlap of a string w, with $w = zu = u\bar{z}$. Then the following hold.*

(a) *z and $\bar{z}$ are primitive, $z = xy$ and $\bar{z} = yx$ for some string x and some nonempty string y, and $u = z^m x$ for some $m \geq 0$.*

(b) *x is the minimum conjugator of z and $\bar{z}$ in the sense that if $zv = v\bar{z}$ then $v = z^k x$ for some $k \geq 0$.*

(c) *$OVL(w) = \{z^k x \mid 1 \leq k \leq m\} \cup \{v \mid v$ is a proper prefix of z and a proper suffix of $\bar{z}\}$.*

Proof See also lemma 3.2 of [15].

(a) From Lemma 1(ii), there are strings x and y, with $y \neq e$, such that $z = xy$, $\bar{z} = yx$, and $u = z^m x$ for some $m \geq 0$. Let p be the primitive root of z, with $z = p^k$. Then $\bar{z} = q^k$, where q is the primitive root of $\bar{z}$, and $pu = uq$. Hence $w = p(p^{k-1}u) = (p^{k-1}u)q$, so, since u is the longest overlap of w, $k = 1$ and z and $\bar{z}$ are primitive.

(b) Suppose that $zv = v\bar{z}$. From Lemma 1(ii), $z = rs$, $\bar{z} = sr$ and $v = z^k r$ for some $k \geq 0$, and some r, s with $s \neq e$. Then $z = rs = xy$ and $\bar{z} = sr = yx$ with s and y nonempty, and z and $\bar{z}$ primitive, so (from Lemma 1(i)) $x = r$, $y = s$ and $v = z^k x$, as desired.

(c) Clearly, every string in the set on the right-hand side is an overlap of w; the point is that any overlap of w that is at least as long as z has the form $z^k x$ for some k. Suppose $v \in OVL(w)$ and $|v| \geq |z|$. Since v is no longer than $u = ov(w)$, $|z| \leq |v| \leq |z^m x| < |z^{m+1}|$. Since v is a prefix of $w = z^{m+1} x$, we have $v = z^k r$ for some k, $1 \leq k \leq m$, and $z = rs$ with $s \neq e$. Then sr is a suffix of $v = (rs)^k r$ and, since v is a suffix of w, it follows that sr is a suffix of w. But $\bar{z}$ is the suffix of w of length $|\bar{z}| = |z| = |rs|$, so $\bar{z} = sr$. At this point, Lemma 1(i) implies that $r = x$, and so $v = z^k x$ with $1 \leq k \leq m$. $\qquad\square$

A *Thue system* on an alphabet A is a set of ordered pairs of strings over A. A Thue system $T \subseteq A^* \times A^*$ gives rise to the following binary relations on A^*.

1. The one-step transformation relation $\longleftrightarrow$ is defined by:

$$x \longleftrightarrow y \text{ if } x = rus, \ y = rvs \text{ with } (u, v) \in T \text{ or } (v, u) \in T.$$

2. The Thue congruence $\overset{*}{\longleftrightarrow}$ is the reflexive, transitive closure of the one-step transformation relation.

3. The one-step reduction relation $\longrightarrow$ is defined by:

$$x \longrightarrow y \text{ if } x = rux, \ y = rvs \text{ where } (u, v) \in T.$$

4. The reduction relation $\overset{*}{\longrightarrow}$ is the transitive, reflexive closure of the one-step reduction relation $\longrightarrow$, and the proper reduction relation $\overset{+}{\longrightarrow}$ is the transitive closure of $\longrightarrow$.

To emphasize the notion that reduction consists of replacing an occurrence of the left-hand side of a rule by the right-hand side, the rules (u, v) of a Thue system will be written as $(u \rightarrow v)$. To simplify the notation, the set braces will be dropped in displaying one-rule Thue systems. See the monograph by Jantzen [9] for basic notions about Thue systems; the definitions needed here are given in the following paragraphs.

A system T is *length-reducing* if $|u| > |v|$ for each rule $(u \rightarrow v) \in T$. String x is *irreducible* if there is no string y such that $x \longrightarrow y$ (that is, if no factor of x is the left-hand side of a rule).

For a string $x \in A^*$, the congruence class of x (modulo the Thue system T) is the set $[x] = \{y \in A^* \mid x \overset{*}{\longleftrightarrow} y\}$. The Thue congruence determines, in the usual way, a *quotient monoid* $M_T = A^* / \overset{*}{\longleftrightarrow}$.

The Thue system is *confluent* if whenever $z \overset{*}{\longrightarrow} x$ and $z \overset{*}{\longrightarrow} y$ there is some w such that $x \overset{*}{\longrightarrow} w$ and $y \overset{*}{\longrightarrow} w$. It is *Noetherian* if there is no infinite chain of reductions $x_1 \longrightarrow x_2 \longrightarrow x_3 \longrightarrow \ldots$. A system that is both confluent and Noetherian is called a *complete* (or canonical) system. For a complete system, every string has a unique irreducible descendant: for every x, there is a unique string x_0 such that x_0 is irreducible and $x \overset{*}{\longrightarrow} x_0$.

A Thue system is *locally confluent* if whenever $z \longrightarrow x$ and $z \longrightarrow y$ there is some w such that $x \overset{*}{\longrightarrow} w$ and $y \overset{*}{\longrightarrow} w$. A Thue system is *very strongly locally confluent* if whenever $z \longrightarrow x$ and $z \longrightarrow y$, either $x = y$ or there is some w such that $x \longrightarrow w$ and $y \longrightarrow w$. The term "locally confluent" is a standard one, but "very strongly locally confluent" is *ad hoc*.

Proposition *If a Thue system is Noetherian and locally confluent, then it is confluent. If a Thue system is very strongly locally confluent, then it is confluent.*

Proofs of the first statement in this proposition and a stronger version of the second statement can be found in the paper by Huet [8]. The second statement is what is used here, and it can be proved by (integer) induction on the number of transformation steps. The first statement is not used here because the characterization of confluence for one-rule Thue systems need not presuppose the systems to be Noetherian.

3 Characterization Theorem and Discussion

The main result of the paper is the following.

Theorem *A one-rule Thue system $(u \rightarrow v)$ is confluent if and only if either*

(a) $u = z^k v$, with z primitive and $k \geq 1$, and
   $OVL(u) = OVL(v) \cup \{z^i v \mid 0 \leq i \leq k - 1\}$; or

(b) $OVL(u) \subseteq OVL(v)$.

The part of this characterization given by (a) was proved by Otto and Wrathall [15], and the part given by (b), by Kurth [13]. Both parts are restated here in different notation and the proof given below is a simplification of the two proofs. The reason for the two parts is that (a) applies when the right-hand side v of the rule is no longer than the longest proper overlap of the left-hand side u, and (b) applies otherwise.

As an example of application of the theorem, if u has no nontrivial overlap (i.e., if $OVL(u) = \{e\}$), then every system $(u \to v)$ is confluent. When v is nonempty, part (b) applies, since $OVL(u) = \{e\} \subseteq OVL(v)$. When v is empty, part (a) applies with $z = u$ and $k = 1$, since $v = ov(u)$ and a string with no nontrivial overlap must be primitive.

When part (a) applies, the system $(u \to v)$ is length-reducing and hence Noetherian, so if it is confluent then it is complete. Part (b) applies even to non-Noetherian systems: for example (using the remark in the previous paragraph), if x has no nontrivial overlap, then the system $(x \to xx)$ is confluent, but clearly not Noetherian. The characterization of confluence in the theorem is independent of the Noetherian property; in fact, the proof reveals that a one-rule Thue system, whether Noetherian or not, is confluent if and only if it is locally confluent.

It is implicit in the statement of part (a) that v is an overlap of u, since $v = z^0 v$. The requirement that z be primitive is to ensure that all the overlaps of u are captured by the expression given. An equivalent statement would be: $u = wv$ and $OVL(u) = OVL(v) \cup \{z^i v \mid 0 \le i \le k - 1\}$ where z is the root of w with $w = z^k$. In each part, the required relationship between $OVL(u)$ and $OVL(v)$ ensures that all critical pairs arising from the rule can be joined.

The following fact proved by Book served as the starting point for these investigations.

Corollary 1 [3]. *A one-rule Thue system of the form $(w \to e)$ is confluent if and only if the root of w has no nontrivial overlap.* □

When the right-hand side v is empty, part (b) cannot be true, so the theorem reduces to the statement "$(u \to e)$ is confluent if and only if $u = z^k$ with z primitive and $k \ge 1$, and $OVL(u) = \{z^i \mid 0 \le i \le k - 1\}$" which is easily seen to be equivalent to the statement in Corollary 1.

For (finite) length-reducing Thue systems, there is a polynomial-time algorithm to test for confluence [5,10], and confluence is a decidable property for finite Noetherian systems, although undecidable for arbitrary finite systems (see, e.g., [9]). The conditions given in the theorem are sufficiently simple that they can be tested very quickly.

Corollary 2 *There is a linear-time algorithm to solve the following problem (for a fixed alphabet A):*

Instance. A one-rule Thue system $T = \{(u \to v)\}$ on A^*
Question. Is T confluent?

□

The following equivalent statements perhaps make the linear-time procedure more obvious:

(a) For z equal to the prefix of u of length $\pi(u)$ and $u = z^{m+1}x$, either (i) $v = z^k x$ for some k, $1 \le k \le m$, or (ii) $v = x$ and u has no overlap of length strictly between $|x|$ and $|z|$.

(b) $ov(u)$ is both a prefix and a suffix of v.

(The equivalence of (a) here and condition (a) of the Theorem follows easily from Lemma 2.) Efficient string-matching algorithms [7,12] can be used to mark all the overlaps of u and so find its longest overlap; this is essentially what calculation of the "failure function" of u achieves. From that information, the period $\pi(u)$, strings z and x, and exponent m can be easily computed (see also [1]). The extra work to be done for the tests in (a) and (b) clearly requires only an amount of time linear in $|uv|$.

4   Proof of Characterization Theorem

In the proof, arrows directed from right to left denote the transpose of the corresponding relation: for example, $x \longleftarrow y$ means $y \longrightarrow x$. If the left-hand side of the rule is empty, then the system is easily seen to be (very strongly locally) confluent, and (b) holds since $OVL(e)$ is the empty set; assume, therefore, that $u \neq e$.

A. First, suppose that either (a) holds or (b) holds.

1. If $u = pq = qr$ with p, q, r nonempty, then either $vr = pv$ or there is some w such that $vr \longrightarrow w \longleftarrow pv$.

To see this, note first that q is a nonempty overlap of u. If $q \in OVL(v)$, then $v = v_1 q = qv_2$ for some v_1, v_2, so $vr = v_1 qr = v_1 u \longrightarrow v_1 v = vv_2 \longleftarrow uv_2 = pqv_2 = pv$. This shows that the statement holds in case (b), and also in case (a) unless $q = z^i v$ for some $i (0 \leq i \leq k - 1)$ where $u = z^k v$. In that case, $u = z^k v = pq = pz^i v$ so $p = z^{k-i}$, and $u = z^k v = qr = z^i vr$, so $pv = z^{k-i} v = vr$.

2. It follows from part 1 that the system $(u \rightarrow v)$ is very strongly locally confluent; hence (by the Proposition) it is confluent.

The proof that it is very strongly locally confluent follows the usual pattern, since the fact in part 1 says that the critical pairs of the system join quickly. Suppose $w \longrightarrow x$ and $w \longrightarrow y$; we must show that either $x = y$ or there is some w' such that $x \longrightarrow w'$ and $y \longrightarrow w'$. Since $w \longrightarrow x$, $w = x_1 u x_2$ and $x = x_1 v x_2$ for some strings x_1, x_2. Similarly $w = y_1 u y_2$ and $y = y_1 v y_2$ for some strings y_1, y_2, where, by symmetry, we may assume that $|x_1| \leq |y_1|$. If $|x_1 u| \leq |y_1|$, then (from $z = x_1 u x_2 = y_1 u y_2$) we have $y_1 = x_1 us$ and $x_2 = suy_2$ for some s, so $x = x_1 v x_2 = x_1 vsuy_2 \longrightarrow x_1 vsvy_2$ and $y = y_1 vy_2 = x_1 usvy_2 \longrightarrow x_1 vsvy_2$. If $|x_1| < |y_1| < |x_1 u|$, then $y_1 = x_1 p$, $u = pq = qr$ and $x_2 = ry_2$ with p, q and r nonempty, so that $x = x_1 v x_2 = x_1 vry_2$ and $y = y_1 vy_2 = x_1 pvy_2$. Using part 1, either $vr = pv$ or there is some t such that $vr \longrightarrow t \longleftarrow pv$; in the first case, $x = y$, and in the second, $x = x_1 vry_2 \longrightarrow x_1 ty_2 \longleftarrow x_1 pvy_2 = y$. Finally, if $|x_1| = |y_1|$, then $x_1 = y_1$ and $x_2 = y_2$, so $x = y$.

B. Now suppose that the system $(u \rightarrow v)$ is confluent. Let t be the longest overlap of u, with $u = zt = t\bar{z}$. If $|v| \leq |t|$ then (a) holds (part 1, below); and if $|v| > |t|$ then (b) holds (part 3).

1. Suppose $|v| \leq |t|$. As noted in Lemma 2, z and $\bar{z}$ are primitive, $z = xy$ and $\bar{z} = yx$ for some string x and some nonempty string y, $u = z^{m+1} x$ for some $m \geq 0$, and $t = z^m x$. Also, $OVL(u) = OVL(t) \cup \{t\} = \{z^i x \mid 1 \leq i \leq m\} \cup \{s \mid s \text{ is a proper prefix of } z \text{ and a proper suffix of } \bar{z}\}$.

Consider the string $u\bar{z} = zu$. We have the pair of reductions $v\bar{z} \longleftarrow u\bar{z} = zu \longrightarrow zv$, so since the system is confluent, $v\bar{z}$ and zv have a common descendant.

If either $v\bar{z}$ or zv is reducible, then it must have u as a factor, but $|v\bar{z}| = |zv| \leq |zt| = |u|$, so in this case $v\bar{z} = u = t\bar{z}$ or $zv = u = zt$, and, in any event, $v = t$. Thus (a) holds with $k = 1$: $u = zv$, z is primitive, and $OVL(u) = OVL(t) \cup \{t\} = OVL(v) \cup \{v\}$.

If both $v\overline{z}$ and zv are irreducible, then $v\overline{z} = zv$. Since x is the minimum conjugator of z and $\overline{z}$, $v = z^n x$ for some $n \geq 0$; also, $n \leq m$ since $|v| \leq |t| < |z^{m+1}|$. Thus, $u = z^{m+1}x = z^k v$ for $k = m-n+1 \geq 1$, with z primitive, and certainly $OVL(v) \cup \{z^i v \mid 0 \leq i \leq k-1\}$ is a subset of $OVL(u)$. For the reverse inclusion, it is enough to argue that every overlap of u that is longer than v has the form $z^i v$ for some i, $0 \leq i \leq k-1$. This is certainly true for an overlap s that is at least as long as z: in that case $s = z^j x$ for some j, $1 \leq j \leq m$, and in fact $j > n$ when $|s| > |v| = |z^n x|$, so $s = z^{j-n}v$ with $1 \leq j - n \leq m - n = k - 1$. The only remaining possibility is that $s \in OVL(u)$ and $|v| < |s| < |z|$ (so that necessarily $n = 0$ and $v = x$); but this cannot in fact occur if the system $(u \to v)$ is confluent.

Suppose otherwise, that $s \in OVL(u)$, $|x| < |s| < |z|$, and the system $(u \to x)$ is confluent. Since $s \in OVL(u)$ and $|s| < |z|$, $z = sq$ and $\overline{z} = ps$ for some strings p, q with $|q| = |p| > 0$. Then $xqt \longleftarrow uqt = t\overline{z}qt = tpzt = tpu \longrightarrow tpx$ and hence, since the system is confluent, xqt and tpx have a common descendant. Since $|tpx| = |xqt| < |sqt| = |u|$, both xqt and tpx are irreducible, so $xqt = tpx$. Now, tpx has the prefix $tp = x\overline{z}^m p = x(ps)^m p = xp(sp)^m$, so xp is a prefix of $tpx = xqt$ and therefore $p = q$. But then $z = xy = sp$ and $\overline{z} = yx = ps$ with y, $p \neq e$ and $|s| > |x|$, contradicting Lemma 1(i).

2. As a preliminary step for the other case, if $|v| > |t|$, then t is a prefix of v (where, again, $t = ov(u)$ with $u = zt = t\overline{z}$ and we assume the system $(u \to v)$ to be confluent).

If v is longer than t but t is <u>not</u> a prefix of v, then there exist strings s, t_1, v_1 and distinct letters a, b such that $t = sat_1$ and $v = sbv_1$. Since $u = t\overline{z}$, sa is a prefix of u; also, since $u = zt = zsat_1$, zs is a prefix of u and hence (since z is nonempty) sa is a prefix of zs.

Define a binary relation $<_{a,b}$ on strings by: $x <_{a,b} y$ if, for some r, ra is a prefix of x and rb is a prefix of y.

(i) The relation $<_{a,b}$ is a strict partial ordering: irreflexive, nonsymmetric and transitive. Also, the proper reduction relation is increasing relative to that ordering: if $w_1 \xrightarrow{+} w_2$ then $w_1 <_{a,b} w_2$.

(ii) If $zv \xrightarrow{*} w$, then sa is a prefix of w (and u is not a prefix of w).

(iii) If $v\overline{z} \xrightarrow{*} w$, then sb is a prefix of w.

Therefore, the system $(u \to v)$ cannot be confluent: $v\overline{z} \longleftarrow u\overline{z} = zu \longrightarrow zv$, but $v\overline{z}$ and zv can have no common descendant, since sa is a prefix of every descendant of zv and sb is a prefix of every descendant of $v\overline{z}$, and $a \neq b$.

To see that the properties in (i) hold, note first that $x <_{a,b} x$ cannot be true, since if ra and rb are both prefixes of x, then $a = b$. If $x <_{a,b} y$ and $y <_{a,b} w$ then $x = pax_1$, $y = pby_1 = qay_2$ and $w = qbw_1$ for some strings p, q, x_1, y_1, y_2, w_1. Since $y = pby_1 = qay_2$ and $a \neq b$, either pb is a prefix of q or qa is a prefix of p. In the first case, pa is a prefix of x and pb is a prefix of w; in the second case, qa is a prefix of x and qb is a prefix of w. Thus $x <_{a,b} w$. The non-symmetry of $<_{a,b}$ follows from the other two properties.

For the further remark in (i), a simple induction argument will suffice once it is noted that if $w_1 \longrightarrow w_2$ then, for some x, xu is a prefix of w_1 and xv is a prefix of w_2, and hence xsa is a prefix of w_1 and xsb is a prefix of w_2.

For (ii), we can also use induction, on the number of reduction steps from zv to w. First, sa, being a prefix of zs, is a prefix of zv; and u cannot be a prefix of zv since $u = zt$ and t is not a prefix of v. Suppose that $zv \xrightarrow{*} w_1 \longrightarrow w_2$. If u were a prefix of w_2 then the prefix zsa of u would be a prefix of w_2 and hence (since zsb is a prefix of zv) $w_2 <_{a,b} zv$

would hold; but this contradicts (i) since $zv \xrightarrow{+} w_2$. Therefore, u is not a prefix of w_2. From the induction hypothesis, sa, but not u, is a prefix of w_1. Since $w_1 \longrightarrow w_2$, there is some x such that xu is a prefix of w_1 and xv is a prefix of w_2, and x is not empty (otherwise, u would be a prefix of w_1). Then sa and xs are prefixes of w_1 with $|sa| \leq |xs|$ so sa is a prefix of xs and hence of xv and of w_2.

For (iii), first note that sb is a prefix of $v\bar{z}$. Continuing by induction, if $v\bar{z} \xrightarrow{*} w_1 \longrightarrow w_2$, then sb is a prefix of w_1 and, for some x, xu is a prefix of w_1 and xv is a prefix of w_2. Since both sb and xsa are prefixes of w_1, x cannot be empty, so sb is a prefix of xs and hence of xv and of w_2.

3. Finally, suppose that the system $(u \to v)$ is confluent and $|v| > |t|$. From part 2, the longest overlap t (hence, every overlap) of u is a prefix of v. The system $(u^R \to v^R)$ must also be confluent (where the superscript R denotes reversal of the string), so applying part 2 to $(u^R \to v^R)$, every overlap of u^R is a prefix of v^R. Since the overlaps of u^R are precisely the reversals of overlaps of u, every overlap of u is a suffix of v. Thus every overlap of u is also an overlap of v. $\qquad \square$

5 Existence of Equivalent Complete Systems

If a given one-rule Thue system is not confluent, might there be an equivalent confluent and Noetherian system? Two Thue systems are termed *equivalent* if they are defined on the same alphabet and give rise to the same congruence on the free monoid generated by that alphabet. Equivalent Thue systems might or might not determine the same reduction relation; they do determine isomorphic quotient monoids. Thue systems on different alphabets might also determine isomorphic quotient monoids.

A confluent and length-reducing system equivalent to the given one would be particularly desirable, since the length of any reduction sequence can be no more than the length of the string with which it starts. In particular, for a finite system, the amount of work to produce, from x, the unique irreducible descendant of x will be bounded by (approximately) the length of x [2]. An infinite confluent and length-reducing system in which the rules follow some simple pattern might also be convenient for calculations.

A finite, confluent and Noetherian system will also serve to find normal forms (that is, unique irreducible descendants), although possibly without any prior knowlege of the amount of time required. For any system T, there is an infinite, equivalent, complete system S: for example, S could contain rules to change each string into the lexicographically first string among the shortest strings in its congruence class modulo T.

Keeping the alphabet fixed, we have three questions for a given non-confluent Thue system: (1) Is there an equivalent finite, confluent, and length-reducing system? (2) Is there an equivalent confluent and length-reducing system (finite or infinite)? (3) Is there an equivalent finite complete system? A positive answer to the first question clearly implies positive answers to the other two. Also, the existence of an infinite, confluent and length-reducing system with a certain property precludes the existence of an equivalent finite, confluent and length-reducing system: if T is infinite, confluent and length-reducing and every proper factor of the left-hand side of a rule in T is irreducible modulo T, then there is no finite, confluent, length-reducing system equivalent to T (see, e.g., example (2) below). Note further that if all the rules of a system are length-preserving, then there can be no equivalent length-reducing system (examples (4) and (5)).

Consider now a non-confluent one-rule system $(u \to v)$.

When v is no longer than the longest overlap of u, there can be no equivalent confluent and length-reducing system, neither a finite nor an infinite one [15]. There might or might not be an equivalent finite complete system. For example, the non-confluent system $(aba \rightarrow e)$ is equivalent to the two-rule complete system $\{(aab \rightarrow e), (ba \rightarrow ab)\}$. However, there is no finite complete system equivalent to the system $(abba \rightarrow e)$ [9].

When the right-hand side v is longer than the longest overlap of the left-hand side u, any of the five possible combinations of positive and negative answers to the three questions above can occur, as shown by the following examples.

(1) $(aba \rightarrow aabb)$ is equivalent to $(aabb \rightarrow aba)$, which is length-reducing and confluent.

(2) $(acaba \rightarrow abac)$ is equivalent to the finite, complete system $(abac \rightarrow acaba)$ and also to the infinite, confluent and length-reducing system $\{(axcaba \rightarrow axbac) \mid x \in (bac^+)^*\}$. However, there is no equivalent finite, confluent and length-reducing system.

(3) $(aba \rightarrow ab)$ is equivalent to the infinite, confluent and length-reducing system $\{(ab^na \rightarrow ab^n) \mid n \geq 1\}$. There is no equivalent finite complete system, since (due to the properties of this system) any complete system equivalent to $(aba \rightarrow ab)$ must contain, for each $n \geq 1$, a rule with left-hand side ab^na.

(4) $(aba \rightarrow aab)$ is equivalent to $(aab \rightarrow aba)$, which is a complete system. There is clearly no equivalent length-reducing system.

(5) $(aba \rightarrow bab)$ has no equivalent finite complete system [11], nor any equivalent length-reducing system.

The question remains of predicting the existence of nicely-behaved systems equivalent to a one-rule system $(u \rightarrow v)$ based on the internal structure of the strings u and v.

References

[1] J. Avenhaus and K. Madlener. String matching and algorithmic problems in free groups. *Revista Colombiana de Matematicas*, vol. 14 (1980), 1–16.

[2] R. Book. Confluent and other types of Thue systems. *J. Assoc. Comp. Mach.*, vol. 29 (1982), 171–182.

[3] R. Book. A note on special Thue systems with a single defining relation. *Math. Systems Theory*, vol. 16 (1983), 57–60.

[4] R. Book. Homogeneous Thue systems and the Church-Rosser property. *Discrete Mathematics*, vol. 48 (1984), 137–145.

[5] R. Book and C. O'Dúnlaing. Testing for the Church-Rosser property. *Theoret. Comput. Sci.*, vol. 16 (1981), 223–229.

[6] V. Diekert and C. Wrathall. On confluence of one-rule trace rewriting, in preparation.

[7] Z. Galil and J. Seiferas. Time-space optimal string matching. *J. Comput. Systems Sci.*, vol. 26 (1983), 280–294.

[8]   G. Huet. Confluent reductions: abstract properties and applications to term rewriting systems. *J. Assoc. Comput. Mach.*, vol. 27 (1989), 797–821.

[9]   M. Jantzen. *Confluent String Rewriting*, EATCS Monographs on Theoretical Computer Science 14, Springer-Verlag, 1988.

[10] D. Kapur, M. Krishnamoorthy, R. McNaughton and P. Narendran. An $O(|T|^3)$ algorithm for testing the Church-Rosser property of Thue systems. *Theoret. Comput. Sci.*, vol. 35 (1985), 109–114.

[11] D. Kapur and P. Narendran. A finite Thue system with decidable word problem and without equivalent finite canonical system. *Theoret. Comput. Sci.*, vol. 35 (1985), 337–344.

[12] D. Knuth, J. Morris and V. Pratt. Fast pattern matching in strings. *SIAM J. Comput.*, vol. 6 (1977), 323–350.

[13] W. Kurth. *Termination und Confluenz von Semi-Thue-Systems mit nur einer Regel*, Dissertation, Mathematische-Naturwissenschaftliche Fakultät, Technische Universität Clausthal, Chapter 6, 1990.

[14] M. Lothaire. *Combinatorics on Words*, Addison-Wesley, Reading, Mass., 1983.

[15] F. Otto and C. Wrathall. A note on Thue systems with a single defining relation. *Math. Systems Theory*, vol. 18 (1985), 135–143.

[16] C. Wrathall. Confluence of one-rule Thue systems. Tech. Report 24, Universität Würzburg, January, 1991.

SYSTEMS OF EQUATIONS OVER A FINITE SET OF WORDS AND AUTOMATA THEORY

(Extended abstract)

JUHANI KARHUMÄKI

Department of Mathematics
University of Turku
Turku, Finland

Equations over free monoids, that is to say over finite words, has been studied quite intensively during the last twenty years, cf. [H], [Lo] and [M]. At the same time there seems to be almost no results on equations over finite sets of words, that is to say over finite languages. Both of these theories has, or would have, remarkable applications. Certainly, automata theory is among the most important applications.

As an example of such an application let us consider the problem of testing the equivalence of two deterministic gsm's, cf. [Sa]. The decidability of this problem is a straightforward consequence of the following implication. For any finite words $x,y,u,v,\bar{x},\bar{y},\bar{u}$ and $\bar{v}$ we have

$$(1) \qquad \left. \begin{array}{l} xy = \bar{x}\bar{y} \\ xuy = \bar{x}\bar{u}\bar{y} \\ xvy = \bar{x}\bar{v}\bar{y} \end{array} \right\} \implies xuvy = \bar{x}\bar{u}\bar{v}\bar{y}$$

The proof of (1) is only a few lines.

The goal of this paper is to consider equations over finite sets of words, as well as to apply this approach to some decidability questions in automata theory. We point out that some "word considerations" can be extended to "language considerations", and consequently several known decidability results in automata theory can be extended as well. However, in most cases we have to state such extensions as open problems.

In more details we observe, as in [P] and [S], that the monoid of finite prefixes is free. This has several consequences. It allows to solve a special case of an interesting problem of deciding whether two finite substitutions $\sigma, \tau : \Sigma^* \rightarrow \Delta^*$ are equivalent on a given regular language L , i.e. whether $\sigma(x) = \tau(x)$ holds for all x in L , cf. [CKIII]. Namely, the problem is decidable for so-called prefix substitutions, that is to say for substitutions satisfying the condition "the images of letters are finite prefixes". This also allows a so far rare possibility to define a class of finite "unboundedly nondeterministic" finite transducers for which the equivalence problem is decidable, cf. [CKII]. Finally, an extension of the famous DOL problem is achieved, cf. [CF] or [CKI].

We also show that the implication (1) holds for finite multisets, or equivalently, for noncommuting polynomials (with nonnegative integer coefficients). This is a striking contrast to Lawrence's example, cf. [La], and shows that, although we do not know how to test the equivalence of two finite substitutions on a regular language (cf. above), we can easily decide whether two finite substitutions $\sigma, \tau : \Sigma^* \rightarrow \Delta^*$ are multiplicitly equivalent on a given regular language L , i.e. whether $\sigma(x)$ and $\tau(x)$ are the same multisets for each x in L . This again solves a special case of an important problem asking whether two finite transductions are multiplicitly equivalent. The genral case has been settled recently in [HK].

All in all we consider several submonoids of the monoid of all finite languages, as well as the monoid of finite multisets. We make several observations on equations over these monoids, and as applications conclude some results in automata theory. However, the essential part of this talk is to point out open problems.

REFERENCES

[CF] Culik II, K. and Fris, I., The decidability of the equivalence problem for DOL-systems, Inform. Control 33 (1976) 20-39.

[CKI] Culik II, K. and Karhumäki, J., A new proof of the DOL sequence equivalence problem and its implications, in: G. Rozenberg and A. Salomaa (eds), The Book of L (Springer, Berlin, 1986).

[CKII] Culik II, K. and Karhumäki, J., The equivalence of finite valued
transducers (on HDTOL languages) is decidable, Theoret. Comput.
Sci. 47 (1986) 71-84.

[CKIII] Culik II, K. and Karhumäki, J., Decision problems solved with
the help of the Ehrenfeucht Conjecture, Bull. EATCS 27 (1985)
30-35.

[HK]   Harju, T. and Karhumäki, J., The equivalence problem for finite
multitape automata, manuscript, 1990.

[H]   Hmelevskii, Y.I., Equations in free semigroups, Proc. Steklov
Inst. Math. 107 (1971), Am. Math. Soc. Transl. (1976).

[La]   Lawrence, J., The non-existence of finite test set for set equivalence of finite substitutions, Bull. EATCS 28 (1986) 34-37.

[Lo]   Lothaire, M., Combinatorics on Words,(Addison-Wesley, Reading,
MA, 1983).

[M]   Makanin, G.S., The problem of solvability of equations in a
free semigroup, Mat. Sb. 103 (1976) 147-236.

[P]   Perrin, D., Codes conjegues, Inform. Control 20 (1972) 221-231.

[Sa]   Salomaa, A., Formal Languages, (Academic Press, New York, 1973).

[S]   Shyr, H.J., Free Monoids and Languages, Lecture Notes, Department of Mathematics, Soochow University, Taipei, Taiwan (1979).

New Systems of Defining Relations
of the Braid Group

A.G. Makanina
Moscow State University, Main Building

The *braid group* $\mathcal{B}_{n+1}$ of order $n+1$ is the group determined by generators $\sigma_i, \ldots, \sigma_n$ and the defining relations

$$\sigma_i \sigma_{i+1} \sigma_i = \sigma_{i+1} \sigma_i \sigma_{i+1} \qquad (i = 1, \ldots, n-1)$$
$$\sigma_i \sigma_j = \sigma_j \sigma_i \qquad (i, j = 1, \ldots, n; |i-j| > 1)$$

The homomorphism of the braid group $\mathcal{B}_{n+1}$ to the symmetric group $\mathcal{S}_{n+1}$ which maps each generator σ_i of $\mathcal{B}_{n+1}$ to the transposition $(i, i+1)$ is denoted as μ_{n+1}. The subgroup of $\mathcal{B}_{n+1}$ which is the kernel of the homomorphism μ_{n+1} is called the *coloured braid group* and denotĕd by $\mathcal{N}_{n+1}$.

The braid $\sigma_{j-1} \sigma_{j-2} \ldots \sigma_{i+1} \sigma_i^2 \sigma_{i+1}^{-1} \ldots \sigma_{j-2}^{-1} \sigma_{j-1}^{-1}$ of the group $\mathcal{B}_{n+1}$ is denoted s_{ij} $/1 \le i < j \le n+1/$. W. Burau [1] proved that the elements s_{ij} $/1 \le i < j \le n+1/$ generate the coloured braid group $\mathcal{N}_{n+1}$ and the system of defining relations for them is

$$s_{ij} s_{kl} = s_{kl} s_{ij} \qquad (i < j < k < l \ \vee \ i < k < l < j)$$
$$s_{ij} s_{ik} s_{jk} = s_{ik} s_{jk} s_{ij} \qquad (i < j < k)$$
$$s_{ik} s_{jk} s_{ij} = s_{jk} s_{ij} s_{ik} \qquad (i < j < k)$$
$$s_{ik} s_{jk} s_{jl} s_{jk}^{-1} = s_{jk} s_{jl} s_{jk}^{-1} s_{ik} \qquad (i < j < k < l)$$

Artin [2] proved that the coloured braid group $\mathcal{N}_{n+1}$ can be defined by generatc rs s_{ij} $/1 \le i < j \le n+1/$ and by defining relations

$$s_{rp} s_{ik} s_{rp}^{-1} = s_{ik} \qquad (p < i \ \vee \ k < r)$$
$$s_{kp} s_{ik} s_{kp}^{-1} = s_{ip}^{-1} s_{ik} s_{ip} \qquad (i < k < p)$$
$$s_{rk} s_{ik} s_{rk}^{-1} = s_{ik}^{-1} s_{ir}^{-1} s_{ik} s_{ir} s_{ik} \qquad (i < r < k)$$
$$s_{rp} s_{ik} s_{rp}^{-1} = s_{ip}^{-1} s_{ir}^{-1} s_{ip} s_{ir} s_{ik} s_{ir}^{-1} s_{ip}^{-1} s_{ir} s_{ip} \qquad (i < r < k < p)$$

The braid $\sigma_{j-1} \sigma_{j-2} \ldots \sigma_{i+1} \sigma_i^2 \sigma_{i+1} \ldots \sigma_{j-2} \sigma_{j-1}$ of the group $\mathcal{B}_{n+1}$ is denoted by ρ_{ij} $/1 \le i \le n+1/$. Moreover, $\rho_{ij} \rightleftharpoons 1$ $/1 \le i \le n+1/$. A.A. Markov [3] proved that the elements

$$\rho_{ij} \qquad (1 \le i < j \le n+1) \qquad (1)$$

generate the coloured braid group $\mathcal{N}_{n+1}$, and the system of defining relations for them is

$$\rho_{ij}^{-\epsilon}\rho_{kl}\rho_{ij}^{\epsilon} = \rho_{kl} \qquad (i < k \ \vee \ k \le i < j < l), \quad \epsilon \pm 1 \qquad (2)$$

$$\rho_{ij}^{-1}\rho_{kl}\rho_{ij} = \rho_{il}\rho_{j+1,l}^{-1}\rho_{jl}\rho_{il}^{-1}\rho_{kl}\rho_{jl}^{-1}\rho_{j+1,l} \qquad (i \le k \le j < l) \qquad (3)$$

$$\rho_{ij}\rho_{kl}\rho_{ij}^{-1} = \rho_{j+1,l}\rho_{j,l}^{-1}\rho_{kl}\rho_{il}^{-1}\rho_{jl}\rho_{j+1,l}^{-1}\rho_{il} \qquad (i \le k \le j < l) \qquad (4)$$

Theorem 1. *The coloured braid group* $\mathcal{N}_{n+1}$ *can be defined by generators (1) and by defining relations*

$$\rho_{kl}\rho_{ij} = \rho_{ij}\rho_{kl} \qquad (j < k \ \vee \ k \le i < j < l) \qquad (5)$$

$$\rho_{kl}\rho_{ij}\rho_{j+1,l}^{-1}\rho_{jl} = \rho_{jl}\rho_{j+1,l}^{-1}\rho_{ij}\rho_{kl} \qquad (i \le k \le j < l) \qquad (6)$$

Definition.

$$\tau_{ij} \rightleftharpoons \sigma_1\sigma_{i+1}...\sigma_{j-2}\sigma_{j-1}^2\sigma_{j-2}...\sigma_{i+1}\sigma_i \qquad (1 \le i < j \le n+1)$$

$$\tau_{ij} \rightleftharpoons \sigma_{i-1}\sigma_{i-2}...\sigma_{j+1}\sigma_j^2\sigma_{j+1}...\sigma_{i-2}\sigma_{i-1} \qquad (1 \le j < i \le n+1)$$

$$\tau_{ii} \rightleftharpoons 1 \qquad (1 \le i \le n+1)$$

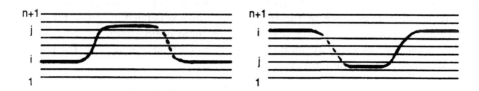

Theorem 2. *The coloured braid group* $\mathcal{N}_{n+1}$ *can be defined by generators*

$$\tau_{ij} \qquad (i,j = 1,...,n+1; i \ne j) \qquad (7)$$

and by the uniform system of defining relations

$$\tau_{lk}\tau_{j,l-1}\tau_{ji}\tau_{jl} = \tau_{jl}\tau_{ji}\tau_{j,l-1}\tau_{lk} \qquad (i \le k \le j < l) \qquad (8)$$

$$\tau_{ij}\tau_{j,i+1} = \tau_{ji}\tau_{i,j-1} \qquad (i < j) \qquad (9)$$

$$\tau_{ij}\tau_{km} = \tau_{km}\tau_{ij} \qquad (i,j < k,m) \ \vee \ (i < k, m \le j) \ \vee \ (i > k, m \ge j) \qquad (10)$$

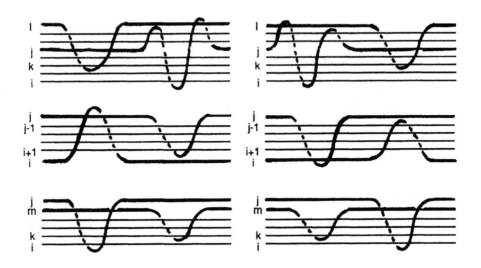

Proof of theorem 1. Let us prove that

$$\rho_{kl}\rho_{ij}\rho_{j+1,l}^{-1}\rho_{jl}\rho_{kl}^{-1}\rho_{ij}^{-1}\rho_{j+1,l}\rho_{jl}^{-1} = 1$$

for $/i \leq k \leq j < l/$ in the group $\mathcal{B}_{n+1}$. Let us rewrite this equality in generators of the braid group $\mathcal{B}_{n+1}$. After that, using correlations of commutation of the braid group $\mathcal{B}_{n+1}$, let us shorten self-inverse elements in the left part of the equality. After some transformations of the resulting word we shall get unit.

Thus, correlation (6) is satisfied in the braid group $\mathcal{B}_{n+1}$ and consequently in the coloured braid group of A.A. Markov. It is evident that correlation (5) is valid in Markov's group. Consequently, the group $\mathcal{N}_{n+1}$ which is defined by generators (1) and by defining relations (2), (3), (4), (5), (6) is a coloured braid group.

Let us prove that equality (3) can be deduced from correlations (5), (6). For this let us rewrite equality (3) in the form of

$$\rho_{kl}\rho_{ij}\rho_{j+1,l}^{-1}\rho_{jl}\rho_{kl}^{-1}\rho_{il}\rho_{jl}^{-1}\rho_{j+1,l}\rho_{il}^{-1}\rho_{ij}^{-1} = 1.$$

Then let us apply correlation (6) for the left part of the equality from left to right and shorten $\rho_{kl}\rho_{kl}^{-1}$. For the received equality

$$\rho_{jl}\rho_{j+1,l}^{-1}\rho_{ij}\rho_{il}\rho_{jl}^{-1}\rho_{j+1,l}\rho_{il}^{-1}\rho_{ij}^{-1} = 1$$

let us apply commutation $\rho_{il}^{-1}\rho_{ij}^{-1} = \rho_{ij}^{-1}\rho_{il}^{-1}$, which is true by correlation (5) when $i < j < l$ and by the definition $\rho_{ij} \rightleftharpoons 1$ when $i = j < l$. The received equality

$$\rho_{jl}\rho_{j+1,l}^{-1}\rho_{ij}\rho_{il}\rho_{jl}^{-1}\rho_{j+1,l}\rho_{ij}^{-1}\rho_{il}^{-1} = 1$$

is correlation (6) when $k = i$.

Let us prove that equality (4) can be deduced from correlations (5), (6). For this purpose we shall rewrite equality (4) in the form of

$$\rho_{ij}^{-1}\rho_{il}^{-1}\rho_{j+1,l}\rho_{jl}^{-1}\rho_{il}\rho_{kl}^{-1}\rho_{jl}\rho_{j+1,l}^{-1}\rho_{ij}\rho_{kl} = 1.$$

Then we shall apply correlation (6) to the left part of equality from right to left and shorten $\rho_{kl}^{-1}\rho_{kl}$. For the received equality

$$\rho_{ij}^{-1}\rho_{il}^{-1}\rho_{j+1,l}\rho_{jl}^{-1}\rho_{il}\rho_{ij}\rho_{j+1,l}^{-1}\rho_{jl} = 1$$

we shall apply commutation

$$\rho_{ij}^{-1}\rho_{il}^{-1} = \rho_{il}^{-1}\rho_{ij}^{-1} \qquad (i \leq j < l).$$

As a result we receive correlation (6) when $i = k$. It is obvious that equality (2) can be deduced from equality (5). Thus, the group which has been defined by generators (1) and defining relations (5), (6) is the coloured braid group.

Proof of theorem 2. The braid $\Delta(i, j)$ for all i, j from the interval $1 \leq i \leq j \leq n$ is defined inductively by equalities

$$\Delta(i, j) \rightleftharpoons \sigma_i \qquad (1 \leq i \leq n)$$
$$\Delta(i, j) \rightleftharpoons \sigma_i\sigma_{i+1}\ldots\sigma_j\Delta(i, j - 1) \qquad (1 \leq i < j \leq n)$$

Moreover, $\Delta(i, i - 1) \rightleftharpoons 1 \qquad (1 \leq i \leq n)$.

Lemma 1. *In the group $\mathcal{B}_{n+1}$ the equalities*

$$\Delta^2(i, j) = \tau_{j+1,i}\Delta^2(i, j - 1) \qquad (1 \leq i \leq j \leq n) \tag{11}$$
$$\Delta^2(i, j) = \tau_{i,j+1}\Delta^2(i + 1, j) \qquad (1 \leq i \leq j \leq n) \tag{12}$$

are valid.

Proof. Since $rev\Delta(i, j) = \Delta(i, j)$ in the group $\mathcal{B}_{n+1}$ [4] we have

$$\Delta(i, j) = \Delta(i, j - 1)\sigma_j\ldots\sigma_{i+1}\sigma_i.$$

For all $k = i + 1, \ldots, j$ we have $\sigma_k\sigma_i\sigma_{i+1}\ldots\sigma_j = \sigma_i\sigma_{i+1}\ldots\sigma_j\sigma_{k-1}$ in the group $\mathcal{B}_{n+1}$. For all $k = i, i + 1, \ldots, j - 1$ we have $\sigma_k\sigma_j\ldots\sigma_{i+1}\sigma_i = \sigma_j\ldots\sigma_{i+1}\sigma_i\sigma_{k+1}$ in the group $\mathcal{B}_{n+1}$. Consequently,

$$\Delta^2(i, j) = \Delta(i, j - 1)\sigma_j\ldots\sigma_{i+1}\sigma_i\sigma_i\sigma_{i+1}\ldots\sigma_j\Delta(i, j - 1)$$
$$= \sigma_j\ldots\sigma_{i+1}\sigma_i\sigma_i\sigma_{i+1}\ldots\sigma_j\Delta^2(i, j - 1) = \tau_{j+1,i}\Delta^2(i, i - 1)$$

and

$$\Delta^2(i, j) = \sigma_i\sigma_{i+1}\ldots\sigma_j\Delta^2(i, j - 1)\sigma_j\ldots\sigma_{i+1}\sigma_i$$
$$= \sigma_i\sigma_{i+1}\ldots\sigma_j\sigma_j\ldots\sigma_{i+1}\sigma_i\Delta^2(i + 1, j) = \tau_{i,j+1}\Delta^2(i + 1, j).$$

Lemma 2. *In group $\mathcal{B}_{n+1}$ the equalities*

$$\tau_{j+1,i} = \rho_{i,j+1} \qquad (1 \le i \le j \le n) \tag{13}$$

$$\tau_{i,j+1} = \rho_{i,j+1}\rho_{ij} \cdots \rho_{i,i+1}(\rho_{i+1,j+1}\rho_{i+1,j} \cdots \rho_{i+1,i+2})^{-1} \qquad (1 \le i \le j \le n) \tag{14}$$

are valid.

Proof. Equality (13) follows from ρ_{ij} and the definition of the τ_{ij}. From equalities (11) and (13) we receive

$$\Delta^2(i,j) = \tau_{j+1,i}\tau_{ji}...\tau_{i+1,i} = \rho_{i,j+1}\rho_{ij}...\rho_{i,i+1}$$

when $/1 \le i \le j \le n/$. From equality (12) we have $\tau_{i,j+1} = \Delta^2(i,j)\Delta^{-2}(i+1,j)$ when $/1 \le i \le j \le n/$.Consequently, equality (14) holds.

Let us denote as $\mathcal{K}_{n+1}$ the group which isdefined by generators (1), (7) and by defining relations (5), (6), (13), (14). The group $\mathcal{K}_{n+1}$ is isomorphic to the group, which is defined by generators (1) and by defining relations (5), (6). By Teitz's rule and by theorem1 it is isomorphic to the coloured braid group.

Lemma 3. *Equality (10) is valid in the group $\mathcal{B}_{n+1}$.*

Proof. For $i, j < k, m$ equality (10) follows from correlations of commutation in the braid group $\mathcal{B}_{n+1}$. For $i < k, m \le j$ and $i > k, m \ge j$ equality (10) follows from equalities

$$\sigma_p\sigma_i\sigma_{i+1} \cdots \sigma_j\sigma_j \cdots \sigma_{i+1}\sigma_i = \sigma_i\sigma_{i+1} \cdots \sigma_j\sigma_j \cdots \sigma_{i+1}\sigma_i\sigma_p \quad (p = i+1, ...j)$$

$$\sigma_p\sigma_j\sigma_{j-1} \cdots \sigma_i\sigma_i \cdots \sigma_{j-1}\sigma_j = \sigma_j\sigma_{j-1} \cdots \sigma_i\sigma_i \cdots \sigma_{j-1}\sigma_j\sigma_p \quad (p = i, ..., j-1)$$

which are true in the group $\mathcal{B}_{n+1}$.

Lemma 4. *Equality (9) is valid in the group $\mathcal{B}_{n+1}$.*

Proof. From equalities (11) and (12)

$$\Delta^2(i,j) = \tau_{j+1,i}\Delta^2(i,j-1) = \tau_{j+1,i}\tau_{ij}\Delta^2(i+1,j-1) \quad (1 \le i < j \le n)$$

$$\Delta^2(i,j) = \tau_{i,j+1}\Delta^2(i+1,j) = \tau_{i,j+1}\tau_{j+1,i+1}\Delta^2(i+1,j-1) \quad (1 \le i < j \le n)$$

follow. Identifying right sides of the equalities, we receive

$$\tau_{j+1,i}\tau_{ij} = \tau_{i,j+1}\tau_{j+1,i+1} \qquad (1 \le i < j \le n),$$

i.e.

$$\tau_{ji}\tau_{i,j-1} = \tau_{ij}\tau_{j,i+1} \qquad (1 \le i < j-1 \le n).$$

According to the definition of τ_{ji}, $\tau_{i+1,i}\tau_{ii} = \tau_{i,i+1}\tau_{i+1,i+1}$ $(1 \le i \le n)$. Combining the two last equalities we receive equality (9).

Lemma 5. *Equality (8) is valid in the group $\mathcal{B}_{n+1}$.*

Proof. By Lemma 4 and Lemma 5

$$\tau_{l,j+1}\tau_{jl} = \tau_{lj}\tau_{j,l-1} \quad (1 \leq j < l \leq n+1)$$
$$\tau_{jl}\tau_{l,j+1} = \tau_{j,l-1}\tau_{lj} \quad (1 \leq j < l \leq n+1)$$

hold in the group $\mathcal{B}_{n+1}$. Consequently,

$$\tau_{l,j+1}^{-1}\tau_{lj} = \tau_{jl}\tau_{j,l-1}^{-1} \quad (1 \leq j < l \leq n+1)$$
$$\tau_{lj}\tau_{l,j+1}^{-1} = \tau_{j,l-1}^{-1}\tau_{jl} \quad (1 \leq j < l \leq n+1)$$

in the group $\mathcal{B}_{n+1}$.
From equality (13) we receive that

$$\rho_{j+1,l}^{-1}\rho_{jl} = \tau_{jl}\tau_{j,l-1}^{-1}, \qquad \rho_{j,l}\rho_{j+1,l}^{-1} = \tau_{j,l-1}^{-1}\tau_{jl} \tag{15}$$

in the group $\mathcal{B}_{n+1}$.
By theorem 1 in the group $\mathcal{B}_{n+1}$ the equality

$$\rho_{kl}\rho_{ij}\rho_{j+1,l}^{-1}\rho_{jl} = \rho_{jl}\rho_{j+1,l}^{-1}\rho_{ij}\rho_{kl} \quad (i \leq k \leq j < l)$$

is valid.
From equalities (13) and (15) we receive

$$\tau_{lk}\tau_{ji}\tau_{jl}\tau_{j,l-1}^{-1} = \tau_{j,l-1}^{-1}\tau_{jl}\tau_{ji}\tau_{lk} \quad (i \leq k \leq j < l)$$

Applying lemma 3 for the last equality we receive equality (8).

Let us denote as $\mathcal{K}'_{n+1}$ the group which is defined by generators (1), (7) and by defining relations (5), (6), (13), (14), (8), (9), (10). By lemmas 5, 4, 3 equalities (8), (9), (10) hold in the group $\mathcal{B}_{n+1}$. Since the braids of equalities (8), (9), (10) are coloured the equalities (8), (9), (10) hold in the coloured braid group, in particular in $\mathcal{K}_{n+1}$. Consequently, by the rule of Tietze, the group $\mathcal{K}'_{n+1}$ is isomorphic to group $\mathcal{K}_{n+1}$, i.e. it is a coloured braid group.

Lemma 6. *The correlations (5) and (6) in group $\mathcal{K}'_{n+1}$ follow from correlations (13), (14), (8), (9), (10).*

Proof. Equality (5) can be deduced from the equalities (10) and (13). Equalities (15) can be deduced from correlations (9), (10) and (13). Equality

$$\tau_{lk}\tau_{ji}\tau_{jl}\tau_{j,l-1}^{-1} = \tau_{j,l-1}^{-1}\tau_{jl}\tau_{ji}\tau_{lk} \quad (i \leq k \leq j < l)$$

we receive easy from correlations (8) and (10). Equality (6) may be deduced from the last equality with the help of the equalities (13) and (15).

Let us denote as $\mathcal{K}''_{n+1}$ the group which is defined by generators (1), (7) and by defining relations (13), (14), (8), (9), (10). According to lemma 6 the group $\mathcal{K}''_{n+1}$ is a coloured braid group.

Lemma 7. *Equality*

$$\tau_{j+1,i+1}\tau_{j,i+1}\cdots\tau_{i+2,i+1}\tau_{i,j+1} = \tau_{j+1,i}\tau_{ji}\cdots\tau_{i+1,i} \quad (1 \le i \le j \le n) \tag{16}$$

can be deduced from the correlations (9), (10).

Proof. Using the correlation (10), the $\tau_{i,j+1}$ of the left part of the equality (16) we shall carry to the beginning of the word. Then we apply $(j-i)$ times correlation (9) and receive the right part of the equality (16):

$$\tau_{j+1,i+1}\tau_{j,i+1}\cdots\tau_{i+2,i+1}\tau_{i,j+1} = \tau_{i,j+1}\tau_{j+1,i+1}\tau_{j,i+1}\cdots\tau_{i+2,i+1} = \tau_{j+1,i}\tau_{ij}\tau_{j,i+1}\cdots\tau_{i+2,i+1}$$

$$= \tau_{j+1,i}\tau_{ji}\tau_{i,j-1}\cdots\tau_{i+2,i+1} = \cdots = \tau_{j+1,i}\tau_{ji}\tau_{j-1,i}\cdots\tau_{i+1,i}.$$

Let us denote as $\mathcal{K}'''_{n+1}$ the group which can be defined by generators (7) and by defining relations (16), (8), (9), (10).

Lemma 8. *The $\mathcal{K}'''_{n+1}$ group is isomorphic to the $\mathcal{K}''_{n+1}$ group.*

Proof. For each pair i,j from $/1 \le i \le j \le n/$ the generator $\rho_{i,j+1}$?occurs? in the correlation (13) one time. Let us exclude these generators in the $\mathcal{K}''_{n+1}$ group by the rule of Tietze. Equality (14) will turn into equality (16), equality (13) will vanish. Since equality (16) can be deduced from the correlations (8), (9), (10), by lemma 7, the $\mathcal{K}'''_{n+1}$ group is isomorphic to the group which is defined by generators (7) and by defining relations (8), (9), (10). Theorem 2 follows from lemma 8.

BIBLIOGRAPHY

1 Burau, W., *Über Zopfinvarianten*, Abh. Math. Sem. Hamburg Univ. 9 (1933), 117-124.
2 Artin, E., *Theory of braids*, Annals of Math. 48 (1947), 101-126.
3 Markov, A.A., Trudy Math. Inst. Steklov. 16 (1945), 1-53 (Russian; English summary).
4 Garside, F.A.: *The braid group and other groups*, Quart. J. Math. Oxford, 20, N 78 (1969), 235-254.

Lecture Notes in Computer Science

For information about Vols. 1–481
please contact your bookseller or Springer-Verlag

Vol. 524: G. Rozenberg (Ed.), Advances in Petri Nets 1991. VIII, 572 pages. 1991.

Vol. 525: O. Günther, H.-J. Schek (Eds.), Advances in Spatial Databases. Proceedings, 1991. XI, 471 pages. 1991.

Vol. 526: T. Ito, A. R. Meyer (Eds.), Theoretical Aspects of Computer Software. Proceedings, 1991. X, 772 pages. 1991.

Vol. 527: J.C.M. Baeten, J. F. Groote (Eds.), CONCUR '91. Proceedings, 1991. VIII, 541 pages. 1991.

Vol. 528: J. Maluszynski, M. Wirsing (Eds.), Programming Language Implementation and Logic Programming. Proceedings, 1991. XI, 433 pages. 1991.

Vol. 529: L. Budach (Ed.), Fundamentals of Computation Theory. Proceedings, 1991. XII, 426 pages. 1991.

Vol. 530: D. H. Pitt, P.-L. Curien, S. Abramsky, A. M. Pitts, A. Poigné, D. E. Rydeheard (Eds.), Category Theory and Computer Science. Proceedings, 1991. VII, 301 pages. 1991.

Vol. 531: E. M. Clarke, R. P. Kurshan (Eds.), Computer-Aided Verification. Proceedings, 1990. XIII, 372 pages. 1991.

Vol. 532: H. Ehrig, H.-J. Kreowski, G. Rozenberg (Eds.), Graph Grammars and Their Application to Computer Science. Proceedings, 1990. X, 703 pages. 1991.

Vol. 533: E. Börger, H. Kleine Büning, M. M. Richter, W. Schönfeld (Eds.), Computer Science Logic. Proceedings, 1990. VIII, 399 pages. 1991.

Vol. 534: H. Ehrig, K. P. Jantke, F. Orejas, H. Reichel (Eds.), Recent Trends in Data Type Specification. Proceedings, 1990. VIII, 379 pages. 1991.

Vol. 535: P. Jorrand, J. Kelemen (Eds.), Fundamentals of Artificial Intelligence Research. Proceedings, 1991. VIII, 255 pages. 1991. (Subseries LNAI).

Vol. 536: J. E. Tomayko, Software Engineering Education. Proceedings, 1991. VIII, 296 pages. 1991.

Vol. 537: A. J. Menezes, S. A. Vanstone (Eds.), Advances in Cryptology – CRYPTO '90. Proceedings. XIII, 644 pages. 1991.

Vol. 538: M. Kojima, N. Megiddo, T. Noma, A. Yoshise. A Unified Approach to Interior Point Algorithms for Linear Complementarity Problems. VIII, 108 pages. 1991.

Vol. 539: H. F. Mattson, T. Mora, T. R. N. Rao (Eds.), Applied Algebra, Algebraic Algorithms and Error-Correcting Codes. Proceedings, 1991. XI, 489 pages. 1991.

Vol. 540: A. Prieto (Ed.), Artificial Neural Networks. Proceedings, 1991. XIII, 476 pages. 1991.

Vol. 541: P. Barahona, L. Moniz Pereira, A. Porto (Eds.), EPIA '91. Proceedings, 1991. VIII, 292 pages. 1991. (Subseries LNAI).

Vol. 543: J. Dix, K. P. Jantke, P. H. Schmitt (Eds.), Nonmonotonic and Inductive Logic. Proceedings, 1990. X, 243 pages. 1991. (Subseries LNAI).

Vol. 544: M. Broy, M. Wirsing (Eds.), Methods of Programming. XII, 268 pages. 1991.

Vol. 545: H. Alblas, B. Melichar (Eds.), Attribute Grammars, Applications and Systems. Proceedings, 1991. IX, 513 pages. 1991.

Vol. 547: D. W. Davies (Ed.), Advances in Cryptology – EUROCRYPT '91. Proceedings, 1991. XII, 556 pages. 1991.

Vol. 548: R. Kruse, P. Siegel (Eds.), Symbolic and Quantitative Approaches to Uncertainty. Proceedings, 1991. XI, 362 pages. 1991.

Vol. 550: A. van Lamsweerde, A. Fugetta (Eds.), ESEC '91. Proceedings, 1991. XII, 515 pages. 1991.

Vol. 551: S. Prehn, W. J. Toetenel (Eds.), VDM '91. Formal Software Development Methods. Volume 1. Proceedings, 1991. XIII, 699 pages. 1991.

Vol. 552: S. Prehn, W. J. Toetenel (Eds.), VDM '91. Formal Software Development Methods. Volume 2. Proceedings, 1991. XIV, 430 pages. 1991.

Vol. 553: H. Bieri, H. Noltemeier (Eds.), Computational Geometry - Methods, Algorithms and Applications '91. Proceedings, 1991. VIII, 320 pages. 1991.

Vol. 554: G. Grahne, The Problem of Incomplete Information in Relational Databases. VIII, 156 pages. 1991.

Vol. 555: H. Maurer (Ed.), New Results and New Trends in Computer Science. Proceedings, 1991. VIII, 403 pages. 1991.

Vol. 556: J.-M. Jacquet, Conclog: A Methodological Approach to Concurrent Logic Programming. XII, 781 pages. 1991.

Vol. 557: W. L. Hsu, R. C. T. Lee (Eds.), ISA '91 Algorithms. Proceedings, 1991. X, 396 pages. 1991.

Vol. 558: J. Hooman, Specification and Compositional Verification of Real-Time Systems. VIII, 235 pages. 1991.

Vol. 559: G. Butler, Fundamental Algorithms for Permutation Groups. XII, 238 pages. 1991.

Vol. 560: S. Biswas, K. V. Nori (Eds.), Foundations of Software Technology and Theoretical Computer Science. Proceedings, 1991. X, 420 pages. 1991.

Vol. 561: C. Ding, G. Xiao, W. Shan, The Stability Theory of Stream Ciphers. IX, 187 pages. 1991.

Vol. 562: R. Breu, Algebraic Specification Techniques in Object Oriented Programming Environments. XI, 228 pages. 1991.

Vol. 563: A. Karshmer, J. Nehmer (Eds.), Operating Systems of the 90s and Beyond. Proceedings, 1991. X, 285 pages. 1991.

Vol. 564: I. Herman, The Use of Projective Geometry in Computer Graphics. VIII, 146 pages. 1992.

Vol. 565: J. D. Becker, I. Eisele, F. W. Mündemann (Eds.), Parallelism, Learning, Evolution. Proceedings, 1989. VIII, 525 pages. 1991. (Subseries LNAI).

Vol. 566: C. Delobel, M. Kifer, Y. Masunaga (Eds.), Deductive and Object-Oriented Databases. Proceedings, 1991. XV, 581 pages. 1991.

Vol. 567: H. Boley, M. M. Richter (Eds.), Processing Declarative Kowledge. Proceedings, 1991. XII, 427 pages. 1991. (Subseries LNAI).

Vol. 568: H.-J. Bürckert, A Resolution Principle for a Logic with Restricted Quantifiers. X, 116 pages. 1991. (Subseries LNAI).

Vol. 569: A. Beaumont, G. Gupta (Eds.), Parallel Execution of Logic Programs. Proceedings, 1991. VII, 195 pages. 1991.

Vol. 570: G. Schmidt, R. Berghammer (Eds.), Graph-Theoretic Concepts in Computer Science. Proceedings, 1991. VIII, 253 pages. 1992.

Vol. 571: J. Vytopil (Ed.), Formal Techniques in Real-Time and Fault-Tolerant Systems. Proceedings, 1992. IX, 620 pages. 1991.

Vol. 572: K. U. Schulz (Ed.), Word Equations and Related Topics. Proceedings, 1990. VII, 256 pages. 1992.